Environmental Science

Series editors: R. Allan · U. Förstner · W. Salomons

Springer
*Berlin
Heidelberg
New York
Barcelona
Hong Kong
London
Milan
Paris
Singapore
Tokyo*

Jürgen Schmidt (Ed.)

Soil Erosion

Application of Physically Based Models

With 119 Figures and 85 Tables

Springer

Editor

Prof. Dr. Jürgen Schmidt
TU Bergakademie Freiberg
Fachgebiet Boden- und Gewässerschutz
Agricolastraße 22
D-09599 Freiberg
Germany

ISSN 1431-6250
ISBN 3-540-66764-4 Springer-Verlag Berlin Heidelberg New York

Library of Congress Cataloging-in-Publication Data
Soil erosion : application of physically based models / Jürgen Schmidt (ed.).
p. cm. – (Environmental science)
Includes bibliographical references.
ISBN 3-540-66764-4 (hc : alk. paper)
1. Soil erosion–Mathematical models.
I. Schmidt, Jürgen, 1953- II. Environmental science (Berlin, Germany)
S627.M36 S65 2000
631.4'5'015118--dc21   00-027989
                                                                                                       CIP

This work is subject to copyright. All rights are reserved, whether the whole or part of the material is concerned, specifically the rights of translation, reprinting, reuse of illustrations, recitation, broadcasting, reproduction on microfilms or in any other way, and storage in data banks. Duplication of this publication or parts thereof is permitted only under the provisions of the German Copyright Law of September 9, 1965, in its current version, and permission for use must always be obtained from Springer-Verlag. Violations are liable for prosecution under the German Copyright Law.

Springer-Verlag is a company in the BertelsmannSpringer publishing group.
© Springer-Verlag Berlin Heidelberg 2000
Printed in Germany

The use of general descriptive names, registered names, trademarks, etc. in this publication does not imply, even in the absence of a specific statement, that such names are exempt from the relevant protective laws and regulations and therefore free for general use.

Cover Design: Struve & Partner, Heidelberg
Dataconversion: Büro Stasch, Bayreuth

SPIN: 10575560 – 32/3136/xz – 5 4 3 2 1 0 – Printed on acid-free paper

Preface

Soil erosion at the present extent is mainly a result of human activities and not a product of natural processes. Without human impact, the earth's soil surface would be almost completely covered by permanent vegetation with the exception of extreme climatic environments, such as deserts, polar or high mountainous areas. The main natural hazards which may cause erosion under natural conditions would be natural fires, storms, volcanic eruptions, or meteorite impacts. Since such hazards would have only local and temporary effects on the vegetation cover, one can assume that – especially in the regions of temperate climate – the amount of soil loss as a result of natural processes would be negligible.

The use of soils by man – in particular for agriculture – constrains to remove the natural vegetation cover and to replace it by crops. Thus, the protection of the soils from the direct impact of wind and water is, at least, temporarily suspended. Accelerated erosion caused by water and wind is the inevitable consequence.

Erosion leads to the irreversible degradation of soils and to the loss of their ecological and economic functions. Once the soil has been lost, it cannot be compensated by natural soil restoration within reasonable time periods. In addition, erosion usually causes further off-site damages by depositing the transported material on adjacent sites. Moreover, eroded sediments and sediment-bound chemicals may enter the surface water system, resulting in long-term eutrophication and toxification.

Erosion is usually regarded as a slow and almost imperceptible process which occurs in a large number of isolated erosion events. In fact many difficulties are associated with the monitoring and surveying of erosion processes. In most cases direct measurements of soil loss are limited to small experimental plots on which the relevant hydraulic conditions of erosion cannot be completely reproduced. For the same reasons, plot measurements cannot be directly transferred to natural slopes and watersheds without taking the differing hydraulic conditions into account.

Nevertheless, the first mathematical approach to describe soil erosion by water, the UNIVERSAL SOIL LOSS EQUATION (USLE) by Wischmeier and Smith (1965), was derived by correlating the amount of soil loss gained from experimental plots with various topographic, climate, soil, and land use parameters. More recently developed soil erosion models mainly use physically based approaches which allow adequate representation and quantitative estimation of erosion (soil detachment and transport) and deposition. Such models and their practical application are the main subject of this book.

Table 0.1 provides an overview of the various soil erosion models which are described in this book. The models are sorted by the year of their first publication. The

Table 0.1. Overview of the soil erosion models described in this book

Name	Author	Chapter
AGNPS	Young et al. (1987)	3
SMODERP	Holy et al. (1989)	8
WEPP	Lane and Nearing (1989)	1, 11
EUROSEM/KINEROS	Morgan et al. (1992); Woolhiser et al. (1990)	10, 11, 13
EROSION 2D/3D	Schmidt (1991); von Werner (1995)	5, 6, 7, 9, 11
PEPP-HILLFLOW	Schramm (1994); Bronstert (1994)	4, 12
LISEM	De Roo et al. (1996)	2

table shows that the first physically based soil erosion models were developed about 15 years ago. This book was particularly motivated by the fact that, in the meantime, the results of numerous practical applications of these models have become available.

This book is divided into three parts. It mainly focuses on the papers in Part I in which nine different examples for practical model applications are described. Part II consists of three papers that deal with the validation of physically based soil erosion models. Finally, the two papers in Part III provide information on current developments of recent modelling approaches.

References

Bronstert A (1994) Modellierung der Abflußbildung und der Bodenwasserdynamik von Hängen. Mitt Inst Hydrologie und Wasserwirtschaft, Univ Karlsruhe, H 45

De Roo APJ, Wesseling CG, Ritsema CJ (1996) LISEM: a single event physically-based hydrologic and soil erosion model for drainage basins. Hydrological Processes, Part I:1107-1117, Part II:1119-1126

Holy M, Váška J, Vrána K (1989) SMODERP – Simulation Model for Determination of Surface Runoff and Prediction of Erosion Processes. User's Guide. Faculty of Civil Engineering CTU Prague, pp 24

Lane LJ, Nearing MA (eds) (1989) USDA – Water erosion prediction project: hillslope profile model documentation. NSERL Report 2 (USDA-ARS National Soil Erosion Laboratory), West Lafayette, Indiana, USA

Morgan RPC, Quinton JN, Rickson RJ (1992) EUROSEM documentation manual, Version 1. Silsoe College, Silsoe

Schmidt J (1991) A mathematical model to simulate rainfall erosion. In: Bork HR, De Ploey J, Schick AP (eds) Erosion, transport and deposition processes – theories and models. Catena, Supplement 19:101-109

Schramm M (1994) Ein Erosionsmodell mit zeitlich und räumlich veränderlicher Rillengeometrie. Mitt Inst Wasserbau und Kulturtechnik, Univ Karlsruhe, H 190

Werner M von (1995) GIS-orientierte Methoden der digitalen Reliefanalyse zur Modellierung von Bodenerosion in kleinen Einzugsgebieten. PhD thesis, Department of Geography, Berlin Free University

Wischmeier WH, Smith DD (1965) Predicting rainfall-erosion losses from cropland east of the Rocky Mountains. Agr Handbook 282 (USDA), Washington DC

Woolhiser DA, Smith RE, Goddrich DC (1990) KINEROS, a kinematic runoff and erosion model, documentation and user manual. US Department of Agriculture – Agricultural Research Service 77, Washington DC

Young RA, Onstad CA, Bosch DD, Anderson WP (1987) AGNPS, Agricultural Non-Point-Source Pollution Model – a watershed analysis tool. United States Department of Agriculture, Conservation Research Report, 35-80

Acknowledgements

The editor thanks the contributors to this book for their willingness of collaboration and their patience, as well as he thanks the Springer Verlag for publishing this book in its program.

Last, but not least the editor thanks Mrs. U. Krause, Mr. A. Schröder and Mrs. G. Etzold and Mr. A. Stasch for their active assistance in reviewing the text and the technical realization of this book.

Freiberg, June 2000

Jürgen Schmidt

Contents

Part I Practical Model Applications 1

1 The Influence of Global Greenhouse-Gas Emissions on Future Rates of Soil Erosion: a Case Study from Brazil Using WEPP-CO$_2$ 3
1.1 Background 3
   1.1.1 Greenhouse Gases, Climate Change and Soil Erosion 3
   1.1.2 Potential and Actual Increases in Future Soil Erosion 4
   1.1.3 The Aim of this Study 6
1.2 Data and Tools 7
   1.2.1 Emissions Policies 7
   1.2.2 GCM Estimates of Future Climate 7
   1.2.3 WEPP-CO$_2$ 9
1.3 Modeling Future Erosion at Sorriso, Mato Grosso, Brazil 10
   1.3.1 The Area 10
   1.3.2 Modeling Current Erosion Rates 14
   1.3.3 Modeling Future Erosion Rates 15
1.4 Discussion 22
   1.4.1 Implications 22
   1.4.2 Uncertainties 25
1.5 Conclusions 27
   Acknowledgements 28
   Acronyms Used 28
   References 28

2 Applying the LISEM Model for Investigating Flood Prevention and Soil Conservation Scenarios in South-Limburg, the Netherlands 33
2.1 Introduction 33
2.2 Current Policy 33
2.3 LISEM 34
2.4 Scenario Results 36
2.5 Cost-Benefit Analysis 38
2.6 New Policy 39
2.7 Conclusions 40
   Acknowledgements 40
   References 40

3	**Application of Modified AGNPS in German Watersheds**	43
3.1	Objectives	43
3.2	Methodology	43
3.3	Results	47
3.4	Conclusions	56
	References	57

4	**Physically Based Modeling of Surface Runoff and Soil Erosion under Semi-arid Mediterranean Conditions – the Example of Oued Mina, Algeria**	59
4.1	Introduction	59
4.2	Model Planning and Theory	60
	4.2.1 Surface Runoff	60
	4.2.2 Infiltration	63
	4.2.3 Soil Removal, Transportation and Sedimentation	64
	4.2.4 Solid Material Continuity	65
4.3	Determining Parameters	67
	4.3.1 Contribution of Remote Sensing	67
	4.3.2 Soil Mapping	68
	4.3.3 Method of Classification	68
	4.3.4 Soil Classification	70
	4.3.5 Mapping Soil Humidity	72
	4.3.6 Principal Components Analysis	73
	4.3.7 Soil Humidity in the Oued Mina Catchment	74
	4.3.8 Model Results	76
4.4	Conclusions	76
	References	77

5	**Assessing the Impact of Lake Shore Zones on Erosional Sediment Input Using the EROSION-2D Erosion Model**	79
5.1	Introduction	79
5.2	Lake Belau and the Typifiying of its Shore Zone	80
5.3	Modeling	82
5.4	Upscaling	88
5.5	Conclusions	90
	References	92

6	**Modeling the Sediment and Heavy Metal Yields of Drinking Water Reservoirs in the Osterzgebirge Region of Saxony (Germany)**	93
6.1	Introduction	93
6.2	Modeling Principles and Methods	93
6.3	Watersheds	95
6.4	Results	96
	6.4.1 Erosion Simulation	96
6.5	Comparison of Predicted and Measured Sediment Yields	98
	6.5.1 Malter Reservoir	99
	6.5.2 Saidenbach Reservoir	99
6.6	Estimating the Delivery of Particle-Attached Heavy Metals	100

	6.6.1 Methodology 101
	6.6.2 Plot Rainfall Simulations 101
	6.6.3 Saidenbach Reservoir (Hölzelbach Subwatershed) 104
6.7	Strategies for Minimizing the Sediment Delivery 105
6.8	Conclusions 107
	References 107

7 A Multiscale Approach to Predicting Soil Erosion on Cropland Using Empirical and Physically Based Soil Erosion Models in a Geographic Information System 109

7.1	Introduction 109
7.2	Multiscale Investigation and Modeling of Soil Erosion Processes 110
7.3	Assessment of Erosion Hazard and Potential Sediment Yield at the Macro- and Mesoscale Level 112
	7.3.1 Procedures for Areal Estimation of the Potential Erosion Hazard 112
	7.3.2 Identifying the Potential Sites at which Fine Sediments Enter the Surface Water and Estimating the Risk of Sediment Delivery 115
7.4	Areal Prediction of Soil Loss at the Microscale Level 120
	7.4.1 Estimating the Threat to Soil Fertility for individual Plots Employing the Allgemeine Bodenabtragsgleichung (ABAG) 120
	7.4.2 Areally Differentiated Erosion-Deposition Modeling with Physically Based Soil Erosion Models: Application of the Model EROSION-3D within a Geographic Information System 123
7.5	Linking a Soil Moisture Budget Model with EROSION-3D for the Model Assisted Estimation of the Parameter Soil Moisture 129
7.6	Prospects 132
	References 133

8 SMODERP – A Simulation Model of Overland Flow and Erosion Processes 135

8.1	Introduction 135
8.2	Concept of the SMODERP Model 135
	8.2.1 Concept of the Overland Flow Model 135
	8.2.2 Concept of the Erosion Model 138
8.3	Variable Description 139
	8.3.1 Rainfall 139
	8.3.2 Interception 139
	8.3.3 Surface Depression Storage 140
	8.3.4 Manning's Roughness Coefficient for Overland Flow 140
	8.3.5 Infiltration 140
	8.3.6 Rainfall Kinetic Energy 141
	8.3.7 Relative Soil Erodibility 141
	8.3.8 Vegetation Cover and Management (C-Factor) 141
8.4	Model Inputs and Outputs 142
	8.4.1 Input Files 142
	8.4.2 Output Files 144
8.5	Sensitivity Analysis 144

8.6	System Requirements	148
8.7	Use of SMODERP in Soil and Water Conservation	148
8.8	Using the SMODERP Simulation Model for the Design of Conservation Measures as a Part of Landscape Revitalisation Studies	149
	8.8.1 Input Data for the SMODERP Model	149
	8.8.2 Calculation of the Critical Slope Length	150
	8.8.3 Design of Conservation Measures	150
	8.8.4 Conclusion	150
8.9	Protection of the Urban Areas against Surface Runoff and Sediment from a Small Agricultural Watershed (Application of the SMODERP Model)	151
	8.9.1 Description of Current Situation	151
	8.9.2 Methodology	151
	8.9.3 The Solution	152
	8.9.4 Input Data	152
	8.9.5 Alternative Soil Protection Measures	153
	8.9.6 Alternatives	154
	8.9.7 Conclusions	156
8.10	Numerical Simulation of Overland Flow on the Radovesice Waste Dump (North Bohemia)	157
	8.10.1 Introduction	157
	8.10.2 Theoretical Considerations	158
	8.10.3 Methodology	159
	8.10.4 Results and Conclusions	160
	References	161

9 Modeling Overland Flow and Soil Erosion for a Military Training Area in Southern Germany ... 163

9.1	Objective	163
9.2	Overview of Existing Erosion Models	163
	9.2.1 RUSLE/MUSLE	164
	9.2.2 EPIC	164
	9.2.3 AGNPS	165
	9.2.4 CREAMS	165
	9.2.5 GLEAMS	166
	9.2.6 ANSWERS	166
	9.2.7 EROSION-2D and EROSION-3D	167
	9.2.8 KINEROS	168
	9.2.9 OPUS	168
	9.2.10 SPUR	169
	9.2.11 WEPP	169
	9.2.12 EUROSEM	170
9.3	Assessment of Model Applicability	170
9.4	Applying EROSION-3D to a Subcatchment	173
9.5	Simulation of Different Soil Moisture Scenarios	174
9.6	Conclusions	175
	References	178

Part II Model Validation 179

10 A Process-Based Evaluation of a Process-Based Soil-Erosion Model ... 181
10.1 Introduction 181
10.2 EUROSEM 181
10.3 The Test Site 183
10.4 Results 185
    10.4.1 Runoff 185
    10.4.2 Erosion 189
10.5 Discussion 189
    10.5.1 Runoff 189
    10.5.2 Erosion 191
10.6 Conclusions 194
    Acknowledgements 196
    References 196

11 WEPP, EUROSEM, E-2D: Results of Applications at the Plot Scale 199
11.1 Introduction 199
11.2 Objectives and Methodology of the Study 199
11.3 Model Comparison 200
    11.3.1 Input and Output Parameters 201
    11.3.2 Model Structure 204
11.4 Sensitivity Analysis 224
    11.4.1 Definition of Sensitivity Parameter 224
    11.4.2 Results 225
11.5 Model Application 228
    11.5.1 Description of Test Plots 229
    11.5.2 Model Performance 229
    11.5.3 Assessment of Tillage Effects for the Methau Plots 240
11.6 Model Assessment 243
11.7 Summary and Conclusions 245
    Acknowledgements 247
    References 247

12 Simulating Hydrological and Erosional Processes Using the PEPP-HILLFLOW Model-Parameter Determination and Model Application .. 251
12.1 Introduction 251
12.2 PEPP-HILLFLOW Model 251
    12.2.1 Surface Runoff 251
    12.2.2 Transport Capacity 253
    12.2.3 Detachment Capacity 254
    12.2.4 Erosion and Deposition 256
    12.2.5 Sensitivity Analysis PEPP 258
    12.2.6 Infiltration 260
    12.2.7 Sensitivity Analysis HILLFLOW 261
12.3 Determination of the Model Parameters for PEPP 262

	12.3.1 Temporal Variability of the Erosion Resistance	262
	12.3.2 Spatial Variability of the Erosion Resistance	265
	12.3.3 Estimation of Manning's n	268
12.4	Application of the PEPP-HILLFLOW Model	270
	12.4.1 Simulation of Rainfall Experiments	270
	12.4.2 Simulation of Natural Rainfall Events on Slopes	271
	12.4.3 Simulation of Natural Rainfall Events on Small Catchments	273
12.5	Summary and Outlook	276
	References	278

Part III Recent Model Developments 281

13 Dynamics and Scale in Simulating Erosion by Water 283

13.1	Introduction	283
13.2	Basic Features of KINEROS2	284
	13.2.1 Runoff	285
	13.2.2 Sediment Transport	285
	13.2.3 Initial and Boundary Conditions	287
	13.2.4 Solution Characteristics	287
	13.2.5 Treating Soil with Distributed Particle Sizes	288
13.3	Application Examples	289
	13.3.1 Splash vs. Erosion Limiting Conditions	289
	13.3.2 Temporal Changes	291
	13.3.3 Steady Flow and Length Effects	291
13.4	Summary	293
	References	293

14 Modeling Surface Runoff 295

14.1	Introduction	295
14.2	Approaches to Surface Runoff	296
14.3	Some Examples	301
14.4	Conclusions	305
	References	305

Index 307

Contributors

Böhm, Albert
DPW CMTC Hohenfels
Environmental Management Office
Camp Nainhof
Building 34
D-92366 Hohenfels, Germany
albert.boehm@bsbdpw.hohenfels.army.mil

Deinlein, Roland
gibs geologen+ingenieure
Deichslerstraße 25
D-90489 Nürnberg, Germany
rc.deinlein@ngi.de

Dr. De Roo, Ad P.J.
Department of Physical Geography
Utrecht University
POBox 80115
Utrecht, The Netherlands
ad.de-roo@jrc.it

Present address:
European Commission
DG Joint Research Centre
Space Applications Insitute, TP 263,
21020 Ispra (Va), Italy

Dostál Tomáš, MSc, C. Eng.
Department of Irrigation, Drainage and Landscape Engineering
Faculty of Civil Engineering
Czech Technical University in Prague
Thákurova 7
16629 Prague 6, Czech Republic
dostal@fsv.cvut.cz

Dr. Duttmann, Rainer
University of Hannover
Geographical Institute
Department of Physical Geography and Landscape Ecology
Schneiderberg 50
D-30167 Hannover, Germany
duttmann@geog.uni-hannover.de

Dr. Favis-Mortlock, David T.
Queen's University of Belfast
School of Geography
Belfast BT7 1NN
Northern Ireland
d.favis-mortlock@qub.ac.uk

Prof. Dr. Frede, Hans-Georg
Universität Giessen
Institut für Landeskultur
Senckenbergstraße 3
D-35390 Giessen, Germany
Hans-Georg.Frede@agrar.uni-giessen.de

Dr. Gerlinger, Kai
Ingenieurbüro Dr.-Ing. Karl Ludwig
Herrenstraße 14
76133 Karlsruhe, Germany
kai.gerlinger@t-online.de

Dr. Gomer, Dieter
Deutsche Gesellschaft für Technische Zusammenarbeit (GmbH)
Postfach 5180
D-65726 Eschborn, Germany
dieter.gomer@gtz.de

Dr. Grunwald, Sabine
Department of Soil Science
University of Wisconsin-Madison
1525 Observatory Drive
Madison, WI 53706, USA
sgrunwald@facstaff.wisc.edu

Dr. Guerra, Antonio J.T.
LAGESOLOS (Laboratory of Experimental Geomorphology and Soil Erosion)
Department of Geography, Federal University of Rio de Janeiro
Cidade Universitaria – Ilha do Fundao
Rio de Janeiro – RJ
CEP. 21940-590, Brazil
guerra@igeo.ufrj.br

Dr. Hergarten, Stefan
Geodynamics – Physics of the Lithosphere
University of Bonn
Nussallee 8
D-53115 Bonn, Germany
hergarten@geo.uni-bonn.de

Jelinek, Stefan
Ökologie-Zentrum der CAV Kiel
Schauenburger Straße 112
D-24118 Kiel, Germany
Stefan-j@pz-oekosys.uni-kiel.d400.de

Prof. Dr. Mosimann, Thomas
University of Hannover, Geographical Institute
Department of Physical Geography and Landscape Ecology
Schneiderberg 50
D-30167 Hannover, Germany
mosimann@geog.uni-hannover.de

Prof. Dr. Neugebauer, Horst J.
Geodynamics – Physics of the Lithosphere, University of Bonn
Nussallee 8
D-53115 Bonn, Germany
neugb@geo.uni-bonn.de

Prof. Parsons, Anthony
Department of Geography
University of Leicester
University Road
Leicester, LE1 7RH, United Kingdom
AJP16@le.ac.uk

Dr. Paul, Guido
Clemens-August-Straße 47
D-53115 Bonn, Germany

Dr. Quinton, John
Cranfield University, Silsoe
Institute of Water and Environment
Bedfordshire MK45 4DT, United Kingdom
j.quinton@cranfield.ac.uk

Prof. Dr. Schmidt, Jürgen
TU Bergakademie Freiberg, Fachgebiet Boden- und Gewässerschutz
Agricolastraße 22
D-09599 Freiberg, Germany
jschmidt@orion.hrz.tu-freiberg.de

Dr. Schoger, Heinrich
gibs geologen+ingenieure
Deichslerstraße 25
D-90489 Nürnberg, Germany
hschoger@gibs-online.de

Schröder, Axel
Wiesenstraße 17
D-13357 Berlin, Germany
schroeda@rz.uni-potsdam.de

Dr. Smith, Roger
Engineering Research Center, Colorado State University
Fort Collins, CO, 80523-1320, USA
sroger@engr.colostate.edu

Váška Jiøí, Assoc. Prof., PhD, C. Eng.
Department of Irrigation, Drainage and Landscape Engineering
Faculty of Civil Engineering, Czech Technical University in Prague
Thákurova 7
16629 Prague 6, Czech Republic
vaska@fsv.cvut.cz

Vogt, Teodora
Université Louis Pasteur
3 Rue de L'Argonne
67083 Strasbourg, France

Vrána Karel, Assoc. Prof., PhD, C. Eng.
Department of Irrigation, Drainage and Landscape Engineering
Faculty of Civil Engineering, Czech Technical University in Prague
Thákurova 7
16629 Prague 6, Czech Republic
vrana@fsv.cvut.cz

Dr. Wainwright, John
Department of Geography
King's College London
Strand, London, WC2R 2LS, United Kingdom
j.wainwright@kcl.ac.uk

Dr. von Werner, Michael
TU Bergakademie Freiberg
Fachgebiet Boden- und Gewässerschutz
Agricolastraße 22
D-09599 Freiberg, Germany
erosion@geog.fu-berlin.de

Dipl. Geogr. Wickenkamp, Volker
Grosse Pfahlstraße 14
D-30161 Hannover, Germany
wickenkamp@geo-net.de

Part I
Practical Model Applications

Chapter 1

The Influence of Global Greenhouse-Gas Emissions on Future Rates of Soil Erosion: a Case Study from Brazil Using WEPP-CO$_2$

D.T. Favis-Mortlock · A.J.T. Guerra

1.1
Background

1.1.1
Greenhouse Gases, Climate Change and Soil Erosion

Since the mid-1980s, increases in the global concentrations of greenhouse gases have been paralleled by rising international concern over their potential to affect climate. Concentrations of these gases (most importantly, carbon dioxide, methane, nitrous oxide, tropospheric ozone, chlorofluorocarbons and water vapour) have been observed to increase dramatically during the last 100 years or so. This rise results from anthropogenic activity. Emissions of the naturally-occurring gases have increased due to modifications of natural cycles by growing human populations, while some new gases (e.g. chlorofluorocarbons) have been added. Atmospheric concentrations of carbon dioxide, for example, have risen by about 26% since the Industrial Revolution (Fig. 1.1): this results both from increased burning of fossil fuels, and from deforestation (Houghton et al. 1990).

The root of this concern is the potential of these gases to produce so-called greenhouse warming (e.g. Houghton et al. 1990). The theoretical link between carbon dioxide and temperature – radiative forcing – was first proposed by the Swedish scientist Arrhenius in the late 19th century (e.g. Arnell 1996). Observational evidence for the greenhouse effect comes from a variety of sources: for example, palaeo-environmental studies of air trapped in Antarctic ice cores have found close correlations between levels of carbon dioxide and methane, and local temperature (Lorius et al. 1990). Confirmation of a present-day enhanced greenhouse effect, by showing a link between present-day greenhouse gas increases and temperature, is however more contentious. Further discussion of this controversial area of science is beyond the scope of this study. However, Working Group 1 of the Intergovernmental Panel on Climate Change's Second Assessment were able to state that "… the balance of evidence now suggests a discernible human influence on global climate" (Houghton et al. 1996).

While it may reasonably be assumed that increases in greenhouse gas concentrations will result in a rise in global temperature, estimation of the effects on other aspects of climate is far from straightforward. General Circulation Models are the principal tool for this task (e.g. Henderson-Sellers 1994). However, GCMs (a list of acronyms is given at the end) are – like all models – a simplification of the real world; as such, they have numerous deficiencies. While GCMs are now able to satisfactorily replicate large-scale atmospheric features, those regional features which affect precipitation are simulated much less well (Kerr 1997). This is in part because the rather coarse spatial resolution of GCMs (necessitated mainly by computational restrictions) does

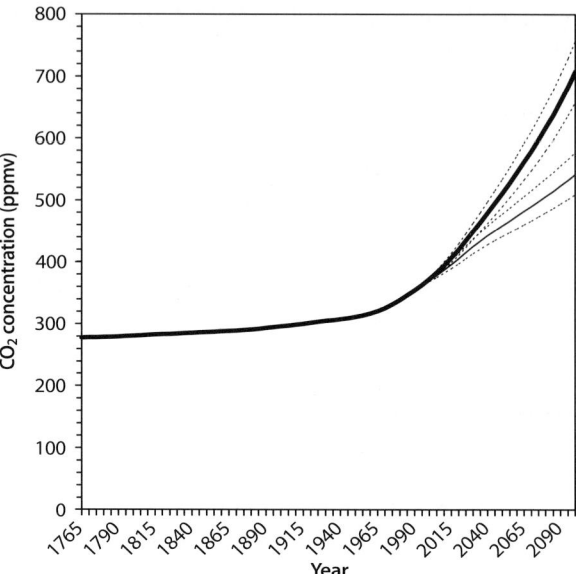

Fig. 1.1. Past and future concentrations of atmospheric CO_2. Two scenarios of estimated future concentrations are shown, IS92a (*thick line*) and IS92d (*thin line*). The *dotted lines* illustrate the uncertainty around these scenarios (data from Climatic Research Unit 1997)

not permit them to explicitly model relatively small-scale atmospheric features such as convectional rainfall. As a result, inter-model agreement for precipitation is poorer than is the case for temperature (e.g. Wilby 1995). This is particularly problematic for consideration of future soil erosion. Changes in rainfall intensity are likely to strongly affect rates of erosion; yet it is the small scale atmospheric features which cannot be modelled by GCMs which are most likely to produce changes in intensity.

Nonetheless, some consensus indications are emerging. For some areas, it appears that future rainfalls are likely to increase at certain times of the year: for example, during the winter months for the southern UK (e.g. United Kingdom Climate Change Impacts Review Group 1991; Arnell 1996). All else being equal, any increase in rainfall – either in quantity or in intensity – at times and in areas where land usage leave the soil surface unprotected will result in increased soil erosion (e.g. Boardman and Favis-Mortlock, in preparation b).

1.1.2
Potential and Actual Increases in Future Soil Erosion

However, the translation of any potential for increased future soil erosion into an actual (and quantified) increase is complicated by at least three additional factors:

- the effects of both climate change and increased atmospheric CO_2 on plant growth
- changes in land use
- the applicability of the erosion model used.

Other more speculative effects, such as transpiration-driven feedback between vegetation change and regional climate, will not be considered here.

1.1.2.1
Plant Growth

Change in temperature, rainfall or solar radiation will affect the growth of plants: these may be considered to be the direct effects of climate change on vegetative growth. However, plants also respond directly to atmospheric CO_2 content: this may be considered an indirect effect of greenhouse-gas driven climate change. Increased concentrations of atmospheric CO_2 were noted to enhance plant growth in the early years of the nineteenth century (Kimball et al. 1993). In many plants increased CO_2 concentrations result in a higher exchange resistance vapour between leaf and ambient air to gases such as CO_2 and water. This reduces water losses by transpiration, and so increases water use efficiency (e.g. Kimball and Idso 1983; Kimball et al. 1993; Wolf 1993). Not all plant species respond similarly, however. Plants such as wheat have a C3 photosynthetic pathway and so are generally more responsive to elevated CO_2 concentrations than plants such as maize with a C4 pathway. In addition, plant responses to interactions between temperature, water stress, nutrient availability and enhanced CO_2 are not yet clear; also the responses of crops to increased CO_2 in the (relatively) competitive ecosystem of a field may be very different from those noted during the isolation of a pot experiment (Tegart et al. 1990; Kimball et al. 1993). Implications for the early stages of growth are poorly known (J.I.L. Morison, personal communication 1994): this is the most critical period for soil erosion, as the young plant begins to cover the ground.

In this way, interactions between direct and indirect effects complicate erosion's response to climate change. The interplay between climate change-induced shifts in the timing and amount of rainfall and the rate of crop growth in a changed and CO_2-enriched climate give rise to complex non-linear responses, as noted in a modeling study by Favis-Mortlock and Boardman (1995).

1.1.2.2
Land Use

In many (perhaps most) situations the erosional system is more sensitive to land use change than it is to climate change. Land use frequently governs erosion to a greater extent than climate (Evans 1996, p. 79; Morgan 1995, p. 5). This was indicated, for example, in an analysis of Canadian lake sediments by De Boer (1997). Modeling studies have also produced a similar result: Favis-Mortlock et al. (1997) carried out a simulation study of past land use and climate change on the UK South Downs and found land-use change to be a stronger determinant of erosion rate than climate change. Another modeling study, this time for probable future conditions at another UK site (Boardman and Favis-Mortlock, in preparation a), obtained a similar result.

Thus in order to estimate future erosion, some forecast of future land use must be made. This is very difficult to do in any objective way, however. Attempts have been made to model future changes in land use (mainly in the developed world) by building upon assumptions regarding future economic conditions and land suitability (e.g. Hossell et al. 1995; Turner et al. 1995). At present though, the only generally feasible approach is extrapolation from current practices (cf. Evans 1996, pp. 89–90). The simplest of all such extrapolations is that present land use will continue unchanged in the future: this has been adopted for the great majority of studies of future erosion (e.g. Boardman et al. 1990; Parry et al. 1991, pp. 36–37; Boardman and Favis-Mortlock 1993; Phillips et al. 1993; Botterweg 1994; Favis-Mortlock et al. 1991; Favis-Mortlock and Boardman 1995; Favis-Mortlock and Savabi 1996; Kallio et al. 1997; Lee et al. 1996; Lee 1998; Favis-Mortlock and Guerra 1999).

1.1.2.3
Model Type

Modeling soil erosion under any novel conditions inevitably involves some extrapolation. This is particularly true of climate change studies. The main danger associated with such extrapolation is that of using relationships within the model beyond the range for which they are valid: this will be reduced if physically-based, rather than empirical, models are used (Favis-Mortlock et al. 1996; Favis-Mortlock and Savabi 1996).

1.1.3
The Aim of this Study

Notwithstanding these complicating factors, there are four conceptual cause-and-effect 'layers' which link emissions of greenhouses gases and future soil erosion (Table 1.1). For the top layer, future rates of greenhouse gas emissions, it is clear that these will largely be the result of policy decisions made at a national and international level (e.g. Intergovernmental Panel on Climate Change 1991; Fankhauser and Tol 1996). Thus, following the links through from emissions, via greenhouse warming and climate change, to rainfall-driven erosion, it may be said that future rates of soil erosion will also be affected by such policies, albeit to an unknown extent.

The aim of this study is a first attempt at quantification of that relationship. It builds upon earlier work by Favis-Mortlock and Guerra (1999).

The layers of Table 1.1 roughly correspond to the methodological steps of this study. The first three layers are modelled by use of MAGICC and SCENGEN (see below), which in turn call upon the stored results of previous GCM simulations. A stochastic weather

Table 1.1. The four 'layers' linking emissions policies and future soil erosion

'Layer'	'Driver'	Commonly-used modeling approaches
Future emissions of greenhouse gases	Emissions policies	Scenario-based extrapolations of current trends, e.g. IS9a and IS92d
Increased concentrations of greenhouse gases producing global warming	Atmospheric dynamics	GCMs
Precipitation changes in a warmed climate system	Atmospheric dynamics	GCMs
The response of the erosional system to changed climate	Future local climate; future land use; processes of soil erosion by water	Techniques for downscaling GCM output such as stochastic weather generators; assumptions regarding future land use; use of a process-based soil erosion model e.g. WEPP

generator, CLIGEN, provides the link between the regional focus of the third layer and the necessarily local focus of the fourth layer; while the fourth layer itself is modelled using the erosion model WEPP-CO_2.

1.2
Data and Tools

1.2.1
Emissions Policies

Two emissions policies are considered here, IS92a and IS92d (Fig. 1.1). These were defined by Working Group III of the IPCC (Leggett et al. 1992). IS92a represents a 'business as usual' scenario, which assumes that few or no steps are taken to limit greenhouse gas emissions. IS92d is a 'mitigation' scenario, which assumes progressively increasing controls to reduce the growth of emissions.

Future CO_2 emissions under these scenarios are shown in Fig. 1.2. Carbon dioxide has been estimated to contribute about 55–60% of the anthropogenic greenhouse effect (e.g. Houghton et al. 1990).

1.2.2
GCM Estimates of Future Climate

GCMs are complex models which require the largest computers, and produce considerable quantities of output. This makes it difficult for the impacts modeller to work

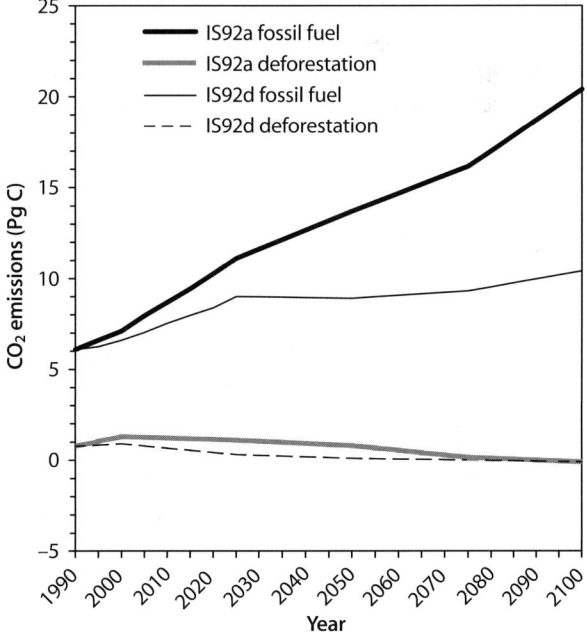

Fig. 1.2. Future global CO_2 emissions resulting from two scenarios of greenhouse gas emissions, IS92a and IS92d (data from Climatic Research Unit 1997)

directly with GCM data. The recent development of software tools such as MAGICC (Hulme et al. 1995a; Climatic Research Unit 1997) and SCENGEN (Hulme et al. 1995b, 1997) has made this much easier. These linked models aim to "allow the user full scope to generate global and regional scenarios of climate change, based on GCM experimental results of their own choosing".

MAGICC is a set of coupled gas-cycle, climate and ice-melt models that produce an estimate of the global-mean temperature change resulting from a given emissions scenario. SCENGEN builds upon MAGICC's output to construct geographically explicit future climate change scenarios. SCENGEN achieves this by taking the spatial patterns for change in climatic parameters produced by a range of GCMs (ten equilibrium and four transient in SCENGEN version 2.1a) and interpolating these onto a common 5°latitude × 5°longitude global grid. A simple model within SCENGEN then rescales these patterns in response to MAGICC's global-mean warming estimates, and with respect to assumptions regarding overall climate sensitivity, and with respect to different baseline climatologies. The simple and global climate models used in or by MAGICC and SCENGEN have all been used or reported by the IPCC, most of them in the IPCC Second Assessment Report (Houghton et al. 1996). Version 2.3 of MAGICC has been designed to replicate exactly the results in the IPCC Second Assessment Report.

While such tools are convenient, their estimates can only be as reliable as the GCMs on which their assumptions are based. A GCM validation study is described by Hulme et al. (1995b) which compared present-day global precipitation estimates from the GCMs used in SCENGEN against the observed Legates and Willmott (1990) climatology, following the principle that "the closer these two climates agree, the more confidence one may have in the performance of the GCM under conditions of climate change". Table 1.2 gives details of three recent GCM experiments: correlations for these range between 0.77 and 0.67 (Hulme et al. 1997), with the UK Meteorological Office

Table 1.2. Correlations between GCM-simulated control (present day) precipitation and the observed climatology of Legates and Wilmott (1990), for three models (from Favis-Mortlock and Guerra 1999)

GCM	Description	Location	Run type	Run date	Reference	Mean correlation
HADCM2	11-layer atmospheric GCM, coupled to a 20-layer ocean model	Hadley Centre, Meteorological Office, UK	Transient	1994–95	Mitchell et al. (1995); Houghton et al. (1996)	0.77
CSIRO9 Mk2	9-layer atmospheric GCM, coupled to a mixed-layer ocean model	Commonwealth and Scientific Industrial Research Organsation, Australia	Equilibrium	1995	Dix and Hunt (1995)	0.71
ECHAM3TR	19-layer atmospheric GCM, coupled to an 11-layer isopycnal ocean model	Max Planck Institute for Meteorology, Germany	Transient	1995	Houghton et al. (1996)	0.67

HADCM2 GCM (the Hadley Centre Unified Model: Mitchell et al. 1995; Houghton et al. 1996) performing best.

These results for GCM-simulated global precipitation may appear reasonable. However a good deal of caution is still needed if these are to be used for erosion modeling. Since GCMs simulate precipitation less well than temperature, climate impacts modellers frequently use seasonally averaged values of GCM-simulated precipitation change in preference to monthly values in order to smooth out any spurious peaks (Mike Hulme, personal communication 1998). When cumulative precipitation is required (e.g. in simulations of crop yield) this technique may be appropriate. It is less useful for erosion modeling however: a large increase in rainfall during a month when the soil's surface is unprotected will produce a very different response to the same increase averaged over a longer period, during which time cover will have changed.

The spatial and temporal scales of GCM output present further problems for the erosion modeller. As discussed previously, there is a good deal of uncertainty as to how climate (particularly precipitation) may change at the regional scale, i.e. for areas smaller than a GCM grid square (Wilby 1995). This uncertainty is enhanced when dealing with areas as small as a catchment or even a single field, which are the spatial scales of present-day process-based erosion models. The information which GCMs are able to provide on local changes in rainfall duration and intensity is even less certain (Kerr 1997). As a result, the strategy adopted in virtually all studies of future erosion is to hold rainfall durations and intensities unchanged, i.e. to retain present-day values. While almost certainly unrealistic (Waggoner 1990), this approach has also been adopted here.

In addition, none of the GCMs in the current version of SCENGEN consider the effects of sulphate aerosols (Hulme et al. 1997), yet these are likely to be of considerable importance in determining regional climate (Kerr 1997).

1.2.3
WEPP-CO_2

The WEPP soil erosion model is based on fundamental relationships for infiltration, surface runoff, soil consolidation and erosion mechanics (Lane et al. 1992). A version of the Green and Ampt equation, modified to consider ponding, is used for infiltration. Soil detachment, transport and deposition processes are represented using a steady-state continuity equation which considers rill and inter-rill processes, under both sediment- and transport-limited conditions (Lane et al. 1992). As a result, WEPP has the ability to estimate both erosion and deposition on complex slope forms (Flanagan and Nearing 1995). Within WEPP, soil crusting is considered by a relationship which assumes an exponential decrease of hydraulic conductivity with total rainfall since tillage (Alberts et al. 1995). WEPP's plant growth submodel is based upon that of EPIC (Arnold et al. 1995). Its water balance submodel is derived from that of SWRRB (Williams et al. 1985).

An experimental version of WEPP has been produced which incorporates the direct effects of increased atmospheric CO_2 on plant growth. The model uses relationships for the effects of CO_2 on plant growth which were originally developed by Stockle et al. (1992) for the EPIC model (Williams et al. 1990). WEPP-CO_2 is based upon version 95.001 of WEPP, and has been used in climate change impact studies by Favis-Mortlock and Savabi (1996) and Favis-Mortlock and Guerra (1999).

1.3
Modeling Future Erosion at Sorriso, Mato Grosso, Brazil

1.3.1
The Area

As long ago as 1939, Jacks and Whyte (1939, p. 59) noted that extensive deforestation was resulting in erosion problems in South America as large areas were cleared for cultivation, to provide fuel, or for grazing land. The case study focuses on the Mato Grosso area of Brazil, where there is a notable present-day erosion problem.

The rural municipality of Sorriso was founded in 1986 and covers an area of 10 048 km². It is located in Mato Grosso State (12° S, 56° W), in the Centre West Region of Brazil (Fig. 1.3). Situated on the Central Plateau, it has a mean altitude of 350 m with slope lengths ranging between 500 m and 3 000 m (Almeida and Guerra 1994). Temperature varies little throughout the year (Table 1.3) and mean annual rainfall is around 1 550 mm, with rains concentrated between November and April (Table 1.4). The soils

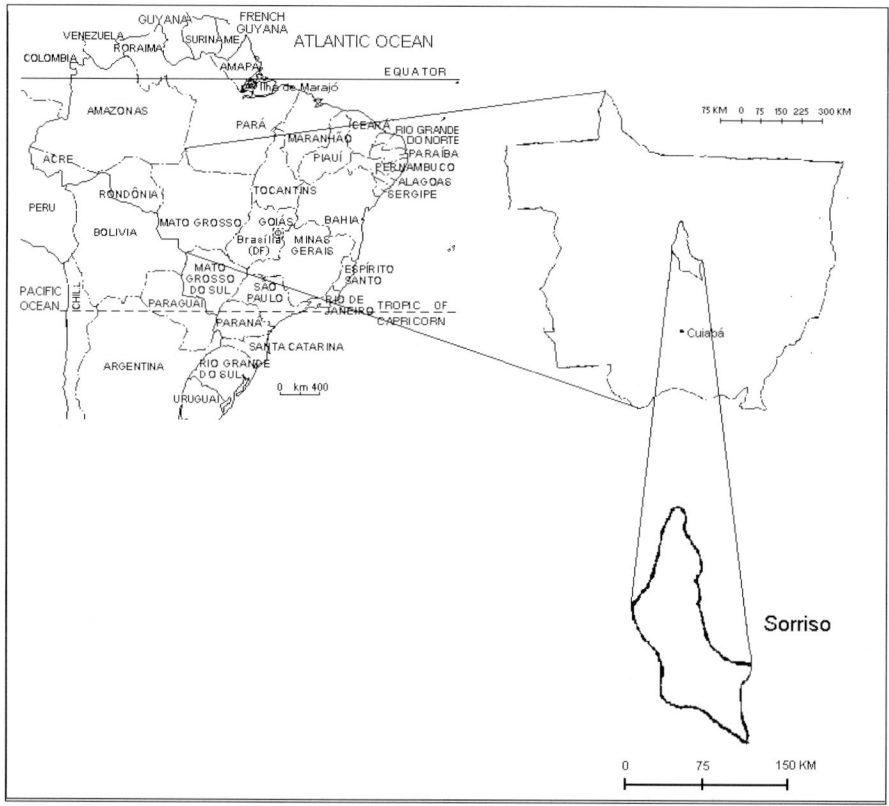

Fig. 1.3. Map of Brazil showing the study area in the Mato Grosso (from Favis-Mortlock and Guerra 1999)

Table 1.3. Mean monthly minimum and maximum temperatures (°C) for Sorriso 1977–1991 (from Favis-Mortlock and Guerra 1999)

	J	F	M	A	M	J	J	A	S	O	N	D
Min	21.0	20.8	20.9	20.6	18.8	16.1	15.1	16.2	19.0	20.3	20.6	20.8
Max	30.4	30.6	30.9	31.5	31.4	31.4	32.0	33.1	32.9	32.2	30.8	30.0

Table 1.4. Monthly rainfall (mm) for Sorriso 1977–1991. Dashes indicate a month with more than 10% of daily values missing (from Favis-Mortlock and Guerra 1999)

Year	J	F	M	A	M	J	J	A	S	O	N	D	Total
1977	–	–	–	–	–	–	–	–	–	–	273.2	324.0	
1978	312.8	268.0	200.8	155.4	68.0	52.2	7.4	0.0	86.4	200.4	92.9	275.8	1 720.1
1979	201.0	210.1	56.0	54.1	64.0	0.0	0.0	6.0	17.0	35.0	141.7	281.6	1 066.5
1980	429.6	448.6	125.2	195.3	94.0	6.3	0.0	0.0	–	118.5	187.9	266.8	
1981	–	235.8	420.5	82.3	13.0	0.0	0.0	0.0	0.0	–	–	–	
1982	–	361.2	492.1	383.7	6.0	0.0	–	0.0	52.5	77.5	–	–	
1983	–	83.5	–	–	–	–	0.6	0.0	–	87.2	41.1	185.6	
1984	42.5	198.9	76.8	93.4	25.9	0.0	0.0	16.8	9.1	96.0	155.3	108.5	823.2
1985	–	86.5	253.5	159.8	8.8	0.0	0.0	0.0	141.0	81.6	53.6	190.3	
1986	239.4	300.5	207.6	59.8	12.9	0.0	7.0	61.0	21.0	81.4	–	–	
1987	163.9	282.9	131.2	49.8	19.0	0.0	0.0	0.6	0.2	54.5	272.0	167.5	1 141.6
1988	145.4	295.9	283.1	48.3	0.0	0.2	0.0	0.0	0.0	152.5	388.9	388.6	1 702.9
1989	330.1	315.0	437.9	100.2	53.2	25.0	0.4	52.0	81.5	181.0	85.3	516.2	2 177.8
1990	325.9	407.5	172.0	145.1	70.1	0.3	2.0	8.0	75.1	133.2	86.9	329.7	1 755.8
1991	519.4	253.5	334.0	229.1	18.9	0.0	0.0	0.0	112.0	141.7	251.7	199.6	2 059.9
Mean	271.0	267.7	245.4	135.1	34.9	6.5	1.3	10.3	49.7	110.8	169.2	269.5	1 571.5

in the study area are mainly sandy, silty, sandy-loam and silty loam, with fine sand reaching 50% and silt up to 40%, while clay content shows values under 20% for most soils (Almeida 1997). Organic matter content is also low, usually under 2.5%. The soils are therefore very erodible. They result from the weathering of the Cretaceous sediments deposited by large fluvial systems, together with large lakes (Almeida and Guerra 1994). After this sedimentation, several erosive phases took place, forming a number of distinctive surfaces, some of which constitute the *chapadas*, which are part of the plateau, made of sedimentary rocks (limestones, sandstones, conglomerates, etc.).

Natural vegetation (where it still exists) is mainly *cerrado*, which is a kind of savannah. The first step in clearing the *cerrado* is to cut down the vegetation: after this the farmers burn the plant residue, which is left on the soil. This process decreases drastically the soil organic matter content and kills off much of the soil biota. This in turn causes a depletion of soil nutrients.

Land management, together with the environmental characteristics discussed above, play an important role in determining soil erosion and land degradation in this area (Guerra and Favis-Mortlock 1998). After deforestation and burning, soya bean is the main crop grown in this area. It is cultivated under a monoculture system, and is planted in October. Thus cropping patterns which leave the land largely bare during the months of October to December, at the beginning of the high-rainfall period in the southern summer (Table 1.4). Conditions then are ideal for splash erosion and the formation of crusts, resulting in soil sealing and decreased infiltration. The farmers also use very heavy machinery which compacts the soil, further decreasing infiltration. The resultant runoff causes rill and gully erosion (Fig. 1.4 and 1.5). It also transports agricultural chemicals, polluting the rivers and increasing siltation. Since the ten years or so since the municipality was founded, some soils cultivated up and down the slope have completely lost their A horizon. The final result of these processes has been the formation of more than ten gullies in the study area. They vary from a few hundred metres long and a few metres width and depth, to more than 1 500 m long, 50 m wide and 10 m deep.

Fig. 1.4. Rill on a soya bean field in the Sorriso area, showing crusting

Fig. 1.5. An actively growing ephemeral gully, again on soya beans in Sorriso

Fig. 1.6. Soya bean production in Brazil 1974–1994 (from Favis-Mortlock and Guerra 1999)

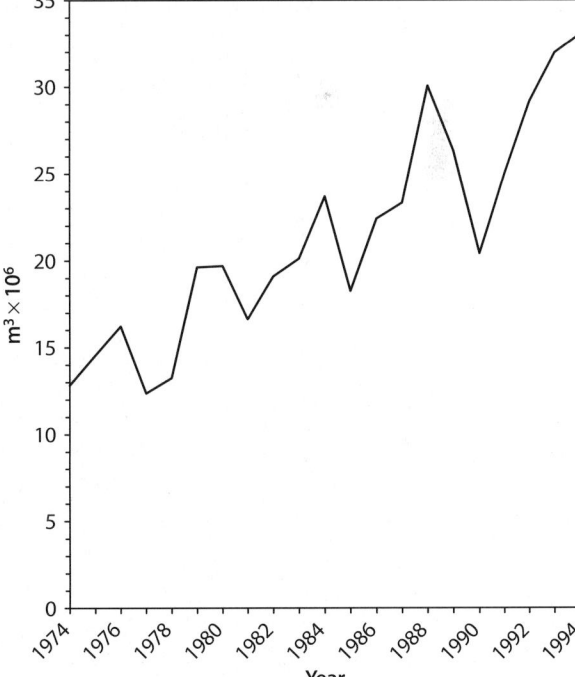

Although farmers have begun to make use of terracing and other conservation techniques, the problem still remains. The combination of clearance of large areas, heavy rainfall over a short period while crop cover is low, inherently erodible soils, and the use of heavy machinery all combine to create a severe present-day erosion problem.

Soya bean production in Brazil continues to increase (Fig. 1.6): in Sorriso *cerrado* clearing still takes place at a rate of about 3% per year (Almeida 1997). For this study, it is assumed that soya beans will still be grown in the area in the middle of the twenty-first century.

1.3.2
Modeling Current Erosion Rates

Favis-Mortlock and Guerra (1999) obtained data for a 48 ha hillslope field in the study area (Table 1.5). Field monitoring and aerial photography had been used to measure sediment yield from this field over a four year period (Almeida 1997). Using data for the field's sandy-loam latosol (Table 1.6), the standard WEPP estimation procedures (Flanagan and Nearing 1995) were used to estimate values for the model's three erodibility parameters, and for the effective hydraulic conductivity parameter (cf. Favis-Mortlock and Savabi 1996; Favis-Mortlock 1998b). Soya bean data from the WEPP databases was used, with management data as in Table 1.7. It was possible to use WEPP parameters for US agricultural implements without modification, such much of the agricultural machinery used in the Mato Grosso is American in origin (cf. Favis-Mortlock 1994).

Using statistics obtained by analysis of daily weather data for the period 1977–1991 (Tables 1.3 and 1.4), an input file was constructed for WEPP's stochastic daily weather

Table 1.5. Sorriso site data (from Almeida 1997)

Field length (m)	800
Field width (m)	600
Slope angle (%)	7
Slope morphology	concave-convex
Average annual sediment yield for 4 years (t ha^{-1})	12

Table 1.6. Soil data for the sandy-loam latosol at the Sorriso site (from Almeida 1997)

Depth to bottom of horizon (mm)	Horizon name	Sand content (%)	Clay content (%)	Organic matter content (%)	Cation exchange capacity (meq/100 g)	Rock fragments (%)
100	A1	77.1	10.2	3.27	8.2	0
350	AB	76.1	15.4	1.89	3.2	0
600	BA	70.7	22.7	1.03	3.4	0
1100	Bw1	69.2	23.5	0.68	2.3	0
1600	Bw2	70.0	22.4	0.51	2.0	0

Table 1.7. Management data for the Sorriso site (from Almeida 1997)

Date	Operation
9 Sep	Chisel plough
10 Sep	Lime via anhydrous applicator
11 Sep	Tandem disk
20 Sep	Field cultivator
7 Oct	Drill soya beans
7 Mar	Harvest

generator, CLIGEN (Nicks et al. 1995). CLIGEN was then used to generate 100 years of current-climate synthetic daily weather data (cf. Favis-Mortlock 1995; Favis-Mortlock and Boardman 1995; Favis-Mortlock and Savabi 1996). Current atmospheric CO_2 content was specified as 350 ppmv.

When run with this data. WEPP-CO_2 underestimated current-climate mean annual sediment yield. The model calculated a value of 6.86 t ha^{-1} yr^{-1}. The value measured during the four years of observation is 12 t ha^{-1} yr^{-1}. Considering that the model was not calibrated in any way, this is a satisfactory result (cf. Favis-Mortlock 1994, 1998a). The observed value is an average of measurements made over only a small number of years. In addition, difficulties in measuring depositional volumes may mean that this value is something of an overestimate.

No attempt was made to improve the fit to the measured value by calibration. Calibration is undesirable in global change studies, which inevitably involve some extrapolation from present conditions (see previous discussion). In addition, note that both this study and that of Favis-Mortlock and Guerra (1999) aim to produce a relative, rather than absolute, result (cf. Barfield et al. 1991 for example).

1.3.3
Modeling Future Erosion Rates

However, calibration was unavoidable in the next step carried out by Favis-Mortlock and Guerra (1999). The crop growth submodel of WEPP-CO_2 was calibrated against measured data (Kimball 1983; Baker and Allen 1993) for change in soya bean yield resulting from increased atmospheric CO_2 content ranging from 350 to 660 ppmv (Stockle et al. 1992; Kimball et al. 1993). The procedure followed was similar to that performed previously for wheat by Favis-Mortlock and Savabi (1996) and for maize by Boardman and Favis-Mortlock (in preparation a).

Before constructing the climate change scenarios, some analysis was made of the reliability of the GCM forecasts. It is of course not possible to validate estimates of future climate by comparison with observed data. However, one measure of confidence is the extent to which different GCMs agree. In Fig. 1.7 and 1.8, predictions of change in monthly temperature and rainfall for the grid square enclosing Sorriso by about 2050 are shown. These are from three GCMs (Table 1.2), for each of the two emissions scenarios, and were generated by SCENGEN assuming a mid-range climate sensitivity. Changes are with respect to climate during 1961–90.

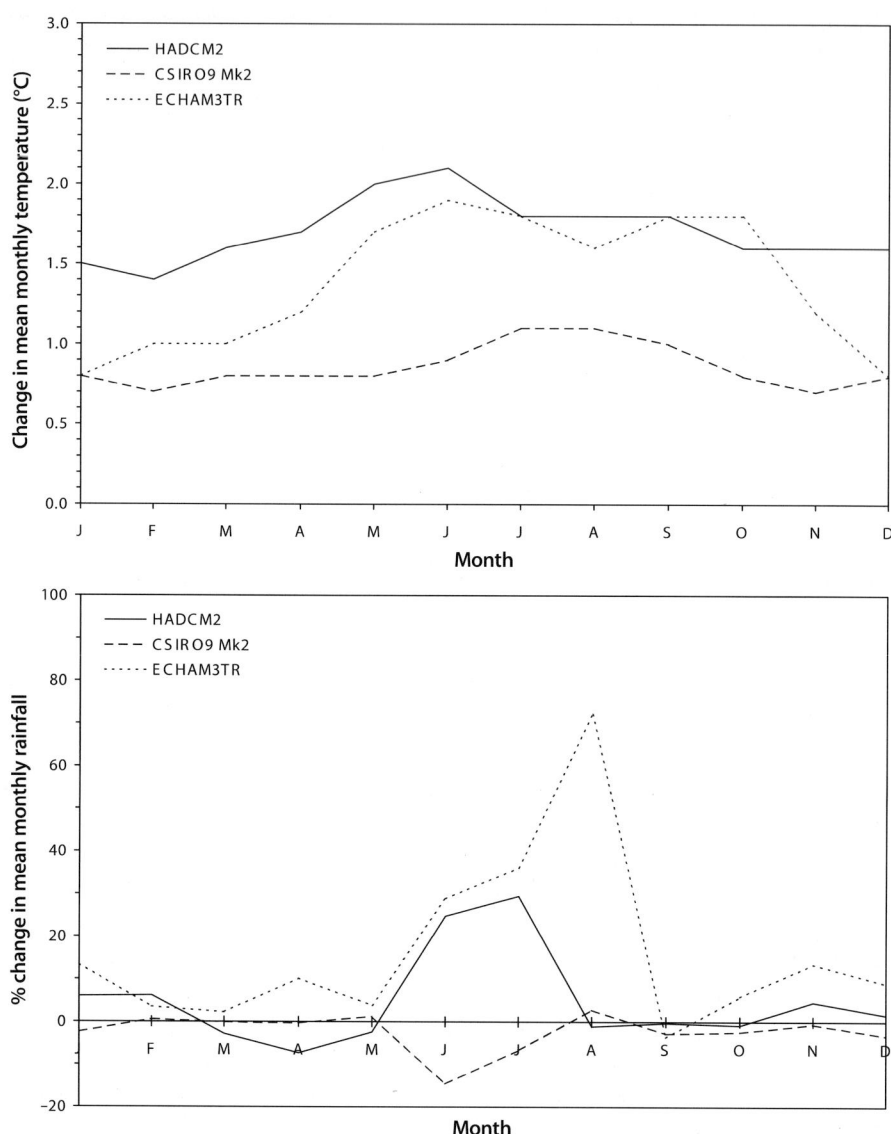

Fig. 1.7. Change in monthly temperature (*upper*) and monthly rainfall (*lower*) for the Mato Grosso area of Brazil in 2050, simulated by three GCMs (see Table 1.2 for details). Emissions scenario IS92d ('mitigation') is assumed

Greater agreement is shown between the GCMs for temperature change than rainfall change. The German ECHAM3TR (Houghton et al. 1996) is generally the 'wettest' and the Australian CSIRO9 Mk2 (Dix and Hunt 1995) the 'driest'; HADCM2 is intermediate. The temporal pattern of rainfall change however differs noticeably between the three. HADCM2 and ECHAM3TR show large percentage increases in the dry

Chapter 1 · The Influence of Global Greenhouse-Gas Emissions

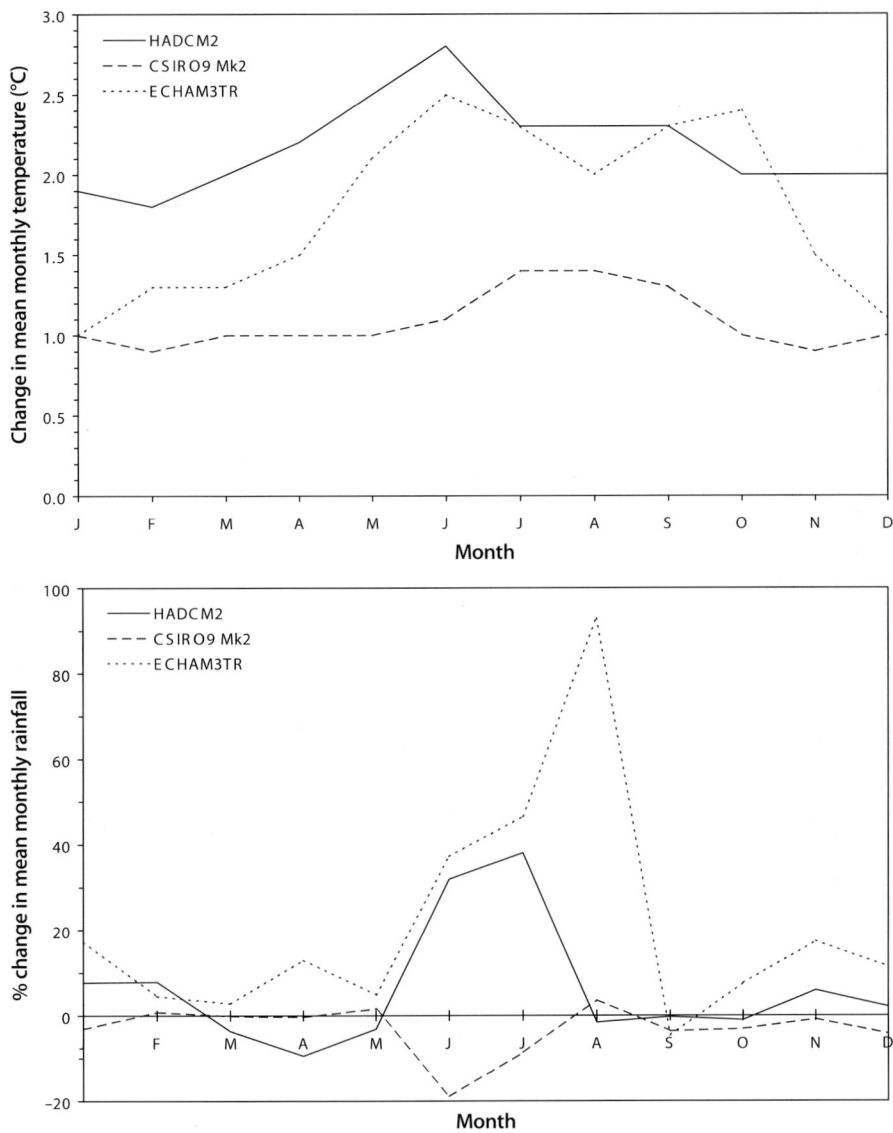

Fig. 1.8. As Fig. 1.7, but assuming emissions scenario IS92a ('business as usual') (from Favis-Mortlock and Guerra 1999)

months of June to August (cf. Table 1.4), but CSIRO9 Mk2 shows a decrease in June and July. Given the low rainfalls in these months, the differences will have little importance for erosion. More importantly, ECHAM3TR and HADCM2 show increases during all or most of the wetter summer months, while CSIRO9 Mk2 does not. Caution is clearly needed before accepting any GCM's predictions of change in future rainfall. The

Table 1.8 a. Mean annual sediment yield (t ha^{-1}) at Sorriso simulated by WEPP-CO$_2$, for 1990 and 2050 under emissions scenario IS92d. All values are 100-year averages

Emissions scenario	–	IS92d					
GCM	–	HADCM2		CSIRO9 Mk2		ECHAM3TR	
Climate scenario	1990	2050	% change	2050	% change	2050	% change
J	0.71	1.02	+45	0.62	−13	1.18	+66
F	0.49	0.78	+60	0.45	−8	0.66	+36
M	0.76	0.88	+17	0.67	−11	0.94	+25
A	0.35	0.33	−5	0.30	−14	0.51	+46
M	0.06	0.06	+3	0.06	−9	0.08	+21
J	0.00	0.00	0	0.00	0	0.00	0
J	0.00	0.00	0	0.00	0	0.00	0
A	0.00	0.00	0	0.00	0	0.00	0
S	0.08	0.08	0	0.07	−12	0.08	−11
O	0.68	0.66	−4	0.62	−9	0.80	+18
N	1.15	1.26	+9	1.06	−8	1.60	+39
D	1.92	2.04	+6	1.70	−12	2.51	+30
Annual total	6.21	7.12	+15	5.54	−11	8.36	+35

Table 1.8 b. Mean annual sediment yield (t ha^{-1}) at Sorriso simulated by WEPP-CO$_2$, for 1990 and 2050 under emissions scenario IS92a. All values are 100-year averages

Emissions scenario	–	IS92a					
GCM	–	HADCM2		CSIRO9 Mk2		ECHAM3TR	
Climate scenario	1990	2050	% change	2050	% change	2050	% change
J	0.71	1.23	+74	0.64	−10	1.48	+109
F	0.49	0.99	+103	0.49	0	0.83	+71
M	0.76	1.03	+37	0.72	−4	1.15	+52
A	0.35	0.36	+3	0.33	−7	0.66	+87
M	0.06	0.07	+16	0.06	−3	0.09	+46
J	0.00	0.00	0	0.00	0	0.00	0
J	0.00	0.00	0	0.00	0	0.00	0
A	0.00	0.00	0	0.00	0	0.00	0
S	0.08	0.09	+1	0.07	−14	0.07	−13
O	0.68	0.65	−4	0.60	−12	0.84	+24
N	1.15	1.33	+15	1.05	−9	1.78	+55
D	1.92	2.11	+10	1.66	−1	2.74	+42
Annual total	6.21	7.87	+27	5.62	−9	9.65	+55

100 year time series of current-climate daily weather which had been generated by CLIGEN was then perturbed (cf. Boardman and Favis-Mortlock 1995; Favis-Mortlock and Savabi 1996; Favis-Mortlock and Guerra 1999) by applying the GCM-generated values for change in monthly precipitation and mean temperature in Fig. 1.7 and 1.8. Changes in mean temperature were applied equally to T_{max} and T_{min}.

All other climate parameters (e.g. daily rainfall time-to-peak and peak intensity, and solar radiation) were held unchanged from present conditions. For simplicity, an atmospheric CO_2 content of 488 ppm was assumed for 2050 for both the IS92a and IS92d runs.

WEPP-CO_2's long-term average results when run with this data are shown in Table 1.8. Under the 'business as usual' scenario IS92a, values for the change in average annual sediment yield range between +55% (ECHAM3TR) and –9% (CSIRO9 Mk2), with HADCM2 predicting an intermediate increase of 27%. Monthly increases of over 100% are predicted in January and February by ECHAM3TR and HADCM2 respectively, with smaller increases are predicted by these two GCMs for most other months in the wet period. Under the 'mitigation' scenario IS92d, patterns for annual and monthly sediment yield are generally similar to their IS92a counterparts, but of lower magnitude.

Shifts in the distributions underlying these average values are illustrated in Fig. 1.9 and 1.10. These plots aim to highlight changes in present and future erosion risk by showing the probability of any given value for annual sediment yield being equalled or exceeded. The probability of a particular value of annual sediment yield (e.g. 5 t ha^{-1} or 10 t ha^{-1}) being equalled or exceeded can therefore be ascertained: results are given in Table 1.9. For example, under present conditions an annual sedi-

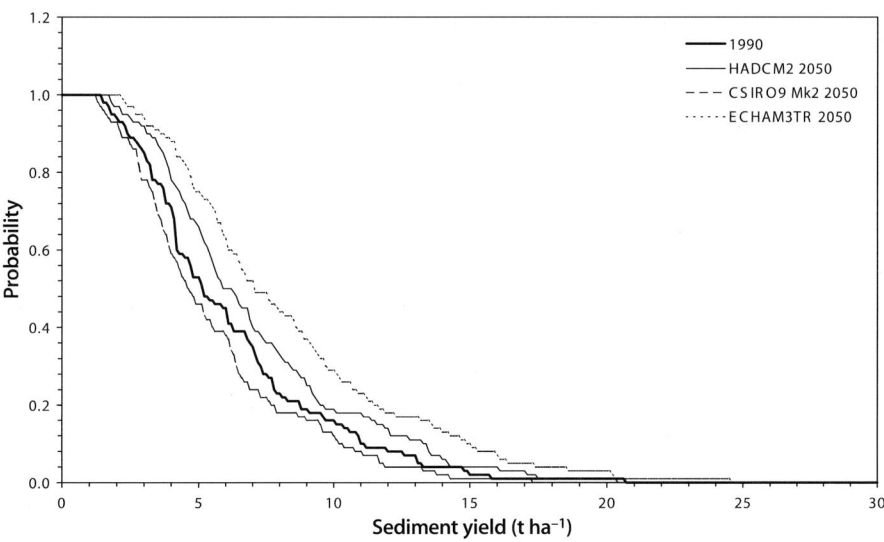

Fig. 1.9. Exceedance probabilities of present and future annual sediment yields estimated by WEPP. The future simulations assume emissions scenario IS92d ('mitigation')

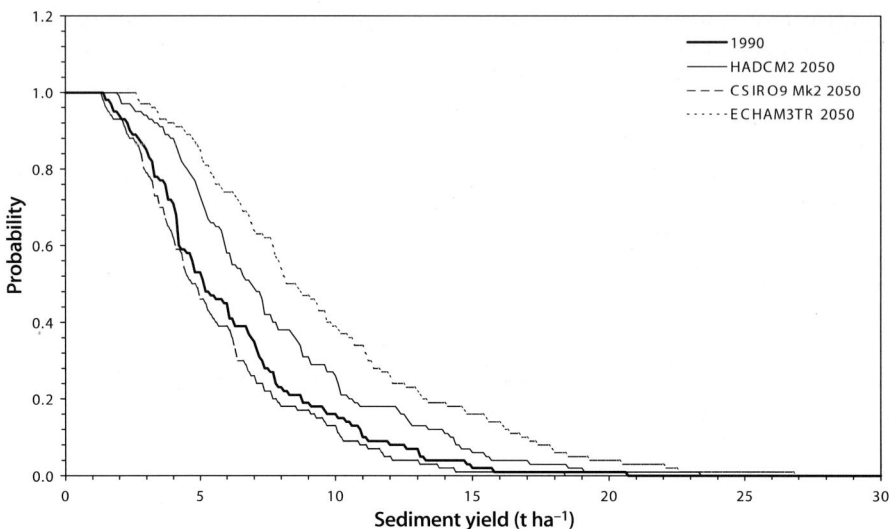

Fig. 1.10. As Fig. 1.9, but assuming emissions scenario IS92a ('business as usual') for the future simulations (from Favis-Mortlock and Guerra 1999)

Table 1.9. Probability (%) of a given annual sediment yield being equalled or exceeded in the WEPP-CO_2 simulations

Probability (%)	Annual sediment yield (t ha^{-1})						
Emissions scen.	–	IS92d			IS92a		
GCM/climate	Present	HADCM2	CSIRO9 Mk2	ECHAM3TR	HADCM2	CSIRO9 Mk2	ECHAM3TR
5	51	66	46	74	71	48	82
10	17	19	14	29	26	14	39
15	2	4	1	10	6	1	17

Table 1.10. Mean change in simulated annual rainfall and sediment yield (%) at Sorriso for 'wet' and 'dry' years. See text for explanation

Emissions scenario	IS92d						IS92a					
GCM	HADCM2		CSIRO9 Mk2		ECHAM3TR		HADCM2		CSIRO9 Mk2		ECHAM3TR	
Year type	wet	dry	wet	dry	wet	dry	wet	dry	wet	dry	wet	dry
Rainfall	1.9	1.7	−1.4	−1.1	8.6	7.4	2.4	2.1	−1.8	−1.5	11.0	9.5
Sediment yield	18.3	12.0	−14.4	−7.8	43.2	28.1	33.4	21.6	−12.5	−7.0	68.7	45.0

ment yield of 5 t ha^{-1} is exceeded in about 51% of the 100 years simulated. Under the IS92a scenario and using the HADCM2 climate data, the same yield will in future be exceeded in about 71% of years. The probability of large events (i.e. events in the upper 'tail' of the distribution) is a good deal higher for ECHAM3TR than for HADCM2 under both emissions scenarios. For both GCMs, future probabilities are higher than those for present conditions. In general, it appears that the more extreme the event, the more its probability increases under future conditions (cf. Katz and Brown 1992). An annual sediment yield of 15 t ha^{-1} occurs in only 2% of years under present conditions, but (for HADCM2) this probability doubles under IS92d, and increases threefold under IS92a. The increase in risk is even more marked for the ECHAM3TR climate data.

Table 1.10 shows the percentage changes in simulated rainfall and sediment yield between present and future conditions for those years of each simulation when present-day rainfall was above or below the average for the whole simulation. These values vary little for rainfall. However, the change in sediment yield in 'wet' years is consistently greater than for 'dry' years for both emissions scenarios and for all GCMs. This non-linear response of the erosional system to climate change was noted by Favis-Mortlock and Boardman (1995). This non-linearity can also be seen in Fig. 1.13 and 1.14. These compare rainfall and sediment yield for individual years of the 100-year simulations. In general, future sediment yield can be seen to increase most in the wetter years of the simulation time series.

Figures 1.11 and 1.12 illustrate changes in the downslope patterns of average annual erosion and deposition for the two scenarios and three GCMs. In all cases, maximum erosion occurs a short distance downslope of the half-way point, i.e. approximately at

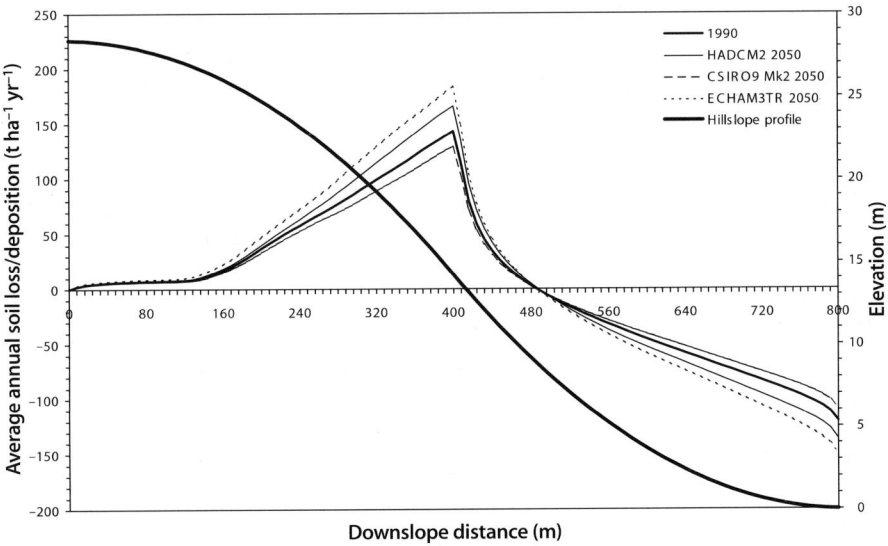

Fig. 1.11. Changes in the downslope pattern of average annual erosion (positive values) and deposition (negative values) predicted by WEPP. Vertical exaggeration of the hillslope profile is approximately 10 : 1. Emissions scenario IS92d ('mitigation') is assumed for the future simulations

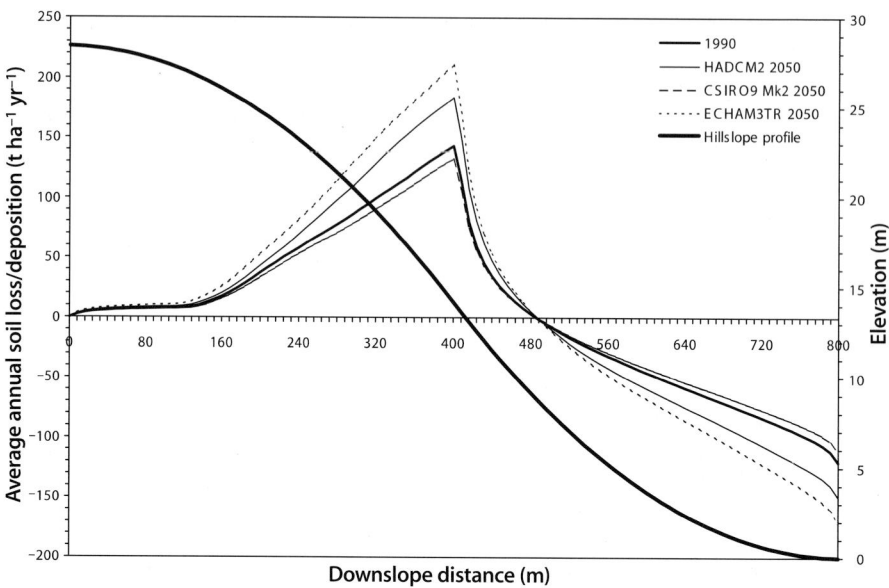

Fig. 1.12. As Fig. 1.11, but assuming emissions scenario IS92a ('business as usual') for the future simulations (from Favis-Mortlock and Guerra 1999)

the slope's point of inflexion. Soil loss values here are around 150 t ha^{-1} yr^{-1} for the present. Under IS92a's 'business as usual' scenario this rises to about 180 t ha^{-1} yr^{-1} (HADCM2) and about 210 t ha^{-1} yr^{-1} (ECHAM3TR). For the IS92d 'mitigation' scenario, the HADCM2 data gives about 170 t ha^{-1} yr^{-1} and ECHAM3TR about 190 t ha^{-1} yr^{-1}. Maximum deposition is at the foot of the slope. This is about 120 t ha^{-1} yr^{-1} at present, and is estimated to rise to about 150 t ha^{-1} yr^{-1} (HADCM2) and 170 t ha^{-1} yr^{-1} (ECHAM3TR). For IS92d, the values are about 140 t ha^{-1} yr^{-1} and 150 t ha^{-1} yr^{-1}. Thus as noted by Favis-Mortlock and Savabi (1996), the greatest changes in erosion or deposition are predicted to occur on those parts of the hillslope at which erosion or deposition rates are currently greatest.

1.4
Discussion

1.4.1
Implications

Irrespective of emissions scenario, erosion is forecast to increase at the Sorriso site according to the HADCM2 and ECHAM3TR GCMs. The increase in erosion rates is considerable for the latter. Increases are greater for the 'business as usual' IS92a emissions scenario. The CSIRO9 Mk2 GCM forecasts decreases in erosion, again irrespective of emissions scenario; although the decrease is less marked for the IS92a scenario.

Table 1.11. The influence of emissions policies on simulated mean annual sediment yield at Sorriso

GCM	Mean annual sediment yield in 2050 assuming IS92d (t ha^{-1})	Mean annual sediment yield in 2050 assuming IS92a (t ha^{-1})	Difference (IS92a – IS92d) as % of 1990 sediment yield
HADCM2	7.12	7.87	12
CSIRO9 Mk2	5.54	5.62	1
ECHAM3TR	8.36	9.65	21

This range of uncertainty, between increase and decrease, reflects the current range of uncertainty in GCM simulations of future precipitation.

Table 1.11 illustrates the influence of emissions scenario on future mean annual sediment yield at Sorriso. Expressed as a percentage of the simulated present-day rate, use of the HADCM2 climate estimates suggest that future erosion will be about 12% higher if the 'business as usual' IS92a emissions scenario is adopted, compared with the 'mitigation' IS92d scenario. The ECHAM3TR GCM climate estimates suggest an even bigger disparity. CSIRO9 Mk2 data suggests only a small (probably insignificant) difference in rates between the two emissions scenarios.

Which GCM's forecasts are most likely? Mitchell et al. (1995) suggest that HADCM2's temperature sensitivity to greenhouse-gas forcing is comparable to that of the real climate system. In addition, HADCM2 has a correlation between present-day precipitation patterns and observed data which is the highest of the fourteen GCMs in SCENGEN 2.1a (cf. Table 1.2). Notwithstanding the many uncertainties involved, HADCM2 may thus be assumed to represent a 'best-guess' future climate. ECHAM3TR and CSIRO9 Mk2 (fortuitously) bracket this.

Based upon this 'best-guess' estimate of future climate, it appears that the currently severe on-site and off-site impacts of erosion in this area will worsen further during the next century. Note that since the hillslope version of WEPP was used here, the predicted increase in erosion is, strictly speaking, likely to apply only to other hillslope fields in the Sorriso area. It is probable, however, that gully erosion in the area will also increase. The increase in erosion risk is not equally spread in time: bigger increases are forecast in wetter years. Those areas of the hillslope profile which currently suffer the greatest erosion or deposition will see the greatest changes in rates of erosion or deposition.

However, the results of global policy decisions regarding greenhouse gas emissions will affect the magnitude of increase in erosion at this site. Pursuance of a 'business as usual' strategy for emissions will see erosion rates around the middle of the twenty-first century which are around 12% higher than erosion rates resulting from a 'mitigation' strategy. The risk of extreme erosion events also increases markedly under the 'business as usual' strategy.

Any increase in future erosion would have serious consequences. Soya beans are an important product and a valuable export in Sorriso, as in many other areas of the Mato Grosso. Decrease in agricultural productivity resulting from increased rates

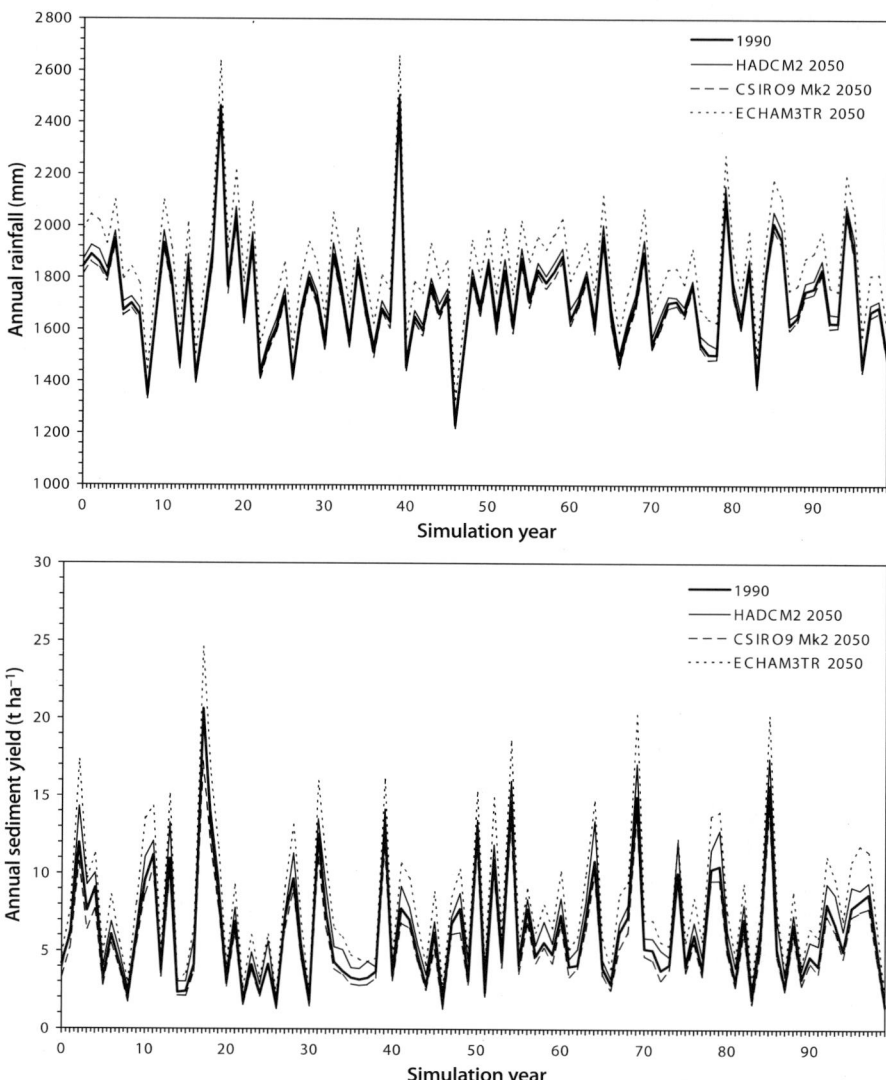

Fig. 1.13. Rainfall (*upper*) and sediment yield (*lower*) for each of the 100 years of the simulations, assuming emissions scenario IS92d ('mitigation')

of soil erosion would thus have serious consequences for the Brazilian economy (Guerra and Favis-Mortlock 1998; Favis-Mortlock and Guerra 1999). Impacts upon the environment would not be confined to the areas that lose their soil. Increased siltation in downstream rivers and lakes, decreased water quality and a probable decrease in biodiversity are other likely consequences (Guerra and Favis-Mortlock 1998; Favis-Mortlock and Guerra 1999). Both economic and environmental impacts would be lessened if global greenhouse-gas emissions policies adopt a 'mitigation' approach.

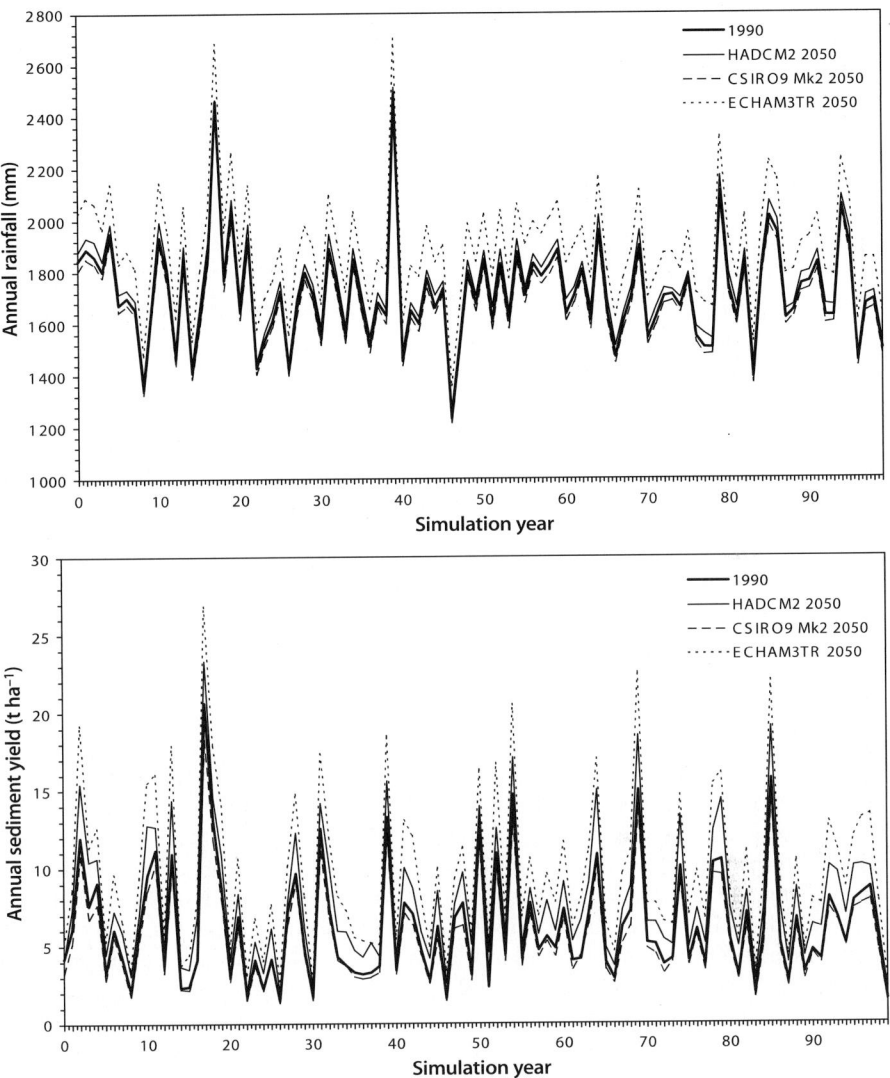

Fig. 1.14. As Fig. 1.13, but assuming emissions scenario IS92a ('business as usual') (from Favis-Mortlock and Guerra 1999)

1.4.2
Uncertainties

There must inevitably be very many uncertainties in any study of future erosion. Following Favis-Mortlock and Guerra (1999), these may be categorised (Table 1.12) as uncertainties resulting from the assumptions made regarding future conditions and those relating to the shortcomings of the erosion model used.

Table 1.12. Some sources of uncertainty for the results from this study with their direction of influence (see text for details, modified from Favis-Mortlock and Guerra 1999)

Source of uncertainty	Origin	Probable direction of influence on erosion rates
WEPP overpredicts on long slopes	model	+ve
WEPP-CO_2's oversimplified handling of plant responses to enhanced CO_2	model	?
CLIGEN underpredicts large rainfall events	model	–ve
Possibly unrealistic future emissions scenarios	assumption	?
Possibly unrealistic future climate/land use scenarios	assumption	?
Rainfall intensities unchanged from present values	assumption	?
Crop management unchanged from present	assumption	+ve
'1990' soil profile used with 2050 climate	assumption	?

1.4.2.1
Problems due to Assumptions

Considerable uncertainty underlies all studies of future erosion as a result of the assumptions made regarding future climate and land use. The need for caution in accepting forecasts of future climate by the three GCMs used here (or any other GCM) has been thoroughly discussed (e.g. Kerr 1997). In addition, there is an extra layer of uncertainty which results from the information on future climate which cannot be supplied by such models. As discussed previously, estimates of changes in precipitation intensity are not available from GCMs: as a result, most impact studies do not change intensity values from present conditions (but see Lee et al. 1996; Boardman and Favis-Mortlock, in preparation a). This is almost certainly unrealistic (Waggoner 1990).

As discussed previously, future land use is a still greater difficulty. It is probably reasonable to assume that soya bean will continue to be grown in Sorriso. Farmers however adapt their management to changes of circumstances, including climatic trends (cf. Easterling et al. 1992a, b; Nicholls 1997). The 'no change' assumption here with regard to management is therefore unrealistic. An example: in the HADCM2 climate there is an increase in temperature and a decrease in rainfall for March to May. These combine to lower soil moisture for these months, so that the WEPP-simulated growing crop suffers increased water stress (Arnold et al. 1995). It therefore grows more slowly, resulting in decreased ground cover and lower yields. For this reason, WEPP predicts increases in erosion during these months (Table 1.8) even though future rainfall decreases then (Fig. 1.7 and 1.8). Adaptive management, such as a change in cultivar or planting date (or an increased use of irrigation), would aim to maximise yields by preventing an increase in water stress during these months.

A final point is that the simulations carried out in this study are equilibrium (i.e. with mean values of climatic parameters held constant) rather than transient (with climatic parameters having time-varying means; Favis-Mortlock and Boardman 1995).

An equilibrium approach was necessary to calculate the exceedance probabilities of Fig. 1.9 and 1.10. It is however unrealistic, because a 1990 soil profile is being used with a 2050 climate. Soils in the Sorriso area will have thinned still further as a result of erosion during the next 50 or so years.

1.4.2.2
Problems due to Model Shortcomings

There should be a decreased danger of invalid extrapolation of the relationships on which the model is based when a physically-based model is used, compared with more empirically-based models (e.g. those incorporating elements from the USLE; Favis-Mortlock et al. 1996; Favis-Mortlock and Savabi 1996). If the model produces a reasonable present-day result without calibration – as it did in this study – then confidence in the model may be further enhanced. However, there remains the possibility that the model may have produced 'the right answer for the wrong reason' (Favis-Mortlock 1994). This could occur if, for example, two or more model shortcomings cancel each other out by biasing results in opposite directions.

A number of authors have discussed the more general shortcomings of WEPP (e.g. Tiscareno-Lopez et al. 1995; Huang et al. 1996; Zhang et al. 1996; Favis-Mortlock 1998a, 1998b). One recently-discovered problem which is particularly relevant to this study is that WEPP tends to overpredict erosion on slope lengths longer than about 50 to 100 m (Mark Nearing, personal communication 1997). This may well have happened for the 800 m slope at Sorriso.

Uncertainty due to WEPP-CO_2's handling of plant responses to enhanced CO_2 is less clear-cut. Plants will probably respond in a more complex way than is modelled by WEPP-CO_2 due to interactions between temperature and enhanced CO_2 which are not yet fully understood, and to the differences between responses to increased CO_2 in the field to those noted in pot experiments (Kimball et al. 1993). Changes in the rate with which the growing crop covers the ground are of particular importance for erosion modeling (Favis-Mortlock and Savabi 1996).

CLIGEN (Nicks et al. 1995) introduces a final modeling problem. While CLIGEN has proved better at reproducing extreme values of present-day daily rainfall for the UK South Downs than EPIC's rainfall generator (Favis-Mortlock 1995, 1998b). However, it appears to be rather less good at reproducing daily rainfall distributions for the present-day tropical climate of the Mato Grosso. Means and standard deviations were well simulated, as were lengths of wet and dry spells, but CLIGEN under-represented the highest values of daily rainfall. This is likely to have resulted in some under-prediction of erosion rates (cf. Favis-Mortlock 1995).

1.5
Conclusions

This study has indicated that future erosion rates may well rise at the Mato Grosso study site, although there is a considerable band of uncertainty. Global greenhouse-gas emissions policies will affect the size of this rise: adoption of a 'mitigation' strategy compared with a 'business as usual' strategy will reduce the magnitude of the increase in erosion rates by a noticeable amount, somewhere in the region of 12% of the current rates. Since any increase in an already-severe erosion problem is likely to have significant economic and environmental effects for the area, there is a clear advantage to be gained for this site at least from the implementation of global strategies for the mitigation of greenhouse-gas emissions.

Finally, this study has illustrated the intimate links between the 'global' and the 'local'. Environmental problems are highly interconnected (e.g. Goudie 1990; Middleton 1995). This study demonstrates that such interconnection links the policy decisions of governments world-wide and future rates of erosion in Brazil. Global environmental problems are indeed something that humanity must learn to solve; but as well as teaching us caution and respect for the balance of natural systems, we may also learn from them something of the deeper truth of the saying that 'no man is an island'.

Acknowledgements

For the weather data used in this and in the earlier paper, we are most grateful to the late Arlin D. Nicks (USDA). We wish also to thank Flavio G. Almeida (GEODINAMICO/Department of Geography – UFF) and Anderson S. Lamin (LAGESOLOS/Department of Geography – UFRJ) for supplying data for the Sorriso site, Antonio S. Silva and Rosangela G. Botelho LAGESOLOS/Department of Geography – UFRJ) for drawing Fig. 1.3, Flavio G. Almeida for Fig. 1.4 and 1.5, Ruth Butterfield (ECU) for information on soya beans, Mike Hulme (Climatic Research Unit) for advice on MAGICC and SCENGEN, Mark Nearing (USDA) for assistance with WEPP, Reza Savabi (USDA) for supplying the CO_2-sensitive version of WEPP, and John Boardman (ECU) for comments on a draft of this paper. The second author acknowledges the receipt of a Research Grant from the Brazilian Research Council (CNPq – Conselho Nacional de Desenvolvvimento Cientifico e Tecnologico). Finally, many thanks to an anonymous referee for some insightful comments.

This paper is a contribution to the Soil Erosion Network of GCTE, which is a Core Research Project of the International Geosphere-Biosphere Programme.

Acronyms Used

- *CLIGEN*   Climate Generator
- *EPIC*   Erosion-Productivity Impact Calculator
- *GCM*   General Circulation Model
- *MAGICC*   Model for the Assessment Of Greenhouse-Gas Induced Climate Change
- *SCENGEN*   Scenario Generator
- *SWRRB*   Simulator for Water Resources in Rural Basins
- *WEPP*   Water Erosion Prediction Project

References

Alberts EE, Nearing MA, Weltz MA, Risse LM, Pierson FB, Xhang XC, Laflen JM, Simanton JR (1995) Soil component. In: Flanagan DC and Nearing MA (eds) USDA-Water Erosion Prediction Project: Hillslope profile and watershed model documentation. NSERL Report 10:7.1–7.20, USDA-ARS, West Lafayette, Indiana, USA

Almeida FG (1997) A estrutura fundiária como uma variável a ser considerada no processo de erosão dos solos - Sorriso - MT. Unpubl PhD thes, Fed Univ Rio de Janeiro, Brazil, 222 pp

Almeida FG, Guerra AJT (1994) Erosão dos solos e impacto ambiental na microbacia do rio Lira – Sorriso – MT. Proceedings 1st Brazilian Symposium Environmental Sci, Rio de Janeiro, Brazil, 1010–1021

Arnell N (1996) Global Warming, River flows and water resources. Wiley, Chichester, UK, 224 pp

Arnold JR, Weltz MA, Alberts EE, Flanagan DC (1995) Plant growth component. In: Flanagan DC and Nearing MA (eds) USDA-Water Erosion Prediction Project: Hillslope profile and watershed model documentation. NSERL Report 10:8.1–8.41, USDA-ARS, West Lafayette, Indiana, USA

Baker JT, Allen Jr LH (1993) Contrasting crop species responses to CO_2 and temperature: rice, soybean and citrus. Vegetatio 104/105:239–260

Barfield BJ, Haan CT, Storm DE (1991) Why model? In: Beasley DB, Knisel WG, Rice AP (eds) Proceedings of the CREAMS/GLEAMS Symposium. Agricultural Engineering Dept, University of Georgia, Athens, Georgia, USA, 3–8

Boardman J, Favis-Mortlock DT (1993) Climate change and soil erosion in Britain. Geographical Journal 159(2):179–183

Boardman J, Favis-Mortlock DT (in prep a) Climate change versus land use change: implications for UK soil erosion. Proceedings of BGRG Annual Conference 'Geomorphology and Climate Change', Dundee, September 1997

Boardman J, Favis-Mortlock DT (eds) (in prep b) Climate change and soil erosion. Imperial College Press, London, UK

Boardman J, Evans R, Favis-Mortlock DT, Harris TM (1990) Climate change and soil erosion on agricultural land in England and Wales. Land Degradation and Rehabilitation 2(2):95–106

Botterweg P (1994) Modeling the effects of climate change on runoff and erosion in central southern Norway. In: Rickson RJ (ed) Conserving soil resources – European perspectives. CAB International, Wallingford, UK, 273–285

Climatic Research Unit (1997) Documentation accompanying MAGICC 2.3, Climatic Research Unit, University of East Anglia, Norwich, UK

De Boer D (1997) Changing contributions of suspended sediment sources in small basins resulting from European settlement on the Canadian prairies. Earth Surface Processes and Landforms 22(7):623–639

Dix MR, Hunt BG (1995) Climatic Modeling – Doubling of CO_2 levels and beyond. Final report to the Federal Department of the Environment, Sport and Territories, CSIRO Division of Atmospheric Research, Melbourne, Australia, 28 pp

Easterling III WE, McKenney MS, Rosenberg NJ, Lemon KM (1992a) Simulations of crop response to climate change: effects with present technology and no adjustments (the 'dumb farmer' scenario). Agricultural and Forest Meteorology 59:53–73

Easterling III WE, Rosenberg NJ, Lemon KM, McKenney MS (1992b) Simulations of crop responses to climate change: effects with present technology and currently available adjustments (the 'smart farmer' scenario). Agricultural and Forest Meteorology 59:75–102

Evans R (1996) Soil erosion and its impacts in England and Wales. Friends of the Earth, London, UK, 121 pp

Fankhauser S, Tol RSJ (1996) Climate change costs: recent advancements in the economic assessment. Energy Policy 24(7):665–673

Favis-Mortlock DT (1994) Use and abuse of soil erosion models in Southern England. Unpublished PhD Thesis, University of Brighton, UK, 310 pp

Favis-Mortlock DT (1995) The use of synthetic weather for soil erosion modeling. In: McGregor DFM, Thompson DA (eds) Geomorphology and land management in a changing environment. Wiley, Chichester, UK, 265–282

Favis-Mortlock DT (1998a) Validation of field-scale soil erosion models using common datasets. In: Boardman J, Favis-Mortlock DT (eds) Modeling soil erosion by water. Springer-Verlag NATO-ASI Global Change Series I-55, Heidelberg, Germany, 89–128

Favis-Mortlock DT (1998b) Evaluation of field-scale erosion models on the UK South Downs. In: Boardman J, Favis-Mortlock DT (eds) Modeling soil erosion by water. Springer-Verlag NATO-ASI Global Change Series I-55, Heidelberg, Germany, 141–158

Favis-Mortlock DT, Boardman J (1995) Nonlinear responses of soil erosion to climate change: a modeling study on the UK South Downs. Catena 25(1–4):365–387

Favis-Mortlock DT, Guerra AJT (1999) The implications of GCM estimates of rainfall for future erosion risk: a case study from Brazil. Catena 37(3/4):329–354

Favis-Mortlock DT, Savabi MR (1996) Shifts in rates and spatial distributions of soil erosion and deposition under climate change. In: Anderson MG, Brooks SM (eds) Advances in Hillslope Processes. Wiley, Chichester, UK, 529–560

Favis-Mortlock DT, Evans R, Boardman J, Harris TM (1991) Climate change, winter wheat yield and soil erosion on the English South Downs. Agricultural Systems 37(4):415–433

Favis-Mortlock DT, Quinton JN, Dickinson WT (1996) The GCTE validation of soil erosion models for global change studies. Journal of Soil and Water Conservation 51(5):397–403

Favis-Mortlock DT, Boardman J, Bell M (1997) Modeling long-term anthropogenic erosion of a loess cover: South Downs, UK. The Holocene 7(1):79–89

Flanagan DC, Nearing MA (1995) USDA-Water Erosion Prediction Project: Hillslope profile and watershed model documentation. NSERL Report No. 10, USDA-ARS National Soil Erosion Research Laboratory, West Lafayette, Indiana, USA

Goudie AS (1990) The human impact on the natural environment. 3rd edn., Blackwell, Oxford, UK, 388 pp

Guerra AJT, Favis-Mortlock DT (1998) Land degradation in Brazil. Geogr Rev 12(2):18–23

Henderson-Sellers A (1994) Numerical modeling of global climates. In: Roberts N (ed) The Changing Global Environment. Blackwell, Oxford, UK, 99–124

Hossell JE, Jones PJ, Rehman T, Tranter RB, Marsh JS, Parry ML, Bunce RGH (1995) Potential effects of climate change on agricultural land use in England and Wales. Environmental Change Unit Research Report 8, Oxford, UK, 82 pp

Houghton JT, Jenkins GJ, Ephraums JJ (eds) (1990) Climate change: the IPCC Scientific Assessment. Cambridge University Press, Cambridge, UK

Houghton JT, Meira Filho LG, Callendar BA, Harris N, Kattenberg A, Maskell K (eds) (1996) Climate change 1995: the science of climate change. Contribution of working group I to the Second Assessment Report of the Intergovernmental Panel on Climate Change. Cambridge University Press, Cambridge, UK, 584 pp

Huang C, Bradford JM, Laflen JM (1996) Evaluation of the detachment-transport coupling concept in the WEPP rill erosion component. Soil Sci Soc Am J 60(3):734–739

Hulme M, Raper SCB, Wigley TML (1995a) An integrated framework to address climate change (ESCAPE) and further developments of the global and regional climate models (MAGICC). Energy Policy 23:347–355

Hulme M, Jiang T, Wigley T (1995b) SCENGEN, a climate change scenario generator: User manual. Climatic Research Unit, University of East Anglia, Norwich, UK, 38 pp

Hulme M, Wigley TML, Brown O (1997) Documentation accompanying SCENGEN 2.1a. Climatic Research Unit, University of East Anglia, Norwich, UK

Intergovernmental Panel on Climate Change (1991) Climate change: the IPCC response strategies. Island Press, Washington DC, USA

Jacks GV, Whyte RO (1939) The rape of the earth: a world survey of soil erosion. Faber and Faber Ltd, London, UK, 313 pp

Kallio K, Rekolainen S, Ekholm P, Granlud K, Laine Y, Johnsson H, Hoffman M (1997) Impacts of climate change on agricultural nutrient losses in Finland. Boreal Environ Res 2:33–52

Katz RW, Brown BG (1992) Extreme events in a changing climate: variability is more important than averages. Climatic Change 21:289–302

Kerr RA (1997) Greenhouse forecasting still cloudy. Science 276:1040–1042

Kimball BA (1983) Carbon dioxide and agricultural yield: an assemblage and analysis of 770 prior observations. Agronomy Journal 75:779–788

Kimball BA, Idso SB (1983) Increasing atmospheric CO_2: effects on crop yield, water use and climate. Agricultural Water Management 7:55–72

Kimball BA, Mauney JR, Nakayama FS, Idso SB (1993) Effects of elevated CO_2 and climate variables on plants. J of Soil and Water Conservation 48(1): 9–14

Lane LJ, Nearing MA, Laflen JM, Foster GR, Nichols MH (1992) Description of the US Department of Agriculture water erosion prediction project (WEPP) model. In: Parsons AJ, Abrahams AD (eds) Overland Flow. UCL Press, London, UK, pp. 377–391

Lee JJ (1998) Cross-scale aspects of EPA erosion studies. In: Boardman J, Favis-Mortlock DT (eds) Modeling soil erosion by water. Springer-Verlag NATO-ASI Global Change Series I-55:191–200, Heidelberg, Germany

Lee JJ, Phillips DL, Dodson RF (1996) Sensitivity of the US corn belt to climate change and elevated CO_2: II. Soil erosion and organic carbon. Agricultural Systems 52:503–521

Legates R, Willmott CJ (1990) Mean seasonal and spatial variability in gauge-corrected, global precipitation. International Journal of Climatology 10:111–128

Leggett J, Pepper WJ, Swart RJ (1992) Emissions scenarios for the IPCC: an update. In: Houghton JT, Callander BA, Varney SK (eds) Climate change 1992: the Supplementary Report to the IPCC Scientific Assessment. Cambridge Univ Press, Cambridge, UK, 75–95

Lorius C, Jouzel J, Raynaud D, Hansen J, Le Treut H (1990) The ice-core record: climate sensitivity and future greenhouse warming. Nature 347:139–145

Middleton N (1995) The global casino. Edward Arnold, London, UK, 332 pp

Mitchell JFB, Johns TC, Gregory JM, Tett S (1995) Climate response to increasing levels of greenhouse gases and sulphate aerosols. Nature 376:501–504

Morgan RPC (1995) Soil erosion and conservation. Longman Scientific and Technical, Harlow, UK, Second Edition, 198 pp
Nicholls N (1997) Increased Australian wheat yield due to recent climate trends. Nature 387:484–485
Nicks AD, Lane LJ, Gander GA (1995) Weather generator. In: Flanagan DC, Nearing MA (eds) USDA-Water Erosion Prediction Project: Hillslope profile and watershed model documentation. USDA-ARS National Soil Erosion Research Laboratory Report No. 10:2.1–2.22, West Lafayette, Indiana, USA
Parry ML, Blantran de Rozari M, Chong AL, Panich S (eds) (1991) The potential socio-economic effects of climate change in South-East Asia. United Nations Environment Programme, Nairobi, Kenya, 126 pp
Phillips DL, White D, Johnson CB (1993) Implications of climate change scenarios for soil erosion potential in the USA. Land Degradation and Rehabilitation 4:61–72
Stockle CO, Williams JR, Rosenberg N, Jones CA (1992) A method for estimating the direct and climatic effects of rising atmospheric carbon dioxide on growth and yield of crops: part I – modification of the EPIC model for climate change analysis. Agric Systems 38:225–238
Tegart WJ McG, Sheldon GW, Griffiths DC (eds) (1990) Climate change: the IPCC impacts assessment. Australian Government Publishing Service, Canberra, Australia
Tiscareno-Lopez M, Weltz MA, Lopes VL (1995) Assessing uncertainties in WEPP's soil erosion predictions on rangelands. Journal of Soil and Water Conservation 50(5):512–516
Turner II BL, Skole D, Sanderson S, Fischer G, Fresco L, Leemans R (1995) Land-use and Land-cover change: Science/Research Plan. IGBP Report No. 35/HDP Report No. 7, IGBP Secretariat, Stockhom, Sweden, 132 pp
United Kingdom Climate Change Impacts Review Group (1991) The potential effects of climate change in the United Kingdom. HMSO, London, UK, 124 pp
Waggoner PE (1990) Anticipating the frequency distribution of precipitation if climate change alters its mean. Agricultural and Forest Meteorology 47:321–337
Wilby RL (1995) Greenhouse hydrology. Progress in Physical Geography 19(3):351–369
Williams JR, Nicks AD, Arnold JG (1985) Simulator for water resources in rural basins. Journal of Hydrological Engineering 111(6):970–986
Williams JR, Jones CA, Dyke PT (1990) The EPIC model. In: Sharpley AN, Williams JR (eds) EPIC – Erosion/Productivity Impact Calculator. 1. Model documentation. US Department of Agriculture Technical Bulletin 1768: 3–92
Wolf J (1993) Effects of climate change on wheat and maize production potential in the EC. In: Kenny GJ, Harrison PA, Parry ML (eds) The effects of climate change on agricultural and horticultural potential in Europe. Environ Change Unit, Univ Oxford, Oxford, UK, 93–119
Zhang XC, Nearing MA, Risse LM, McGregor KC (1996) Evaluation of runoff and soil loss predictions using natural runoff plot data. Trans Am Soc Agric Engineers 39(3):855–863

Chapter 2

Applying the LISEM Model for Investigating Flood Prevention and Soil Conservation Scenarios in South-Limburg, the Netherlands

A.P.J. De Roo

2.1
Introduction

Soil erosion and surface runoff have always been problems concomitant with intensive agricultural land use in hilly areas. These problems can be exacerbated by soil and geology, as is the case in the hill country of South-Limburg (The Netherlands), where soils developed on loess are especially vulnerable to surface runoff and soil erosion. Since people started clearing the forests, soil erosion processes and human reactions to them have created the characteristic landforms dry valleys, incised ('hollow') roads and lynchets.

Until recently, traditional land use practices could keep soil erosion and surface runoff at acceptable levels. During the last two decades, however, the expansion of urban areas, the increased area of sealed surfaces and the intensification of agriculture, and increased arable agriculture, have caused soil erosion and flooding to increase. Re-allotment schemes have resulted in larger fields, causing surface runoff to be more erosive. Changes in land use also contribute to increasing erosion, total runoff and peak runoff. The area of grassland has decreased in favour of urban areas and, after 1975, in favour of arable land (De Roo 1993). Moreover there has been a change in land use and in the kinds of crops grown in South-Limburg. Between 1960 and 1986, the kinds of crops which give rise to a higher erosion risk, such as maize and sugarbeet, have increased in South-Limburg, replacing cereals such as winter wheat. Runoff with a high sediment load causes obstructed waterways and choked up sewerages, causing damage to roads, gardens and houses.

2.2
Current Policy

Since 1980, awareness of soil erosion problems in South-Limburg has increased. Schouten et al. (1985) were among the first to report the causes and damaging effects of surface runoff and soil erosion: flooding and damage to private properties and infrastructure, loss of fertile topsoil, washing away of seedlings, reduced crop yields, and loss of fertilizers, herbicides and pesticides, locally entering nature areas. Consequently, the governmental institutions got interested because they realized that a large part of South-Limburg is susceptible to soil erosion and the subsequent damages. At that time there was no central government soil erosion policy. Therefore, the government proposed to develop a framework within which local and provincial plans for the rural area could be tested with respect to their effects on soil erosion and flooding.

In the meanwhile, in 1990, the Dutch Agricultural Board (Landbouwschap), representing the farmers, issued an 'Erosion Prevention Ordinance' (Verordening Erosiebestrijding), which intended to sustain the potential production capacity of the arable land in South-Limburg. In itself a positive step, they intended to prevent the Province from taking even more strict measures. For example, arable agriculture within fields with an average slope gradient larger than 18% was not allowed anymore. Several other measures were prescribed for fields with lower gradients.

However, these measures did not seem strict enough to prevent the on-site and off-site effects of flooding and erosion. Local and provincial governments needed a more strict regulation to prevent the frequent flooding and sediment problems in the region. But, data to support this decision-making were not or not sufficiently available. Also, there was a demand for research of accurate soil conservation measures and their most suitable locations thus leading to a reduction of the problem to an acceptable level. Local and provincial policy makers and parties concerned (both farmers organisations and environmental groups) needed a quantitative evaluation of the extent and the magnitude of the soil erosion problems and the possible management strategies in South-Limburg on a regional basis. Therefore, field measurements were necessary. Also, there was a need for quantitative simulation models of surface runoff and soil erosion, which can be used to evaluate alternative strategies for improved land management, not only in the monitored areas, but also in ungauged catchments.

2.3
LISEM

To obtain the necessary data to support decision making, a soil erosion project has been carried out in three small experimental catchments in the loess area of South-Limburg, The Netherlands. The project was funded by the Province of Limburg, the Waterboard 'Roer en Overmaas', the Ministry of Agriculture, and 14 municipalities of South-Limburg. The departments of Physical Geography of the Universities of Utrecht and Amsterdam, and the Soil Physics division of the Winand Staring Centre in Wageningen cooperated in this project (De Roo et al. 1995).

Within this scope, as a tool for planning and conservation purposes, a new physically-based hydrological and soil erosion model needed to be developed and tested: the LImburg Soil Erosion Model (LISEM). LISEM is a physically-based hydrological and soil erosion model, which can be used for planning and conservation purposes. LISEM simulates runoff and sediment transport in catchments caused by individual rainfall events. The simulation timestep is free to choose from as low as 1 second up to 15 minutes. LISEM can be applied on small fields and in catchments up to 1 000 km^2. Processes incorporated in the model (Fig. 2.1) are rainfall, interception, surface storage in micro-depressions, infiltration, vertical movement of water in the soil, overland flow, channel flow, detachment by rainfall and through fall, detachment by overland flow, and transport capacity of the flow. Also, the influence of tractor wheelings, small paved roads (smaller than the pixel size), field strips and grassed waterways on the hydrological and soil erosion processes is taken into account. For a detailed description of the processes incorporated in the model is referred to De Roo et al. (1996a). A sensitivity analysis and validation are presented in De Roo et al. (1996b).

Major conclusions are that the quantitative results of the model could be improved by an improved knowledge of the spatial and temporal variability of soil moisture content and hydraulic conductivity in the catchment. The qualitative results are realistic.

CHAPTER 2 · Applying the LISEM Model for Investigating Flood Prevention

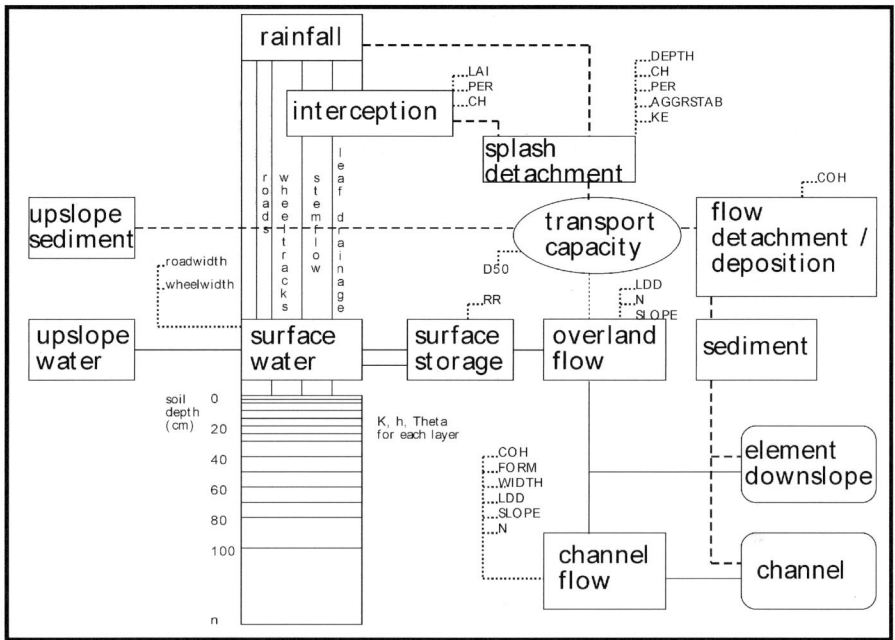

Fig. 2.1. Flow chart of the LISEM model

For infiltration and vertical transport of water in the soil, a LISEM user has the following options: (0) No infiltration; (1) Richards equation for uniform soils; (2) Richards equation for soils and wheel tracks; (3) Richards equation for soils, wheel tracks and crusts; (4) Holtan; (5) one-layer Green-Ampt and (6) two-layer Green-Ampt. The Richards equation is solved separately for uncrusted soils, crusted soils and wheel tracks. In areas without detailed knowledge of soil physical variables, the empirical Holtan/Overton infiltration equation (De Roo 1993) can be used. However, the Holtan equation only simulates Hortonian overland flow, saturation overland flow is not simulated. Using the Richards based sub-model, one can simulate saturated overland flow. In other cases, the user may choose a one or two layer Green and Ampt model. The major differences between LISEM and other erosion models are:

- on a GIS level LISEM uses a raster type representation of the catchment, such as in ANSWERS (Beasley et al. 1980) and EROSION-3D (Schmidt 1991), which allows for a detailed representation of the processes (for example a 10 × 10 m pixel). This is different from other process based models such as EUROSEM (Morgan 1994), KINEROS (Woolhiser et al. 1990) or WEPP (Lane et al. 1992) that use large polygon type elements;
- on a programming level LISEM is constructed with the PCRaster dynamic modeling language (Wesseling et al. 1996) which allows for great flexibility;
- on a process level LISEM comes conceptually close to EUROSEM and WEPP, while e.g. ANSWERS uses many empirical relationships. The possible use within LISEM

of the SWATRE (Belmans et al. 1983) solution of the Richards equation makes it more physically-based. In case of limited data availability, the user also can choose Green and Ampt or the Holtan equation.

In the model, special attention has been given to features that play a crucial role in flooding and erosion problems in South-Limburg: the influence of wheel tracks and small roads. 25% or more of the area of agricultural fields consist of wheel tracks, which have an inferior soil structure and consequently a lower infiltration capacity. Resistance to erosion on the other hand can be larger due to a larger soil cohesion. Since LISEM operates with square raster cells, usually with sizes of 10 or 20 m, small field roads with widths of 3–5 m are difficult to capture. Since these roads can cover up to 2–3% of the total catchment area, and they are situated in such way that they operate as main channels, these roads produce an important part of the surface runoff. Normally 5–10% of the rainfall becomes runoff, so roads can be responsible for 25–50% of this amount!

2.4
Scenario Results

Within the LISEM project, several scenarios were constructed that were considered to lead to a possible reduction of both runoff and erosion. Five scenario groups were formed:

- scenario 0 is the baseline scenario: land use and land management are similar to the situation of 1990, so before the ordinance of the Agricultural Board.
- scenario 1 is the current situation (1993/4), where farmers apply the ordinance.
- scenarios 2A-2D are scenarios of 'improved management', to be carried out by farmers, without further control measures. Considered are mulching, direct sowing, green cover crops in winter.
- scenarios 3A-3C are scenarios of control measures, such as grassed waterways and grass buffer strips, without further management improvements by farmers.
- scenarios 4A-4E are combinations of group 2 and 3 scenarios: integral management scenarios.

Both the effects of frequent and extreme events were evaluated by using rainstorms of a 2 and 25 year return period. Both summer (thunderstorms) and winter (depressional type low intensity rainfall) situations were simulated separately. The input data that were used have been measured in the field and the laboratory. All land use types present in the area have been monitored: grassland, winter wheat, winter barley, sugarbeet, potatoes and maize. Furthermore, on special fields the influences of 'mulching' and direct sowing have been measured. Variables that were monitored both in space and time, whenever relevant, are: soil cover by vegetation, leaf area index, crop height, random roughness, saturated and unsaturated conductivity, water retention curve, soil texture, aggregate stability and soil cohesion. Thus, a large database has been created on the monthly variation of these variables during the growing season.

Within the GIS, a typical land use pattern for mid winter and early summer has been chosen for each catchment. By combining the land use maps and the soils and land use database, maps of the LISEM input variables have been created.

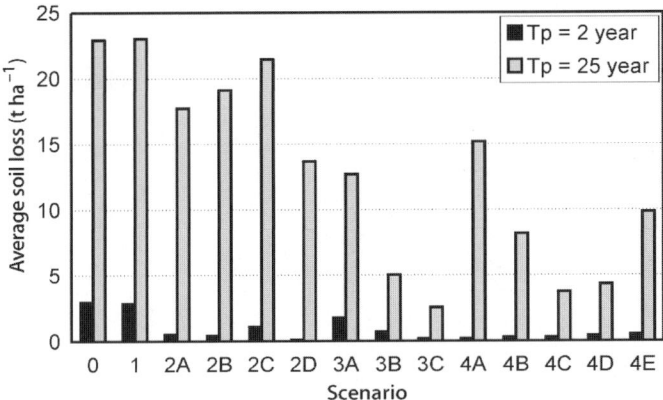

Fig. 2.2. Effects of 14 scenarios on total event soil loss in the St.Gillisstraat drainage basin (Ransdaal, The Netherlands, size is 40 hectares) for summer storms (20 minute duration) with return periods of 2 and 25 year. Scenario 0 is the actual land use in 1990; scenario 1 is the land use in 1993; scenarios 2 are different tillage techniques; scenarios 3 are conservation measures such as field buffer strips and grassed waterways; scenarios 4 are combinations of 2 and 3

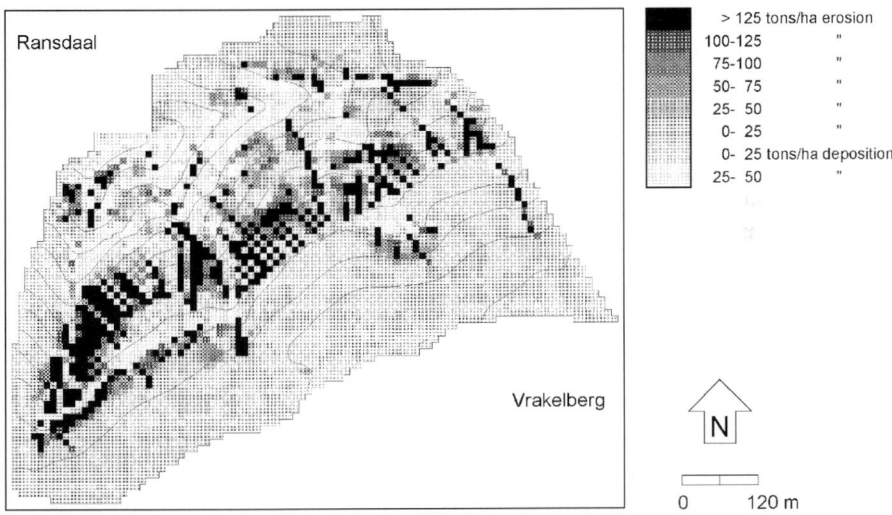

Fig. 2.3. Soil erosion and deposition in the St. Gillisstraat drainage basin (Ransdaal, The Netherlands) for a scenario with field buffer strips and grassed waterways

Using the LISEM model, not only the scenario that leads to the largest reduction in runoff and erosion can be determined (Fig. 2.2 and 2.3), but also the best possible locations for the measures can be determined. Maps of soil erosion and sedimentation of the scenarios can be compared by subtraction. This indicates where possi-

Change in net soil erosion as a consequence of field strips and grassed waterways

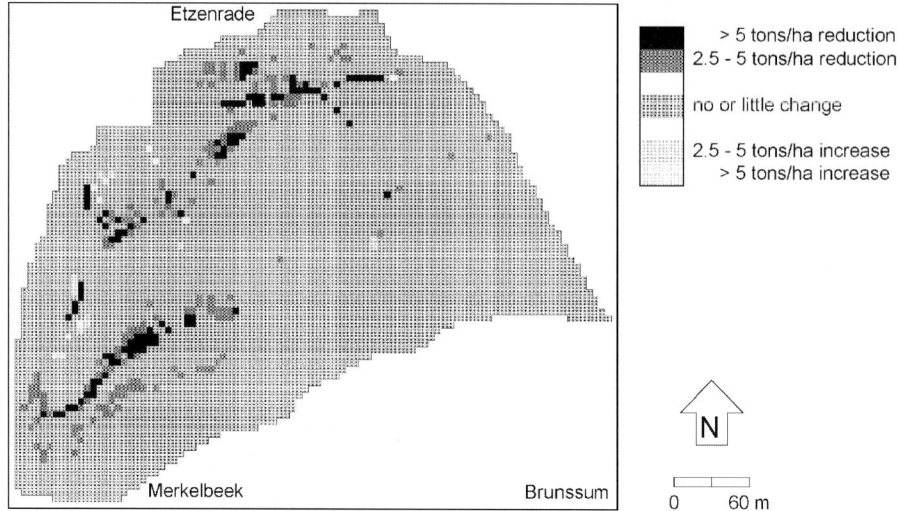

Fig. 2.4. Change of net erosion as a consequence of one of the land use scenario in Fig. 2.3 compared to the actual land use of 1993

ble control measures would have the greatest positive and negative consequences (Fig. 2.4).

Thus, from the simulation results (Fig. 2.2 and 2.3), it can be concluded that scenarios 2D (straw cover), 3C (field strips and grassed waterways) and 4C (mulching, green cover crops in winter, grassed waterways and field strips) would lead to the largest reductions in runoff and soil loss. Given the validation results and the input data uncertainty (De Roo et al. 1996b) one has to be careful with these results. However, qualitatively, the LISEM model yields the same results as the field experiments of individual control measures as carried within the project: scenarios leading to a reduction of discharge and soil loss on field plots also lead to reduced simulated discharge and soil loss with LISEM.

2.5
Cost-Benefit Analysis

The scenario that leads to the largest reduction of runoff and soil loss is not by definition also the 'optimum' scenario. Other factors, such as costs play an important role. Within the LISEM project, an attempt has been made to quantify all costs relating to soil erosion and flooding.

At present, farmers suffer from reduced crop yields and additional costs of sowing whenever a damaging event takes place in winter or spring. On the other hand, the Waterboard plans, constructs and maintains large costly retention basins, to prevent the water and sediment to enter the villages. Also, the damage done has to be repaired and houses, roads and cellars have to be cleaned from the sediment. All these factors

Table 2.1. Total costs and soil loss reductions of several scenarios in the LISEM project. Also the percentage contribution to the total costs of farmers, water-authorities and retention basins (water storage basins to be constructed) are given

Scenario	Total costs (EURO/yr)	Contribution of farmers (%)	Costs payed by water authorities (%)	Costs of retention basins (%)	Reduction soil loss (%) compared to 'actual'
Actual	367	3	0	97	–
2D Straw	454	42	0	58	30
4A Mulch–2%	410	33	0	67	25
4B+Waterways	459	30	14	56	35
4C+FieldStrip2	537	25	30	45	50
4D+FieldStrip5	459	29	16	55	50
4E Mulch–5%	473	19	14	67	30

were incorporated in the cost-benefit analysis. Whenever field or buffer strips are planned, fields become smaller, and farmers have reduced yields and larger labour costs because they have to 'turn' their tractors more often. Also, the land has to be bought by the Waterboard. Crop yields due to mulching and direct sowing change, costs have to be made to buy straw, etc.

Taking all these factors into account, an overall indication of total costs has been given in Table 2.1.

From Table 2.1 can be seen that scenario 4D (mulching >2%, grassed waterways, field strips every 200 m on slopes >5%) reduces soil loss with 50% compared to the actual situation (the 'Actual' scenario). Total costs of this scenario are 25% larger (459 euro yr^{-1} compared to 367 euro yr^{-1}) than now, but now farmers contribute 29% of the costs, and the government 61%, instead of 0 and 100%. The number and size of retention basins can be reduced with 45%!

2.6
New Policy

Based on the results of the LISEM project, local and provincial policy makers and farmers organisations tried to develop a new policy, to further reduce flooding and soil erosion problems, to protect the environment, and to sustain the potential production capacity of the arable land. Compared to the 1990 Ordinance, new regulations have been added, such as:

- the use of wheel track erasers during sowing on fields with a slope gradient between 2 and 18%
- the use of straw or a soil cover crop on fields between 12 and 18% slope gradient

Due to the large political debate, it was decided not to take further general regulations, but to define 20 'bottleneck areas' (Duijsings 1996). These 20 bottleneck areas cover 10% of the total area in South Limburg susceptible to soil erosion, but they comprise about 45% of the locations experiencing problems with flooding and severe erosion. In these areas, problems caused by water and sediment will be reduced to an acceptable level by means of additional measures, to be decided on a case by case basis. Possible additional measures include:

- the use of soil cover crops on fields with slope gradients less than 12%
- construction of grassed waterways in valley bottoms
- construction of grass buffer strips on slopes of 2–10%, about 5 m wide, and at mutual distance of 200 m
- contour farming on slopes <5%
- conversion of arable land into grassland, especially on the steeper slopes
- construction of extra retention basins

2.7
Conclusions

LISEM is a powerful model which simulates hydrological and soil erosion processes during single rainfall events on a catchment scale. Using LISEM it is possible to calculate the effects of land use changes and to explore soil conservation scenarios. Driven with hypothetical rainstorms of known probability of return, LISEM is a valuable tool for planning cost-effective measures to mitigate the effects of runoff and erosion. LISEM produces detailed maps of soil erosion and overland flow that are useful for planners. The integration of LISEM in a raster-based GIS, which holds the many data on the distributions of land attributes, is very useful. Other advantages of LISEM are the use of physically-based mathematical relationships, the ease with which newly developed relationships can be incorporated and the incorporation of information about the spatial variability of land characteristics.

Using the results of the LISEM field project and simulation study, a new soil and water conservation policy has been defined, focussing on the most susceptible areas.

Acknowledgements

The Province of Limburg, the Waterboard 'Roer en Overmaas', the Ministry of Agriculture and 14 municipalities of South-Limburg are greatly acknowledged for funding a large part of the research which lead to the development and validation of the LISEM model. Furthermore, all co-workers in the LISEM project are thanked for their great enthusiasm over the years.

References

Beasley DB, Huggins LF, Monke EJ (1980) ANSWERS: A model for watershed planning. Transactions of the ASAE, 23–4:938–944
Belmans C, Wesseling JG, Feddes RA (1983) Simulation model of the water balance of a cropped soil: SWATRE. Journal of Hydrology 63:271–286
De Roo APJ (1993) Modeling surface runoff and soil erosion in catchments using Geographical Information Systems; Validity and applicability of the 'ANSWERS' model in two catchments in the loess area of South Limburg (The Netherlands) and one in Devon (UK). Netherlands Geographical Studies, Utrecht 157:304 pp

De Roo APJ, Van Dijk PM, Ritsema CJ, Offermans RJE, Cremers NHDT, Kwaad FJPM, Stolte J (1995) Erosienormeringsonderzoek Zuid-Limburg: Veld- en SimulatieStudie. (Soil Erosion Normalisation Project South Limburg, The Netherlands (in Dutch)). Winand Staring Centre, Wageningen, Report 364.1:234 pp

De Roo APJ, Wesseling CG, Ritsema CJ (1996a) LISEM: a single event physically-based hydrologic and soil erosion model for drainage basins. I: Theory, input and output. Hydrological Processes 10:1107–1117

De Roo APJ, Offermans RJE, Cremers NHDT (1996b) LISEM: a single event physically-based hydrologic and soil erosion model for drainage basins. II: Sensitivity analysis, validation and application. Hydrological Processes 10:1119–1126

Duijsings J (1996) To a soil and water conservation policy for the 21st century. Province of Limburg, Maastricht

Lane LJ, Nearing MA, Laflen JM, Foster GR, Nichols MH (1992) Description of the US Department of Agriculture Water Erosion Prediction Project (WEPP) Model. In: Parsons, AJ, Abrahams AD (eds) Overland flow: Hydraulics and erosion mechanics. UCL Press Limited, London, 377–391

Morgan RPC (1994) The European soil erosion model: an update on its structure and research base. In: Rickson RJ (ed) Conserving soil resources: European perspectives. CAB International, Cambridge, 286–299

Schmidt J (1991) A mathematical model to simulate rainfall erosion. In: Bork HR, De Ploey J, Schick AP (eds) Erosion, transport and deposition processes – Theories and models – Heinrich Rohdenburg Memorial Symposium. CATENA Supplement 19:101–109

Schouten CJ, Rang MC, Huigen PMJ (1985) Erosie en wateroverlast in Zuid-Limburg. Landschap 2:118–132

Wesseling CG, Karssenberg D, Burrough PA, Van Deursen WPA (1996) Integrating dynamic environmental models in GIS: The development of a dynamic modeling language. Transactions in GIS 1:40–48.

Woolhiser DA, Smith RE, Goodrich DC (1990) KINEROS: A kinematic runoff and erosion model: documentation and user manual. USDA-ARS, ARS-77

Chapter 3

Application of Modified AGNPS in German Watersheds

S. Grunwald · H.-G. Frede

3.1
Objectives

The objectives have been the application and verification of the AGNPS model (Agricultural Non-Point Source Pollution Model) by Young et al. (1987, 1994) to assess runoff volume, sediment and nutrient yield in medium to large-sized watersheds (>1 km^2) in Germany. The aim was to adapt the model to climate and land use conditions in Germany and to modify some model algorithms to improve description of transport processes (AGNPSm: modified AGNPS).

3.2
Methodology

The event-based water quality model AGNPS Vers. 5.0 was used in this study (Young et al. 1987, 1994). In the original version by USDA-ARS (United States Department of Agriculture – Agricultural Research Service) the empirical Curve-Number-Method was used for runoff volume calculations (SCS 1972). Peak flow rate was computed by the Smith and Williams (1980) algorithm, and Manning's Equation described flow velocity. Soil loss was calculated by a modified Universal Soil Loss Equation (Wischmeier and Smith 1978), which includes the energy-intensity value and a slope-shape factor. Sediment transport and deposition were calculated by a steady-state continuity equation (Foster et al. 1980; Lane 1982), whereby routing was done on a per cell and per particle-size basis. Sediment transport capacity was calculated with a modified Bagnold stream power equation (Bagnold 1966) and, based on the Einstein (1950) approach, each particle class was calculated separately. For the calculation of soluble N and P and nutrients in sediment, analytical approaches from the CREAMS model were used (Frere et al. 1980).

The following modifications were integrated in the AGNPS:

- Lutz (1984) method for runoff volume calculation:

$$RO = (P - Ia) C + \frac{C}{a} \left(e^{-a(P-Ia)} - 1 \right) \tag{3.1}$$

$$Ia = 0.03S \tag{3.2}$$

$$S = 25.4 \left(\frac{10}{C} - 10 \right) \tag{3.3}$$

$$a = C_1 e^{(-C_2/WZ)} e^{(-C_3/q_B)} e^{(-C_4 D)} \tag{3.4}$$

- RO = runoff volume (mm) by Lutz
- P = storm precipitation (mm)
- Ia = initial abstraction (mm)
- C = maximum discharge value (–)
- a = Lutz factor (mm^{-1})
- S = potential maximum retention (mm)
- C_1, C_2, C_3, C_4 = weighting parameters for optimization (–)
- WZ = week value
- q_B = base flow (l s^{-1} km^{-2})
- D = duration of precipitation (h)

- Smith and Williams (1980) algorithm for peak flow calculation:

$$Q_{max} = 3.79 A_{EO}^{0.7} J^{0.16} \left(\frac{RO}{25.4}\right)^{(0.903 A_{EO}^{0.017})} \left(\frac{L^2}{A_{EO}}\right)^{-0.19} \tag{3.5}$$

- Q_{max} = peak flow rate (m^3 s^{-1})
- A_{EO} = drainage area (km^2)
- J = channel slope (%)
- L = maximum flow path (km)

- LS-factor algorithm based on 'stream power theory' by Moore and Burch (1986) (compare Fig. 3.1):

$$f = \frac{A_{pwa}}{b l_{pwl}} \tag{3.6}$$

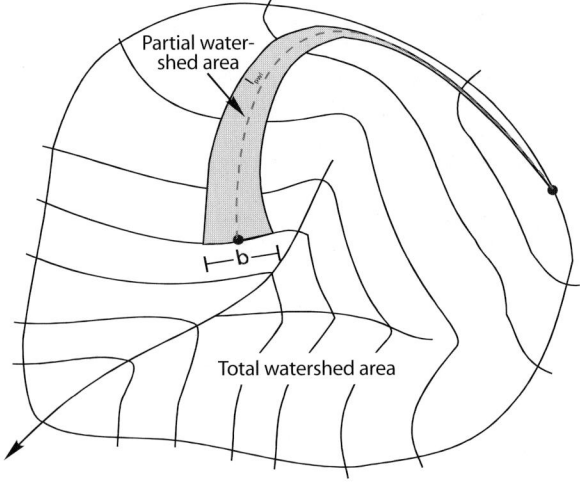

Fig. 3.1. Partial watershed area concept by Moore et al. (1986)

$$Lsp = \left(\frac{fl_{pwl}}{22.14}\right)^{0.4} \tag{3.7}$$

$$Ssp = \left(\frac{\sin s_x}{0.0896}\right)^{1.3} \tag{3.8}$$

- f = shape parameter (–)
- A_{pwa} = partial watershed area (m²)
- b = width of contour element (m)
- l_{pwl} = partial watershed length (m)
- Lsp = L-factor (stream power theory)
- Ssp = S-factor (stream power theory)
- s_x = slope of partial watershed area (°)

- Scouring of particles in channel linked to water flow velocity.
  In Table 3.1, critical flow velocities for different particle classes based on literature are shown.
- Grid-based precipitation input instead of uniform rainfall input for large watersheds.

The modeling concept comprised a Geographic Information System (GIS), an input-interface, the model AGNPSm, and an output-interface (Fig. 3.2). The program Digital Elevation Drainage Network Model (DEDNM) of Garbrecht and Martz (1993) was used to derive the watershed boundary, flow directions, and the drainage network. The GIS (SPANS and IDRISI) was used to store the spatial input data (land use, soil, topography). An interface program written in C linked the spatial data as well as the climate data to AGNPSm. By means of the interface, AGNPSm input variables were calculated by primary and secondary derivation based on spatial data and if-structures. For example, input variables that vary seasonally (e.g. Manning's roughness coefficient) were derived by time-dependent if-structures and land use data. In Table 3.2, the data sources and AGNPS input variables are listed.

The watersheds used in this study denoted G1 and G2, are located on Glonn Creek, in Bavaria, Germany. The watershed G1 is 1.2 km² and G2 1.6 km² in size. The elevation varies between 511–550 m (G1) and 515–560 m (G2), and the average slope is 7% (G1) and 6% (G2). The soils are predominantly loamy-sands, loam, and clay-loam soils with some influence of loess. In the valley bottom, gleyed soils are present. The

Table 3.1. Critical flow velocities for different particle classes (Imhoff 1972; Maue 1988; DVWK 1988)

Flow velocity (m s⁻¹)	Particle classes[a]
0.3	Primary clay
0.6	Primary silt
0.9	Small aggregates
1.2	Large aggregates
1.5	Primary sand

[a] Particle classification: Foster et al. (1980).

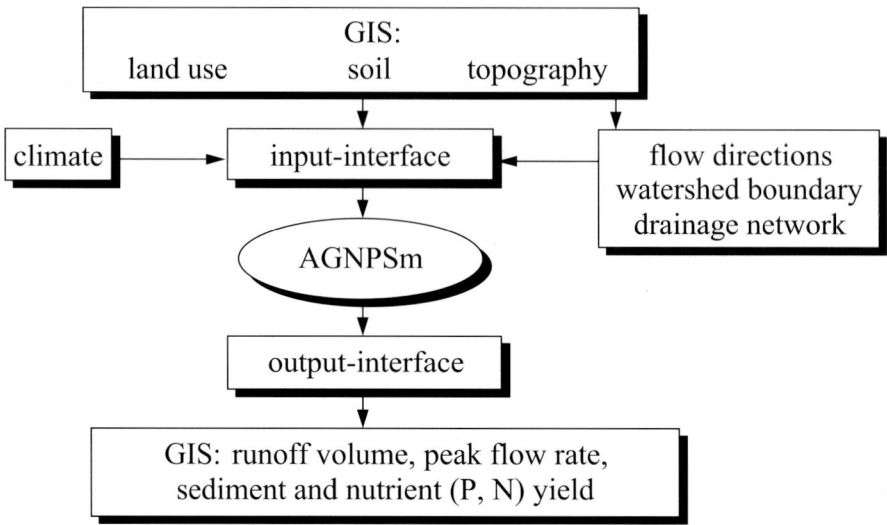

Fig. 3.2. Flow chart of the modeling concept

Table 3.2. Data sources and AGNPS input variables used for the Glonn watershed

Data source	AGNPS input variable
DEM (digitized contours 1 : 5 000) + DEDNM	→ Numerical order of cells → receiving cells (flow directions) → identification of channel cells → slope and slope shape
Topographic map 1 : 5 000	→ Slope length
Mapping of land use + interview of farmers + standard values	→ Manning's n Roughness Coefficient → Surface Condition Constant → Soil Loss Ratio (USLE) → N and P fertilizer amount and availability → organic matter content
Mapping of soil types (1 : 5 000) + soil texture (Reichsbodenschätzung –soil survey)	→ Soil texture → K-factor (USLE)
Analysis of soil N and P	→ Soil P and N
Mapping of land use + soil data information + digitized street and alley system	→ Curve Numbers; maximum discharge values
Mapping of contouring	→ P-factor (USLE)
Data from 2 rain stations + 1 rain recorder	→ Precipitation amount → duration of rainfall event → energy-Intensity-value (USLE)
Rainfall data + measured discharge	→ K-value (% runoff)
Field investigation + DEDNM	→ Channel slope → channel side slope → channel width → channel length

total area in G1 covered by forest is 20.2%, while 79.8% is used as agricultural land, of which 30.9% is corn, 22.5% pasture/meadow, 32.7% grain, 7.3% potatoes, 5.2% forage fodder, and 1.4% waste land. Watershed G2 is dominated by forest (58.8%), while 41.2% is used as agricultural land, of which 60.2% is grain, 21.8% corn, 16.5% potatoes, and 1.5% pasture/meadow. The average yearly precipitation for two rain stations nearby are 830 mm (Mering) and 873 mm (Puch). The number of rainfall/runoff events used in this study were 29 (G1) and 24 (G2). These rainfall events varied between 17 and 89 mm, and measured runoff volume ranged from 1–58 mm (G1) and 0.5–23 mm (G2), respectively. Measured data at the drainage outlet were collected by the Bavarian Water Authority (1984).

Additionally, simulations were carried out in the Salzboede watershed, in the hilly midlands of Germany. This watershed is 81.7 km^2 in size. The elevation range is 190–564 m, and the average slope of the arable land is 9.7%. The watershed is predominantely (46.0%) forest, with 21.5% pasture/meadow, and 23.5% arable land dominated by grain. The soils are predominantely loamy sands, loam or clay-loam soils mixed with loess. The average yearly precipitation is 786 mm. For model verification, 16 measured rainfall/runoff events were used (Rode 1995).

3.3
Results

The prediction results for the G1 watershed are presented in detail, while the results for G2 and Salzboede are summarized. In this study, a grid resolution of 25 m was used, with 1965 cells representing the watershed G1.

Runoff volume by the SCS CN-Method (SCS 1972) underpredicted measured runoff volume. The calculated runoff was very low, due to the magnitude of the initial abstraction (Ia), which was calculated by the equation: $Ia = 0.2S$ (S: potential maximum retention). Especially for small events (<50 mm precipitation amount), the predictions were unreliable with a coefficient of efficiency (Nash and Sutcliffe 1970) of 0.25. Because the results of the SCS CN-Method do not match the conditions of land use, soil, and climate found in watershed G1, a calibration was carried out. For this purpose, the weighting factor for initial abstraction was varied between 0.01 and 0.20. Ten calibration events were used to calculate the coefficient of efficiency E for each weighting factor combination. The highest E was found for a weighting factor of 0.03. The results for the calibrated SCS CN-Method are shown in Table 3.3 and Fig. 3.3. The calibrated runoff volumes for SCS CN-Method show reliable results compared to measured runoff volumes. The E was 0.87 (calibration events) and 0.93 (validation events), respectively. The differences between measured and predicted median of runoff volume for validation events were justifiably low. The results correspond well with literature. For example, Maniak (1992) recommended a reduction of the Ia value to 5% of the water storage capacity in the soil ($Ia = 0.05S$) for German watersheds.

Results for runoff volume calculations by the method of Lutz (1984) are shown in Table 3.4 and Fig. 3.4. C_2 was set according to Lutz (1984), dependent on land use. In calibrating the Lutz method 10 representative rainfall events out of a total of 29 were used. The calibration parameters C_1 and C_3 were varied between 0.02–0.08 and 1.0–6.0, respectively. For each parameter combination, the 10 rainfall events were simulated, and predicted and measured runoff volumes were compared using E.

Table 3.3. Statistics for runoff volume (RO in mm) calculated by calibrated SCS CN-Method (SCS 1972) – ($Ia = 0.03S$); watershed G1

	Calibration events		Validation events	
	Measured	Predicted	Measured	Predicted
Mean	8.1	9.4	9.0	8.6
Median	5.7	6.3	3.8	5.2
S. dev.	9.9	9.5	13.4	10.6
N	10		19	
E	0.87		0.93	

Table 3.4. Statistics for runoff volume (RO in mm) calculated by Lutz Method; watershed G1

	Calibration events		Validation events	
	Measured	Predicted	Measured	Predicted
Mean	8.1	8.1	9.0	8.7
Median	5.7	5.9	3.8	4.3
S. dev.	9.9	9.1	13.4	11.6
N	10		19	
E	0.98		0.96	

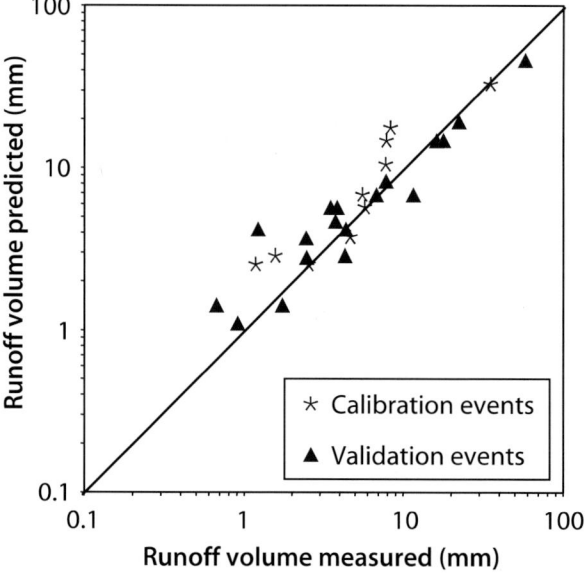

Fig. 3.3. Measured and predicted runoff volume (RO in mm) calculated by calibrated SCS CN-Method ($Ia = 0.03S$); watershed G1

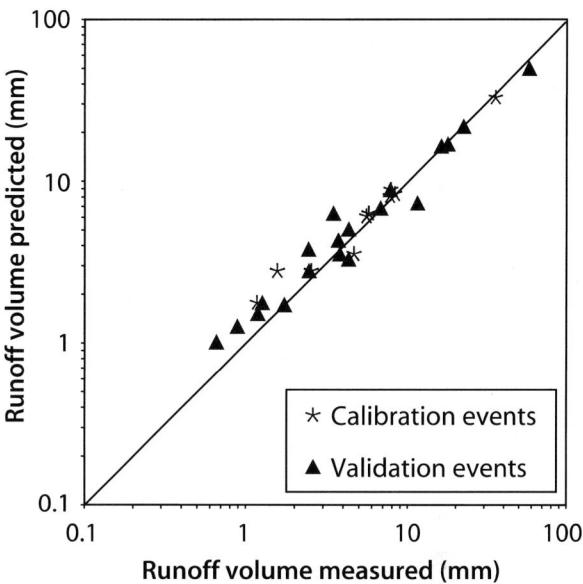

Fig. 3.4. Measured and predicted runoff volume (RO in mm) calculated by Lutz method; watershed G1

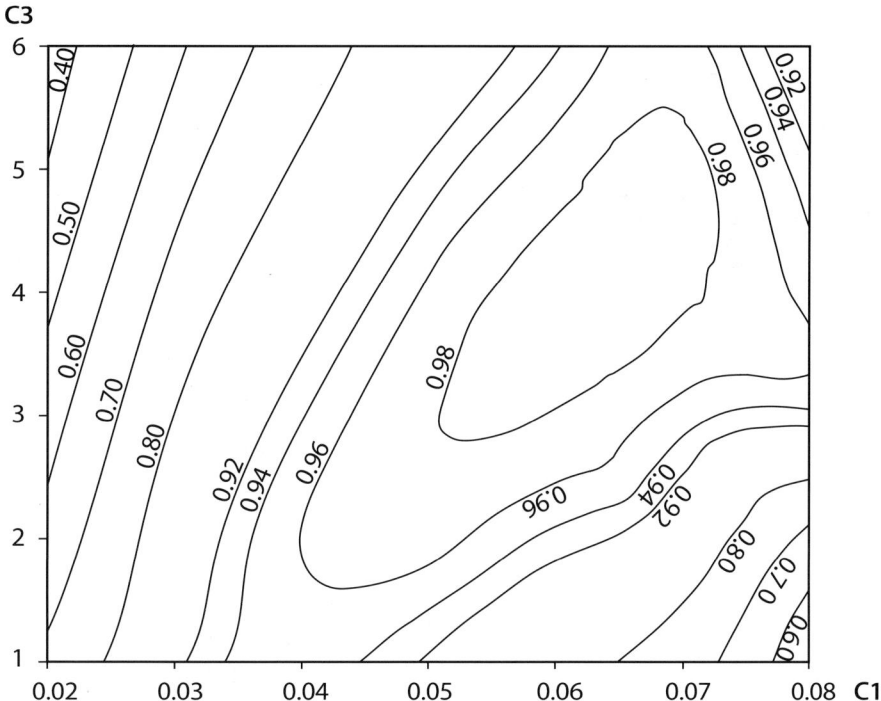

Fig. 3.5. Coefficients of efficiency (Nash and Sutcliffe 1970) for different combinations of Lutz parameters C_1 and C_3

In Fig. 3.5, the isolines of E are shown for different calibration parameter combinations. According to the calibration procedure, a C_1 value of 0.06 and C_3 of 4.0 gave one of the best simulation results for watershed G1.

Even for small events, the measured and predicted values are scattered near the 1 : 1 line in Fig. 3.4. The difference between measured and predicted medians for the validation events of 0.5 mm (Table 3.4) was very low compared to the CN-Method. For all subsequent calculations, the Lutz method was used, as it provides the highest degree of suitability for simulations in the G1 watershed.

Because peak flow calculations by the algorithm of Smith and Williams (1980) overpredicted measured values, a calibration was carried out. Ten calibration events were used, and a calibration factor f, calculated as a function of runoff volume, was integrated into the peak flow algorithm ($f = 0.328e^{(-0.812RO)}$) Eq. 3.5. The results for peak flow predictions are shown in Table 3.5 and Fig. 3.6. For 19 validation events, a coefficient of efficiency of 0.84 was calculated, which is reliably high.

Table 3.5. Statistics for peak flow rate (Q_{max} in $l\,s^{-1}$); watershed G1

	Calibration events		Validation events	
	Measured	Predicted	Measured	Predicted
Mean	435.3	340.4	306.7	334.2
Median	341.4	345.4	312.9	260.9
S. dev.	255.7	176.8	169.9	201.0
N	10		19	
E	0.85		0.84	

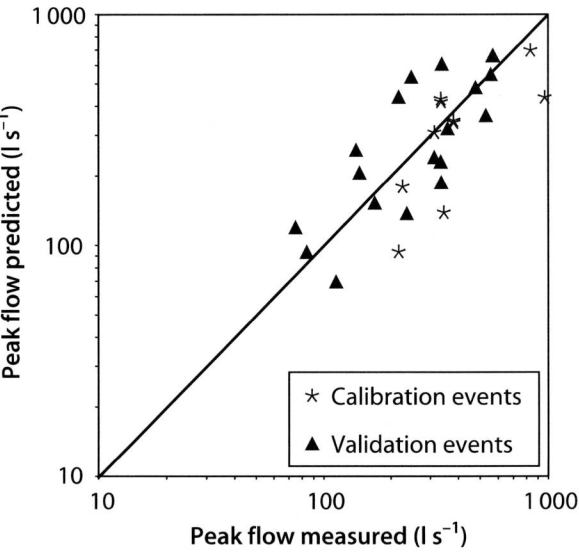

Fig. 3.6. Measured and predicted peak flow rate (Q_{max} in $l\,s^{-1}$); watershed G1

In Table 3.6 and Fig. 3.7, the results for sediment delivery calculated by the Universal Soil Loss Equation (Wischmeier and Smith 1978) are shown. Calibration was carried out to determine which particle class has to be scoured within the channel. Based on calibration for 10 events, the highest E (0.42) was calculated for 'no scouring of particles within channel' (see Table 3.7). It should be indicated that the assumption 'no scouring of particles' contradicts observations in watershed G1, because there was channel erosion when large runoff events occurred. The coefficient of efficiency for the validation events was 0.26, which is very low.

In Table 3.8 and Fig. 3.8, the results for sediment delivery calculated by AGNPSm (LS-factor by method of Moore and Burch 1986) are shown. There, too, the calibration procedure gave the highest E (0.65) for the option 'no scouring of particles within channel' (Table 3.9), and this has been assured for validation. The coefficient of efficiency for validation events was 0.57, which is higher compared with the sediment

Table 3.6. Statistics for sediment delivery (*Sed* in t) calculated by USLE (Wischmeier and Smith 1978); watershed G1

	Calibration events		Validation events	
	Measured	Predicted	Measured	Predicted
Mean	2.93	1.14	3.64	1.23
Median	2.01	0.90	2.10	0.86
S. dev.	2.75	0.91	5.71	1.58
N	10		18	
E	0.42		0.26	

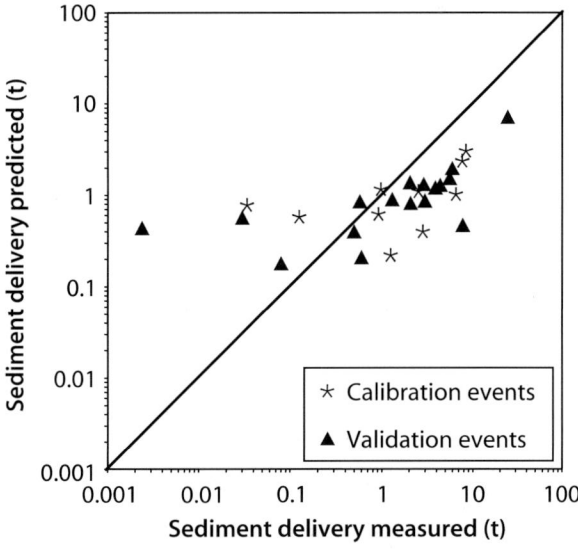

Fig. 3.7. Measured and predicted sediment delivery (*Sed* in t) calculated by USLE (Wischmeier and Smith 1978); watershed G1

Table 3.7. Coefficients of efficiency (Nash and Sutcliffe 1970) for sediment delivery (*Sed* in t) for different combinations of scouring of particles within channel. Sediment delivery was calculated by USLE (Wischmeier and Smith 1978)

	Combinations									
Clay	–	×	×	×	×	×	–	–	–	–
Silt	–	–	×	×	×	×	×	–	–	–
Small aggregates	–	–	–	×	×	×	×	×	–	–
Large aggregates	–	–	–	–	×	×	×	×	×	–
Sand	–	–	–	–	–	×	×	×	×	×
E	0.42	0.30	0.05	–0.03	–0.50	–0.63	0.15	0.21	0.22	0.27

– No scouring of particle within channel;
× Scouring of particle within channel.

Table 3.8. Statistics for sediment delivery (*Sed* in t) calculated by USLE (Wischmeier and Smith 1978), and LS-factor by method of Moore et al. (1986); watershed G1

	Calibration events		Validation events	
	Measured	Predicted	Measured	Predicted
Mean	2.93	1.76	3.64	1.86
Median	2.01	1.49	2.10	1.10
S. dev.	2.75	1.48	5.71	2.71
N	10		18	
E	0.65		0.57	

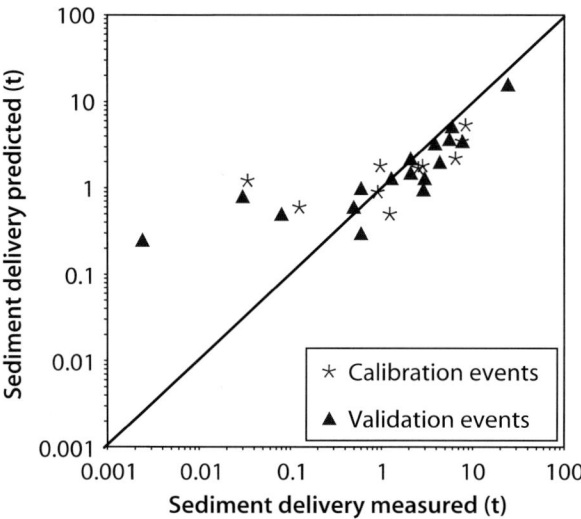

Fig. 3.8. Measured and predicted sediment delivery (*Sed* in t) calculated by USLE (Wischmeier and Smith 1978), and LS-factor by method of Moore et al. (1986); watershed G1

Table 3.9. Coefficients of efficiency (Nash and Sutcliffe 1970) for sediment delivery (*Sed* in t) for different combinations of scouring of particles within channel. Sediment delivery was calculated by AGNPSm (LS-factor of Moore et al. 1986)

	Combinations									
Clay	–	×	×	×	×	×	–	–	–	–
Silt	–	–	×	×	×	×	×	–	–	–
Small aggregates	–	–	–	×	×	×	×	×	–	–
Large aggregates	–	–	–	–	×	×	×	×	×	–
Sand	–	–	–	–	–	×	×	×	×	×
E	0.65	0.55	0.28	0.07	–0.30	–0.44	0.36	0.52	0.60	0.62

– No scouring of particle within channel;
× Scouring of particle within channel.

Table 3.10. Statistics for sediment delivery (*Sed* in t) calculated by USLE (Wischmeier and Smith 1978), LS-factor by the method of Moore et al. (1986), and scouring of particles linked to flow velocity; watershed G1

	Validation events	
	Measured	Predicted
Mean	3.59	2.92
Median	2.08	2.00
S. dev.	4.95	4.11
N	28	
E	0.90	

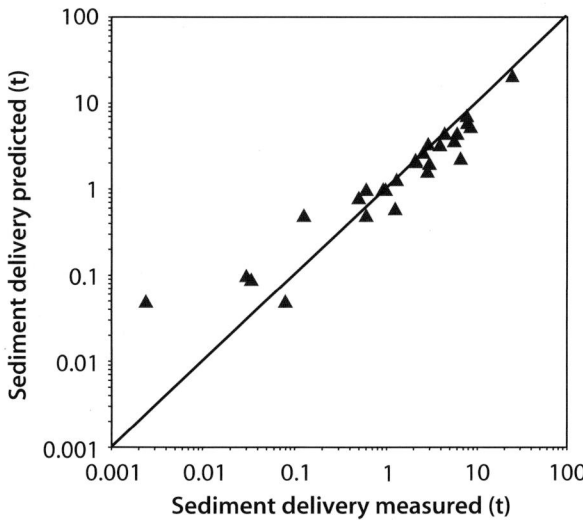

Fig. 3.9. Measured and predicted sediment delivery (*Sed* in t) calculated by USLE (Wischmeier and Smith 1978), LS-factor by the method of Moore et al. (1986), and scouring of particles linked to flow velocity; watershed G1

calculation by AGNPS. The deviation between measured and predicted medians was 1 t. This indicates that the modification using the LS-factor of Moore and Burch (1986) improved sediment delivery calculations. With this modification in the AGNPS model, the description of soil erosion on the field is improved but this does not affect channel erosion.

In Table 3.10, the results for sediment delivery calculated by means of AGNPSm (scouring of particles in channel is linked to flow velocity) are shown. It should be emphasized that no calibration of the sediment routine was necessary. The coefficient of efficiency was 0.90, and the deviation between measured and predicted medians was 0.08 t. These very good results for sediment predictions are plotted in Fig. 3.9. This modification improved the description with respect to channel erosion.

Table 3.11. Statistics for phosphorus in sediment (P_{sed}) and soluble phosphorus (P_{sol} in kg); watershed G1

	Validation events P_{sed}		Validation events P_{sol}	
	Measured	Predicted	Measured	Predicted
Mean	4.24	3.88	3.64	4.24
Median	1.78	1.29	1.57	1.76
S. dev.	5.74	6.00	5.52	6.64
N	28		28	
E	0.71		0.40	

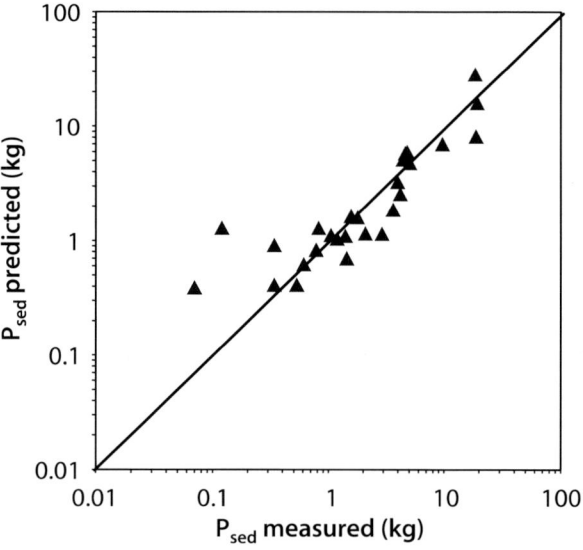

Fig. 3.10. Measured and predicted phosphorus in sediment (P_{sed} in kg); watershed G1

In watershed G1 for nutrient delivery assessment, results for phosphorus in sediment (P_{sed}) as well as for soluble phosphorus (P_{sol}) are shown in Table 3.11 and Fig. 3.10 and 3.11. The reliable results for P_{sed} were based on the excellent results for sediment delivery from AGNPSm. The predictions for P_{sol} were poorer compared to P_{sed}.

In Tables 3.12 and 3.13, a summary for all predictions (AGNPSm) in watersheds G1 and G2 is shown. The predictions were slightly poorer for watershed G2 compared to G1. For runoff the results were reliable for both watersheds. Sediment delivery, phosphorus, and nitrogen in sediment calculated with AGNPSm gave satisfactory results. Very poor results were calculated for soluble nutrients.

The Salzboede watershed was divided into grids of 200 m in width or 2032 grid cells in total. For all events, rainfall rasters were generated by ordinary kriging with the GEO-EAS software package (Englund and Sparks 1988), to use for the grid-based precipitation input to AGNPSm. The Lutz $C1$-factor was fixed at 0.05 and $C3$-factor at 2.0 (see Rode 1995). Peak flow was calculated based on the algorithm of Rode (1995). The re-

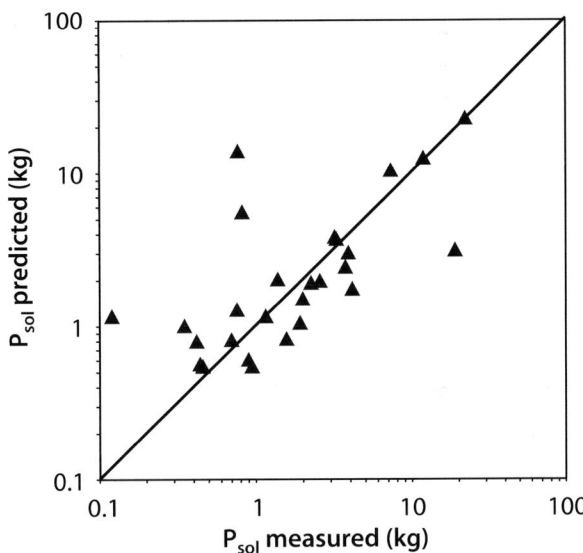

Fig. 3.11. Measured and predicted soluble phosphorus (P_{sol} in kg); watershed G1

Table 3.12. Summary of predictions calculated in watershed G1

	RO (CN)	RO (Lutz)	Q_{max}	Sed	P_{sed}	P_{sol}	N_{sed}	N_{sol}
E (–)	0.93	0.96	0.84	0.90	0.71	0.40	0.79	0.60
D.m.[a]	1.4 mm	0.5 mm	52 l s^{-1}	0.08 t	0.49 kg	0.19 kg	0.06 kg	5.30 kg
N (–)	19[b]	19[b]	19[b]	28	28	28	28	28

[a] Deviation between measured and predicted median;
[b] Validation events.

Table 3.13. Summary of predictions calculated in watershed G2

	RO (CN)	RO (Lutz)	Q_{max}	Sed	P_{sed}	P_{sol}	N_{sed}	N_{sol}
E (–)	0.76	0.83	0.82	0.72	0.64	–1.92	0.40	0.13
D.m.[a]	1.5 mm	0.8 mm	14 l s^{-1}	0.10 t	1.66 kg	0.48 kg	7.85 kg	56.8 kg
N (–)	12[b]	12[b]	12[b]	24	24	24	24	24

[a] Deviation between measured and predicted median;
[b] Validation events.

Table 3.14. Summary of predictions calculated in Salzboede watershed

	RO (Lutz)	Q_{max}	Sed
E (–)	0.87	0.57	0.50
D.m.[a]	0.1 mm	0.15 m^3 s^{-1}	–
N (–)	16[b]	16[b]	7

[a] Deviation between measured and predicted median;
[b] Validation events.

sults in hydrology calculated by AGNPSm are shown in Table 3.14. The median between measured and predicted runoff volume differed only slightly. The coefficient of efficiency for runoff volume was 0.87, for peak flow rate 0.57, and for sediment delivery 0.49.

3.4
Conclusions

Verification of the modified AGNPS model (AGNPSm) was carried out in 3 different watersheds in Germany. The results in hydrology were satisfactory in all watersheds. Sediment delivery was calculated satisfactorily after modifications (LS-factor calculation and scouring of particles linked to flow velocity) were integrated into the AGNPS model. Reliable results were calculated for nutrients in sediment but not for soluble nutrients (P_{sol}). Reasons for the difficulties in nutrient delivery predictions include lack of data on fertilizer application, detailed time of application, and seasonal change in soil nutrients caused by plant uptake, mineralization, vertical wash out, and so forth.

AGNPSm is not a very complex model but uses many empirical algorithms, hence the number of parameters needed is not very high. The advantage of AGNPSm in combination with a GIS and an interface is the possibility it offers to predict runoff volume, sediment, and nutrient yield in medium- to large-sized watersheds. In this study, it was shown that AGNPSm can calculate results in runoff volume, peak flow, and sediment delivery reliably. This tool promises to be useful as a decision support system in further studies.

References

Bagnold RA (1966) An approach to the sediment transport problem from general physics. US Geological Survey Professional Paper 422-I, Washington

Bavarian Water Authority (1984) Nährstoffaustrag aus landwirtschaftlich genutzten Flächen 2/84, 236 pp

DVWK (1988) DVWK-Schriften – Feststofftransport in Fließgewässern – Berechnungsverfahren für die Ingenieurspraxis, Heft 87, Paul Parey, Hamburg Berlin, 135 pp

Einstein HA (1950) The bed-load function for sediment transportation in open channel flows. Tech Bull No. 1026, US Dept Agriculture, Soil Conservation Service, Washington

Englund E, Sparks A (1988) GEO-EAS (Geostatistical Environmental Assessment Software) User's Guide. United States Environmental Protection Agency, Las Vegas, Nevada, EPA/600/4-88/033a

Foster GR, Lane LJ, Nowlin JD (1980) A model to estimate sediment yield from field-sized areas: Selection of parameter values. In: Knisel W (ed) CREAMS: A field scale model for chemicals, runoff and erosion from agricultural management systems. US Dept Agric, Conserv Res Rep 26, Vol 2, Ch 2:193–281

Frere MH, Ross JD, Lane LJ (1980) The nutrient submodel. In: Knisel W (ed) CREAMS: A field-scale model for chemicals, runoff and erosion from agricultural management systems. US Dept Agric, Conserv Res Rep 26:640

Garbrecht J, Martz L (1993) Case application of the automated extraction of drainage network and subwatershed characteristics from digital elevation models by DEDNM. Geographic Information Systems and Water Resources – American Water Resources Association, March, 221–229

Imhoff K (1972) Taschenbuch der Stadtentwässerung. 23. Aufl. R. Oldenburg Verlag, München Wien

Lane LJ (1982) Development of a procedure to estimate runoff and sediment transport in ephemeral streams. In: Recent developments in the explanation and prediction of erosion and sediment yield. Publ. No. 137, Int. Assoc. Hydro Sc, Wallingford, England, 275–282

Lutz W (1984) Berechnung von Hochwasserabflüssen unter Anwendung von Gebiets-kenngrößen. Dissertation, TU-Karlsruhe, 214 pp

Maniak U (1992) Regionalisierung von Parametern für Hochwasserganglinien. In: Kleeberg H-B (Hrsg) Regionalisierung in der Hydrologie. Deutsche Forschungsgemeinschaft VCH Verlagsgemeinschaft, Weinheim, 325–332

Maue G (1988) DVWK-Schriften. Literaturstudie zur Freisetzung von Nährstoffen aus Sedimenten in Fließgewässern. Verlag Paul Parey, 273–344

Moore ID, Burch GJ (1986) Physical basis of the length-slope factor in the universal soil loss equation. Soil Sci Soc Am J 50:1294–1298

Nash JE, Sutcliffe JV (1970) River flow forecasting through conceptual models – Part I: A discussion of principles. J of Hydrology 10:282–290

Rode M (1995) Quantifizierung der Phosphorbelastung von Fließgewässern durch landwirtschaftliche Flächennutzung. Dissertation, Justus-Liebig Universität Giessen, 168 pp

SCS (1972) United States Department of Agriculture – Soil conservation service. National engineering handbook, Sect. 4. Hydrology, 593 pp

Smith RE, Williams JR (1980) Simulation of the surface hydrology. In: Knisel W (ed) CREAMS: A field-scale model for chemicals, runoff and erosion from agricultural management systems. US Dep Agric Conserv Res Report 26:608

Wischmeier W, Smith D (1978) Predicting rainfall erosion losses – A guide to conservation planning. USDA, Handbook No. 537

Young RA, Onstad CA, Bosch DD, Anderson WP (1987) AGNPS: Agricultural Non-Point-Source Pollution Model – A watershed analysis tool. United States Department of Agriculture, Conservation Research Report 35:80

Young RA, Onstad CA, Bosch DD, Anderson WP (1994) Agricultural Non-Point Source Pollution Model, Version 4.03 – AGNPS User's guide (ftp: soils.mrsars.usda.gov)

Chapter 4

Physically Based Modeling of Surface Runoff and Soil Erosion under Semi-arid Mediterranean Conditions – the Example of Oued Mina, Algeria

D. Gomer · T. Vogt

4.1
Introduction

In semi-arid regions such as the Maghreb, the pressure on natural resources such as water and soil are increasing. The mobilisation of water resources is mainly based on the construction of reservoirs.

In the past, these investigations have mostly been made with exclusive profit for the population downstream of reservoirs by providing water for drinking, irrigation and industrial purposes. The watershed upstream was only regarded as provider of water and had to be treated to produce an optimum of water under limited sediment transport.

Soil cultivation for arable farming was long regarded as the principal cause for the sedimentation of the reservoirs, which could supposedly be alleviated by procedures to retard erosion such as terracing and afforestation. Despite many years of expensive measures involving terracing and afforestation which entailed the relocation of people with smallholdings, this strategy was finally acknowledged to be a complete failure and abandoned.

In the case of the Oued Mina example in Algeria, a project involving Algerian-German technical co-operation, new methods and concepts of watershed management under semi-arid Mediterranean conditions have been developed (Fig. 4.1).

The project essentially recognises that the following measures must be implemented which can be summarised as follows:

- Educating decision makers to be more sensitive to the task of harmonising the interests of people living upstream and downstream of dams
- Planning and management strategies for water catchment areas with marly soils in a semi-arid Mediterranean climate
- Investigation and account of the influence that traditional soil cultivation has on the prevention of erosion in catchments with marly soils

The in situ recording and description of runoff and erosion processes in a semi-arid Mediterranean area, based on an engineering approach, was a prerequisite for compiling this new strategy for the harmonised cultivation of water catchment areas. In addition to several years of expensive measurements, conducted on site, which will not be elaborated upon here (Gomer 1994), a distributed, quasi-physical approach was developed and employed to determine surface runoff and soil erosion.

Fig. 4.1. General view in project area

4.2
Model Planning and Theory

4.2.1
Surface Runoff

The modeling of surface runoff is based on a digital grid elevation model. The grid size is variable. For small catchments up to a catchment size of approximately 1 km² a 10 m grid square was selected (at the same time this was the max. definition available for stereoscopic aerial photographs). For catchments up to 100 km² this high geometrical resolution for limiting the data processing time could not be maintained and therefore a 25 m grid square was selected.

The surface runoff in each grid element is regarded as a control volume. The outflow from this control volume runs into a lower grid element, according to the direction of flow.

If, as in Fig. 4.2, a simplified one-dimensional control element on the soil surface is considered with a water level h and an effective lateral inflow S_o, the continuity equation in this one-dimensional case (cf. Moore and Foster 1990) is

$$\frac{\partial h}{\partial t} + \frac{\partial (V_x h)}{\partial x} = \frac{\partial h}{\partial t} + V_x \frac{\partial h}{\partial x} + h \frac{\partial V_x}{\partial x} = S_0 = N_r - f \qquad (4.1)$$

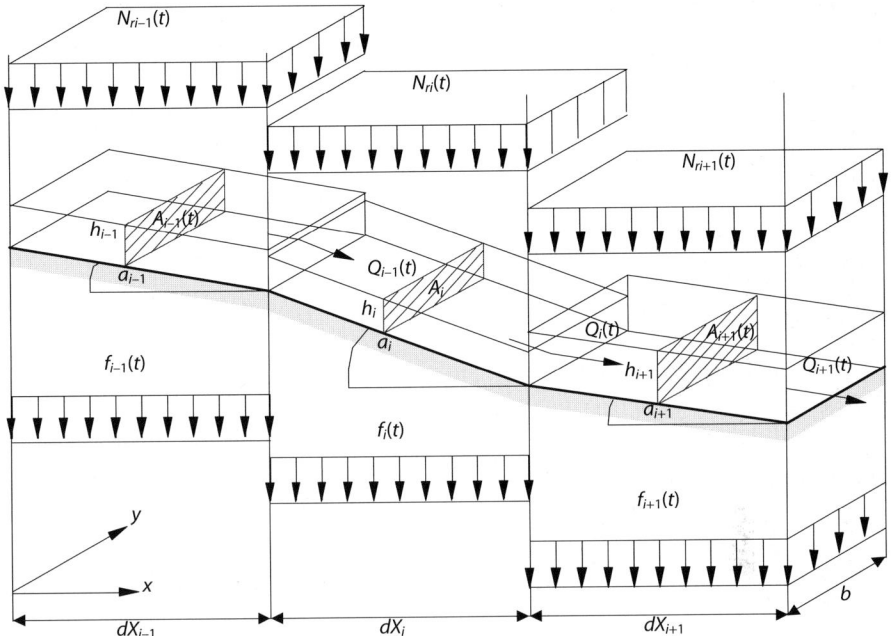

Fig. 4.2. Surface runoff control elements

If the effects of evaporation are disregarded, the effective lateral inflow S_o is formed from the difference in rainfall intensity N_r and the infiltration rate f. This approach, however, is only permissible for short-term appraisal; for observing, for instance, runoff as a result of heavy precipitation. In the case of a sufficiently fine spatial discretion $\partial V_x / \partial x \to 0$ and observation of an grid element with a width b over which flow has occurred, the continuity equation may be applied in the following form

$$\frac{\partial A}{\partial t} + \frac{\partial Q}{\partial x} = S_o b \quad (4.2)$$

In this regard, attention should be paid to the fact that the rainfall intensity N_r applies to a horizontal surface. The infiltration rate f, however, generally applies to inclined surfaces. The following is therefore valid for the continuity equation

$$\frac{\partial A}{\partial t} + \frac{\partial Q}{\partial x} = S_o b = \left(N_r - \frac{f}{\cos \alpha}\right) b \quad (4.3)$$

The attendant motion equation (Eq. 4.4), also known as the second de-Saint-Venant equation, is valid for channels with minimal cross-sectional variation (cf. Moore and Foster 1990). The reduction of the motion equation to the conditions of the kinematic wave presupposes that the local and convective acceleration term and the pressure term can be disregarded which applies to steeply inclined watercourses, without any great loss of accuracy (Woolhiser and Ligget 1967).

$$\frac{1}{A}\frac{\partial Q}{\partial t} + \frac{1}{A}\frac{\partial}{\partial x}\left(\frac{Q^2}{A}\right) + g\frac{\partial h}{\partial x} - g(I_0 - I_f) = 0 \qquad (4.4)$$

Kinematik Wave
Diffusion Wave
Dynamic Wave

Explicit numerical procedures for solving the Eq. 4.4, as opposed to implicit solutions, offer the advantage of being more graphic and simpler to apply. In the case of explicit procedures, however, attention should be paid to the fact, for reasons of stability, that the Courant criterion $\Delta t \leq \Delta x / c$ is adhered to (Woolhiser and Ligget 1967; Schmid 1986). The kinematic wave velocity c, which can be described as the propagation velocity of the wave front, is determined by the quotient $c = \partial Q / \partial A$.

In the case of small time steps, where lateral inflow is disregarded, the continuity equation in discrete form in the one-dimensional case is

$$\Delta A_i \Delta x = \Delta t (Q_{i-1} - Q_i) \qquad (4.5)$$

For Eq. 4.5 the initial conditions $A(x, t=0) \equiv 0$ apply, i.e. no film of water on the surface of the land before precipitation commences. The basic conditions for the uppermost section is $A(x = 0, t) = 0$.

The change in retention capacity $\Delta A_i \times \Delta x$ in section i within a time step Δt is a function of the inflow and runoff Q_{i-1} and Q_i and is not initially known. It is estimated initially from the change in the area through which the flow has taken

$$A_i' = A_i + S_{0i} b \Delta t \qquad (4.6)$$

place, as a result of the lateral inflow S_{oi}.

The outflow from element i can therefore be derived from the well-known Manning-Strickler equation. The following applies to runoff as sheet flow

$$Q_i = \frac{1}{n_i} I_{oi}^{1/2} \frac{A_i^{5/3}}{b_i^{2/3}} \qquad (4.7)$$

and the following to triangular channel runoff

$$Q_i = \frac{1}{n_i} I_{oi}^{1/2} \beta_1 A_i^{4/3} \qquad (4.8)$$

in conjunction with

$$\beta_1 = \left(\frac{m}{8 + 2m^2}\right)^{1/3} \quad \text{and} \quad m = \frac{b}{h} \qquad (4.9)$$

The transition from runoff as sheet flow to channel runoff was described for marly catchments by catchment size (Gomer 1994). During transition from layer to channel runoff it must be borne in mind that this transition does not take place suddenly, thus making it possible for several furrows/rills to occur in a grid element with *rillanz* denoting the parameter for the number of furrows/rills. The following equation is therefore applicable for channel runoff:

$$Q_i = \text{rillanz}_i \frac{1}{n_i} I_{0i}^{1/2} \beta_1 \frac{A_i}{\text{rillanz}_i}^{4/3} \tag{4.10}$$

The area over which flow has taken place is determined from the above estimated value, taking into consideration the inflow and runoff Q_{i-1} and Q_i.

$$A_i = A_i' + \frac{\Delta t}{\Delta x}(Q_{i-1} - Q_i) \tag{4.11}$$

In the case of the Eq. 4.5 to Eq. 4.11 one-dimensional conditional equations relating to *kinematic retention* for channel and sheet flow are therefore available. The surface runoff along the flow channels in a catchment area from its highest to its lowest point can be routed using these conditional equations.

4.2.2
Infiltration

The infiltration characteristics of the soil are described by means of a modified Horton equation (Blum and Gomer 1996). The hydrological model of Horton (1939) and other approaches derived from it are very suitable for describing soil infiltration influenced by surface sealing and crusting, as many authors have already demonstrated. However, instead of the functional dependence between infiltration rate and time chosen by Horton (1939), Gomer (1994), following Roth (1992), Gunnink et al. (1993) and others, contrasted the cumulative kinetic energy with the infiltration rate. For the absolute infiltration rate $f_{(Ekin)}$, is valid, where:

$$f_{(E_{kin})} = (V_{init} - V_{final})e^{-WE_{kin}} + V_{final} \tag{4.12}$$

- $f_{(Ekin)}$ = infiltration rate depending on cum. kinetic energy
- V_{init} = initial infiltration rate
- V_{final} = final infiltration rate
- W = curve-determining factor
- E_{kin} = cumulated kinetic energy of precipitation (Brandt 1989)

Both, the results of multiple small-scale infiltration experiments (1 m²) and of the large-scale rainfall experiments (~ 100 m²) matched very well with an exponential pattern and for identical soils the same parameters are obtained.

Experiments with low antecedent moisture contents demonstrated that in all soils the infiltration rates dropped significantly more slowly than when the antecedent moisture content was high. In the latter case, the final infiltration rate was also reached very rapidly. In contrast to the findings of Gunnink et al. (1993) regarding semi-arid Mediterranean soils in southern France, where the level of precipitation intensity apparently had only a negligible effect on the absolute infiltration rate, here a clear dependence on precipitation intensity could be found.

Gomer (1994) related the infiltration rate f to the rainfall intensity N_r, and with this relative infiltration f_{rel} was able to describe the behaviour of marly soils very accurately with few parameters (Fig. 4.5).

Instead of explaining all parameters dependent on antecedent soil moisture θ_{ini} and rainfall intensity, when matching the actual infiltration rate, it was thus possible to treat the relative initial and final infiltration rates as soil-specific and keep them constant, so that only the curve-determining parameter W exhibited any dependency on antecedent soil moisture θ_{ini}.

$$f_{rel(E_{kin})} = \frac{f_{(E_{kin})}}{N_{r(t)}} = \left(V_{rel,init} - V_{rel,final}\right) e^{-WE_{kin}} + V_{rel,final} \qquad (4.13)$$

where:
- $f_{rel(E_{kin})}$ = relative infiltration rate
  $= f_{(E_{kin})}/N_r$
- $V_{rel,init}$ = relative initial infiltration rate
  $= V_{init}/N_r$
- $V_{rel,final}$ = relative final infiltration rate
  $= V_{final}/N_r$
- W = curve-determining factor
- E_{kin} = cumulated kinetic energy (Brandt 1989)

The curve-determining parameter W, which depends on antecedent soil moisture θ_{ini}, was determined by Gomer (1994) by matching the function to the empirical curve derived from the individual experiments (Fig. 4.6). The curve-determining factor W increased much faster with soil moisture in saline soil types than in soil types suitable to cultivation, with the effect that the final infiltration rate is reached more quickly in saline soils unsuitable to cultivation than in agriculturally used soils.

4.2.3
Soil Removal, Transportation and Sedimentation

Soil removal processes during erosion only arise, according to Schmidt (1991) when external shear forces affecting a grid element overcome the shear resistance of the soil in question. This shear resistance of the soil is made up of the following soil mechanic variables: internal friction, cohesion (if applicable) and gravity. Disregarding the processes during surface weathering and landslides, the external shear forces are determined by the impulse flow of the precipitation and the surface runoff. The conditional equations used to model the erosion of soil particles follow to a large extent the Schmidt model (1991), which is why only a brief description of the most important relationships between the formulae can be given here.

The impulse flow $i_{Q,i}$ from the surface runoff Q_i for grid element i of width b_{raster} and length l_{raster} is calculated at an average flow velocity V_i where

$$i_{Q,i} = \rho_{fluid,i} Q_i V_i \qquad (4.14)$$

and the effective impulse flow $i_{r,i}$ from precipitation r_i is determined by an average slope inclination a_i of the grid element i, taking into consideration the degree of soil covering BBG_i as a function of the average droplet velocity $V_{r,i}$ where

$$i_{r\alpha,i} = \sin \alpha_i \rho_{\text{fluid},i} \cos \alpha_i r_i l_{\text{raster}} b_{\text{raster}} V_{r,i} (1 - BBG_i) \quad (4.15)$$

The average droplet velocity $V_{r,i}$ (m s^{-1}) can be described according to Schramm (1994) as a function of the intensity of precipitation r_i (m s^{-1}) using the following empirical formula

$$V_{r,i} = 4.506 + 0.601 \ln(r_i \cos \alpha_i) \quad (4.16)$$

The transportation of solid matter which occurs when a critical impulse flow is exceeded $m_{\text{crit},i}$, can be determined using an empirical formula devised by Schmidt (1991), where

$$q_s = 1.75 \times 10^{-4} \left(\frac{i_Q + i_{r,\alpha}}{m_{\text{crit}}} - 1 \right) \quad (4.17)$$

The impulse flow m_{crit} is determined in each instance in the case of Schmidt (1991) over a uniform area (1 m²). However, because grid elements of a defined length and width were selected for observation, this must be taken into account accordingly. The transportation of solid material away from a grid element i of width b and length l is calculated by using the relevant critical impulse flow m_{crit} where

$$Q_s = 1.75 \times 10^{-4} lb \left(\frac{i_Q + i_{r,\alpha}}{m_{\text{crit}} lb} - 1 \right) \quad (4.18)$$

In addition to determining the critical impulse flow, the relevant transportation capacity for each grid element must also be known as a function of the general hydrological and sedimentary conditions.

The types of solid material transportation were divided into bed material load and wash load. In each case the maximum possible soil erosion allowed was determined; in the case of bed material load by the maximum transportation capacity, calculated by Engelund and Hansen (1967) and in the case of wash load by establishing a maximum permissible concentration of solid material. The quotient designated the Rouse number z_s (cf. Zanke 1982; Chang 1988) was the criteria used to differentiate between the types of solid material transportation.

4.2.4
Solid Material Continuity

Just as there is a continuity equation for clear water (cf. Eq. 4.2) a corresponding continuity equation for fluids containing suspended matter and sediments can also be formulated (Sloff 1993, etc.). For suspended material in a steeply inclined element the following applies in one-dimensional cases in accordance with Woolhiser et al. (1990).

$$\frac{\partial(AC)}{\partial t} + \frac{\partial(AVC)}{\partial x} + d_s(x,t) = q_s(x,t) \qquad (4.19)$$

During differentiation of this dynamic equation for solid material, it is assumed that the volumetric concentration of suspended matter C_v and the flow velocity V over the runoff cross-section A are averaged out i.e. evenly distributed. If the solid material gathered laterally q_s and deposition d_s are summarised in a parameter Φ as a measurement of the net amount of material gathered, a continuity equation is produced that corresponds to the surface runoff.

$$\frac{\partial(AC)}{\partial t} + \frac{\partial(AVC)}{\partial x} = \Phi(x,t) \qquad (4.20)$$

The use of the kinematic motion formula requires temporally fine discretion, which is why simplistic stationary observation can be applied in the case of the solid material continuity equation.

$$\frac{\partial(AVC)}{\partial x} = \Phi(x,t) \qquad (4.21)$$

If the stationary solid material effects are contemplated for a discrete grid element i, the following applies as a result of $AVC = Q_s$ and $\Delta x \Phi = D_s$ (deposition)

$$Q_s(i) - Q_s(i-1) = D_s(i) \qquad (4.22)$$

According to the effects of solid material transportation Q_s the deposition D_s may assume both positive or negative values. As unrealistic values often arise when establishing a model as a result of linking solid material transportation directly with deposition (or erosion), virtually all known deterministic erosion and deposition models have introduced a so-called transfer coefficient or deposition coefficient C_{dep} (Woolhiser et al. 1990; Schramm 1994).

$$D'_s(i) = C_{dep}(i) D_s(i) \qquad (4.23)$$

This is explained by the fact that simplistic assumptions such as average flow velocity and even distribution of concentrations in the cross-section of the flow in a grid element are not completely reproduced.

It is also known from urban water management that the settling characteristics of granular material and suspended matter vary greatly (Imhoff 1979).

Instead of a general deposition coefficient, deposition coefficients were therefore introduced which were dependant on the transportation process. As in marly areas suspended load is the most important transport mechanism, distinction has been made between suspended bed material load and wash load. The criteria of distinction used is the Rouse number z_s.

If the model of a sand filter bed in a sewage plant is taken, the deposition coefficient C_{dep} as a function of the settling rate w_s, the element length over which flow has occurred Dx, the average flow velocity V and the depth of flow h can be determined. For *bed material load* ($z_s > 0.1$) the deposition coefficient is applied as follows

$$C_{dep} = \frac{\Delta x}{h} \frac{\varpi_s}{V} \leq 1.0 \qquad (4.24)$$

No deposition can be expected during the runoff of very fine suspended material, so that in the case of *wash load* ($z_s < 0.1 \sim 0.06$) the deposition coefficient $C_{dep} = 0$ is produced. In the case of erosion (negative deposition) the coefficient C_{dep} corresponds to a constant transfer rate ($C_{dep} = 1.0$).

4.3
Determining Parameters

To supply reliable results with a distributed surface runoff and soil erosion model, the quality and the exactitude of the spatial distribution of the parameters used within the model are crucial. Besides a well representative digital elevation model and the exact rainfall pattern, the spatial distribution of the physical soil characteristics are most important.

While the physical soil parameters for infiltration and surface runoff are determined by means of small and large-scale rainfall experiments, special methodology was employed to determine the spatial distribution by satellite image interpretation which is dealt with in greater detail below.

4.3.1
Contribution of Remote Sensing

Photo-interpretation has been for a long time a valid help for soil mapping, furthermore infra-red aerial photos give some information on surface wetting status. About a quarter of a century ago satellite recordings came into use. Thus, aerial photographs are generally obtained at a low altitude, whereas the orbits of civil satellites for surveying the Earth's surface are located at a considerable distance (700–800 km for Landsat and SPOT). Optical techniques are therefore ineffective and are replaced by spectral sensors. The spatial resolution is of course less than that of aerial photographs (the Landsat TM pixel is 30 m × 30 m), nevertheless these recordings do have some advantages:

- synoptic view on wide surfaces (less than 3 minutes for recording a Landsat TM scene 185 km × 185 km), that makes the values strictly comparable in every point: this is not the case for aerial photographs, which need several days for the same surface, with changes in lighting, shadows etc. throughout the day;
- high periodicity (18 to 26 days), that permits diachronic observation.

Moreover, the two mid infrared (MIR) bands TM5 and 7 are specially suitable for mapping soil moisture (e.g. Musick and Pelletier 1986): the higher the water content, the stronger infrared absorption is, and the lower reflectance values are.

Satellite data is supplied in a digital form, so that it can be easily processed using all sorts of statistical and image analyses. Statistical processing of physical data avoids the drawback of subjective interpretation. On the another hand, as the data is comprehensive, no interpolation is needed.

4.3.2
Soil Mapping

Soil maps are too often based on criteria irrelevant to evaluate water runoff and soil degradation: precise indications about texture, stoniness, thickness, water holding capacity, etc. are seldom to be found. With regard to the Mina catchment, the map of soil associations at 1 : 100 000 (Semmel and Nierste 1987) worked out from genetic conception, is too general for application requirements. The scale is too small, the part of interpolation too large, and genetic criteria are difficult to relate to hydrological data. The soils are classified according to FAO system and their geographical distribution is assumed to be dependent on geomorphological evolution from the Tertiary age. This is surely a good hypothesis, if evolution and dynamics of the landscape are well known, but no thorough geomorphological study of this region exists. A more detailed map, also genetic, covering the small experimental basins ($\approx$ 300 ha, Schweickle 1993) shows three main soil groups:

- little developed soils on marls and limestones (regosols, alluvial soils, rendzines);
- vertic soils and calcic brown soils (calcic brown vertic soils, soils with sesquioxides, vertic soils);
- alcaline soils (saline soils and solonetz).

The hydrologic behaviour has to be deduced from textural and structural characteristics. The amount of the plant available water content depends on the thickness, which is difficult to appreciate from the map. Anyhow, with regard to hydrologic, hydrodynamic, and agronomic problems, genetic criteria are less pertinent than those based on physical properties. Furthermore, the accuracy of those maps depends on the number of observations and quality of interpolation, but once checked in the field, the accuracy appeared to be unsatisfactory for an information layer to be introduced in a GIS.

This is why a soil classification was experimented using Landsat TM data, which own the double asset of good spatial resolution and wide spectral range.

A quarter scene Landsat TM of 9 January 1990 was chosen, a date which provided a maximum amount of bare soils and represented average moisture conditions, being far from noticeable rainfall episode (December 1989: 50 mm, 5 mm at the beginning of January 1990). Data was processed using a CARTEL software package (Hirsch and Schneider 1983).

4.3.3
Method of Classification

The best inventory technique is classification, which groups in homogeneous classes the multitude of pixels of an image. This has to be made by multivariate statistical processing as, whichever the scanner type is, each pixel contains several values.

Two main methods for classifying satellite data are currently used: unsupervised and supervised. Unsupervised classifications, based only on statistical distances, are no use for a precise inventory of soil cover.

For *supervised classifications*, the routine is a previous accurate field survey aiming to exhaustively identify the types of soil to be introduced as training samples in the classification. Some problems arise from this procedure. First of all, the disparity between visual analysis (even if helped by a radiometer) of the landscape and size of the image pixel (900 m² for TM), an heterogeneous surface that nobody knows exactly how the satellite scanners integrate. Second, the impossibility of getting a wide enough range of ground data to describe all the physical variations seen within one taxon in the study area. Third, the idea that remote sensing is just a convenient way of quickly surveying a well-known and *a priori* well classified reality: what soil scientist look for by remote sensing is the soil legend they are used to. Present-day experiments, employing expert-systems, exogene data, etc. to obtain thematic maps from satellite data, pose the question of the availability of remote sensing in little-known regions, devoid of basic maps, those very areas for which remote sensing should be mostly useful ...

One can consider satellite data from another point of view, bearing in mind that they provide some information which escapes our observation, and are therefore able to enrich our knowledge of the Earth's surface. This is evident for Landsat TM, as amongst 7 bands, 4 record infra-red radiation, unseizable to our sight. Instead of trying to find what we already know, it seems more interesting to discover what else this data can bring us, and instead of starting from ground survey, to previously analyse and process them for posing hypotheses to be checked afterward in the field.

The most known remote sensing images are colour composites, graphic overlay of three bands, that allow visual analyses. Considering a set of 6 bands Landsat TM (the thermal one being different, with a spatial resolution of 120 m × 120 m), these images supply only half the information. Furthermore, visual analyses and comparisons are subjective.

Principal Component Analysis (PCA) is a statistical procedure currently employed in remote sensing to compress data and reduce redundancy. The transformation obtained by PCA ranges the whole data in axes (Principal Components, PCs), aligned along the main directions of variance, orthogonal each other, that seems they are uncorrelated. Over 90% of TM data variance is described by the first three PC. Their colour composite contains near the whole information of all bands, without any redundancy. Areas of different radiometric behaviour (that means different physical properties) are better visualised than by colour composite of the original bands. Training samples are retrieved from the factor colour composite.

Classification methods are based on a multinormality hypothesis. From a thematic point of view this condition cannot be accepted, as taxons little represented in a given area, but thematically significant, could be neglected. Now, for geographical objects, their load is totally independent from their frequency. To overstep this condition, all types of surfaces (= all different colour shades), that means all types of radiometric behaviour, appearing on the factor image are traced, sampled in the original 6-bands file, and put in the classification matrix.

The most performant method of classification we have experimented is stepwise discriminant analysis (BMDP software), which allows very fine discrimination between classes. When a good discrimination (i.e. no overlapping) is obtained, the classification coefficients are applied to the whole file to produce an image, which is to be checked in the field. If spatial distribution of the classes is not random or confused, one can consider that the image represents a natural reality, even if the exact content of classes is unknown. The spectral curve and the place in the graph of the discriminant analysis gives some indications for working hypothesis. By field checking the thematic content of classes is recognised.

In this way, supervision is made *a posteriori* instead of *a priori*. Classes too close are grouped, samples of some taxon that were not taken into account are added, etc. A final classification is then performed. Our experience is that this method leads to a wider spectrum of significant soil types than obtained by a classic method (Vogt 1991). Of course, these classes are synthetic, as they represent global physical behaviour, and their analytical characteristics have to be detailed by field description and laboratory work (texture, retention capacity etc.).

4.3.4
Soil Classification

This was the procedure we used. An initial classification distinguished 20 classes of soils and surfacial formations for field-checking. The questions to be answered were:

1. Are these classes relevant, that means do they really exist in nature, or are they just statistical groups?
2. If they really exist, what is their actual content, i.e. which type of soil do they correspond to?

The image obtained from this classification was used as a basis for field work.

All classes were checked in the field and for each one 2 to 3 pits were excavated. This confirmed that each type distinguished by data classification corresponds to a soil with specific characteristics and physical behaviour. Some classes could be grouped so that 13 soil types and 3 surficial formations (marls, calcretes, sand) were retained (Fig. 4.3):

- *thick clayey soils*: (1) in depressions and north-facing slopes (wetter), (2) on plateaux. Vertic soils (> 80 cm thick, clay > 60%)
- (3) *clayey-silty soils* ($\approx$ 80 cm, clay $\approx$ 40%) generally on a calcrete, equilibred texture in surficial horizons, clayey downwards; (4) *sandy-silty stony soils* ($\approx$ 30 cm thick, sand > 50%), on slopes surrounding the previous class and deriving from its erosion, so that calcrete is attained by ploughing (stoniness; the bigger sand percent comes from disaggregation and dissolution of calcrete)
- (5) *thick red clayey-silty/clayey soils on calcretes* (> 60 cm thick, clay 30–50%), well structured, rich in iron sesquioxides; (6) *stony sandy red soils on calcretes* ($\approx$ 20 cm thick, sand > 50%), linked to runoff erosion (stones and sand deriving from calcrete) of the previous class on slight slope
- (7) *stony clayey soils* (> 100 cm, clay > 50%) without differentiated horizons: they correspond to old solifluction mounds nourished by the marls, slightly salted, with a dense network of desiccation cracks

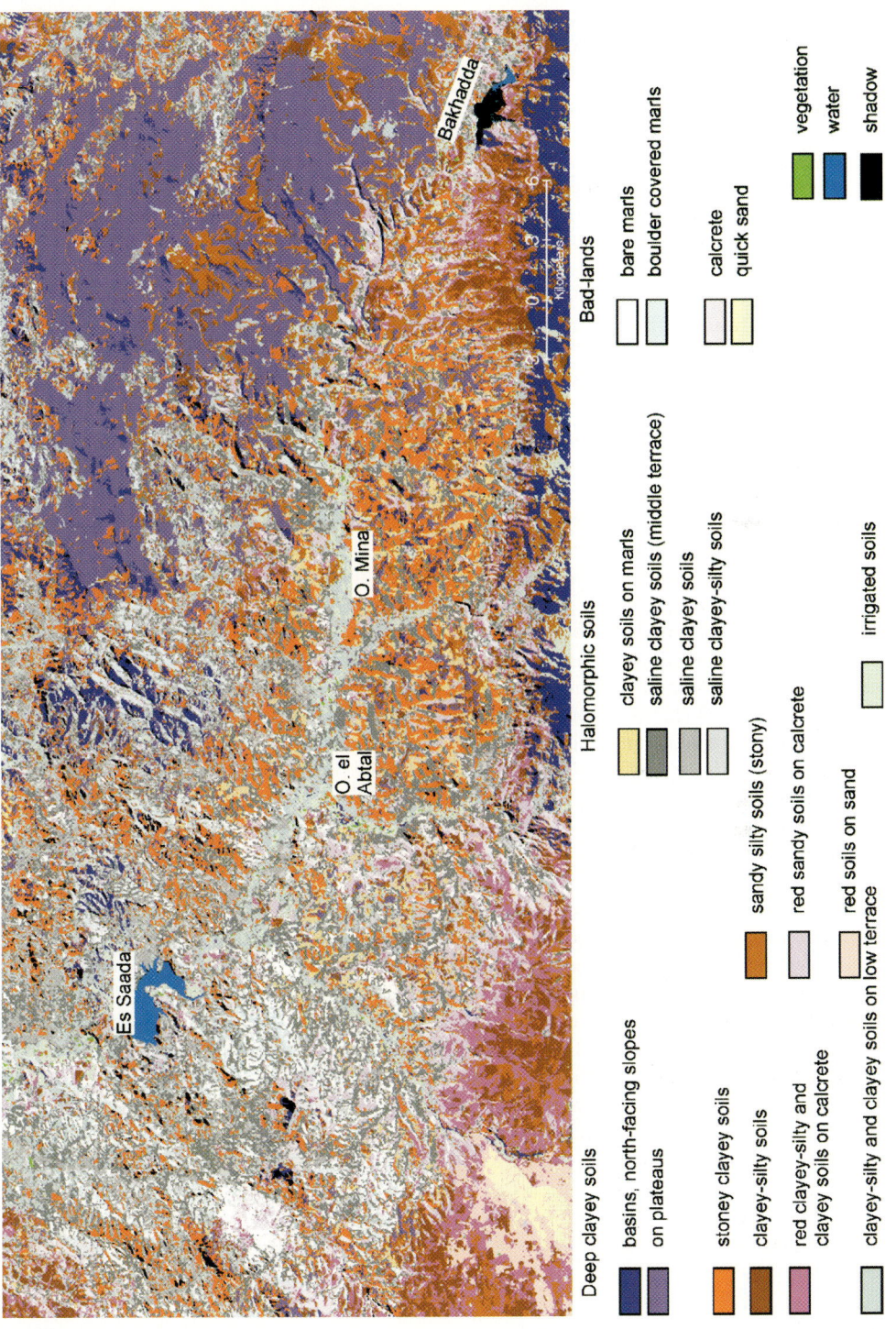

Fig. 4.3. Soils and surficial formations. Zone of Tertiary marls, Oued Mina catchment. Classification of Landsat TM data

- (8) *alluvial soils* of the lower terrace of Oued Mina (90 cm thick or more, fine to balanced texture; clay < 60%, sometimes stones)
- (9) *thick clayey soils on marls* (> 1 m thick, clay ≈ 70%), developed from the underlying marls, they are slightly saline and vertic
- *halomorphic soils*, found in all densely incised areas of the "marl zone", they are thick (90 cm/>1 m), with heavy texture and degraded structure, generally devoid of vegetation, showing three main types: (10) *saline clayey* soils of the middle terrace (clay ≈ 70%), (11) *saline clayey soils* (clay 50%), (12) *clayey-silty saline soils* (clay 30–50%)
- (13) irrigated soils

The classification obtained using remote sensing data is finer: e.g. classes 1–2 and 8 correspond to the vertisols of the soil map due to field survey. The separation is relevant and shows different characteristics, as the first ones contain > 60% clay and no stones. Differences also appear in red soils on calcretes (5 and 6) and between clayey-silty soils and sandy-silty stony soils (3 and 4), which are the same class from a pedogenetic point of view, but not from the hydrological one. Only one class of saline soils, corresponding to class 9 (thick clayey soils on marls) had been recognised by field work, when classes 10 to 12 had been classified as *little developed alluvial soils*: indeed, their topographic characteristics (flat areas) and visual appearance (light, crusty, fine-textured) lead to confusion, whereas in the discriminant graph they are well separated from all other classes.

Field observation had not obtained such a result, because we tend to recognise what we already know. Remote sensing brings new information which enriches our knowledge of the environment, provided we are able to extract and gather them.

This method of classification could be defined as a supervised classification based on non-preconceived, non-subjective sampling. Field checking showed that the accuracy was very fine, in any event better than that of the former soil map.

4.3.5
Mapping Soil Humidity

Soil moisture is function of a set of variables, some interrelated, others uncorrelated: climatic variables, intrinsic soil properties, features of the whole profile ...

Perfecting satellite scanners started from laboratory and field work, in an analytic way isolating the effects of single factors, as soil chemical and mineralogical characteristics, granulometry, roughness, water content, etc. on soil reflectance. Those experiments were performed on small surfaces. The observations made by satellites are synthetic, as spectral bands are broader, scanned surfaces wider and of course heterogeneous. There is a large gap between laboratory and field measurements, and satellite data. If TM5 and 7 in MIR wavelengths have been specially selected for their usefulness in plant and soil moisture evaluation, they supply only a part of the information about humidity provided by the set of TM data. As the water amount increases, reflectances decrease both in the visible and infrared wavelengths. It would be a mental aberration to forget that wetness is a complex phenomenon affecting the whole spectral range.

It follows that the best procedure for remote sensing of soil water content status should be multivariate analysis, instead of simple arithmetical operations as e.g. band rationing.

4.3.6
Principal Components Analysis

As previously stated, PCA ranges the whole data in axes (PCs) aligned along the main directions of variance and orthogonal to each other. By analysing the correlations between TM bands in each PC, say the structure of PCs, some interpretation may be proposed. The first PC, usually showing a positive correlation of all bands, is considered as an equivalent of "brightness", which is responsive to the variance in total reflectance. Another PC responds to high reflection in TM4 (near infrared – NIR), sometimes together TM5 (mid infrared – MIR), and absorption in all other bands, that is characteristic of photosynthetic activity, therefore it is interpreted as an equivalent of "greenness". A third axis shows positive correlation of TM5 and 7, with VIS and NIR in a negative correlation: as this information is related to MIR bands, this axis is considered as an indicator of soil moisture condition (Jensen 1986), that can be termed as "wetness".

Crist and Cicone (1984) obtained simulated TM data by field and laboratory measurements on a set of more than 800 samples of cultures and bare soils. Those data were processed using the Tasseled Cap Transformation (variant of PCA). PC3 shows the positive correlation of MIR bands, with the other wavelengths negatively correlated. The authors demonstrated that this PC is related to differences in soil moisture and is independent of soil surface characteristics, as colour and roughness. The coefficients obtained with this transformation are currently employed for mapping wet areas (e.g. Estes et al. 1991).

However, it appears that these coefficients are too dependent on vegetative cover and not at all useful in the case of bare soils (Vogt and Vogt 1991), that is possibly due to the original sampling. With standardised PCA (as the internal variance of the TM bands is higher than the variance between the bands, the reflectance values are reduced to a standard deviation of 1, that gives all bands equal importance) an axis is obtained which opposes MIR (TM5 and 7) to the other wavelengths and is irrespective of vegetative cover (Vogt 1987; Rimbert and Vogt 1991; Vogt and Lenco 1995). Its structure is identical to that of Tasseled Cap PC3 and can also be interpreted as a "wetness" indicator. The part of variance explained in the factor space as well as factor loadings depends on the state of the landscape: in well-vegetated areas, as temperate regions, or semi-arid environments during or immediately after a rainy season, "greenness" PC (MIR dominant) takes second place after "brightness". In little vegetated landscapes, "wetness" PC is more explicative and takes the second place, "greenness" being little loaded. Therefore, a PCA processed directly on the study area gives a more appropriate result than just applying coefficients obtained from other areas or sets of samples.

Previous works (Vogt and Vogt 1991, 1996; Vogt and Lenco 1995) have shown that the images of PC "wetness" scores well evidentiate differences in soil water content state.

4.3.7
Soil Humidity in the Oued Mina Catchment

In average humidity conditions: For the time the image of 9 January 1990 was processed using PCA. Three factors were obtained, explaining 99.2% of the variance. As usually, the first one shows a positive correlation of all bands and represents more than 93% of total variance. The second PC (4.3% of the variance explained) shows positive correlation of TM5 and 7, in negative correlation with all other bands, and can therefore be considered as PC "wetness". The image of factor-scores shows the moisture being concentrated in lower hillslopes and valley bottoms and the areas near-

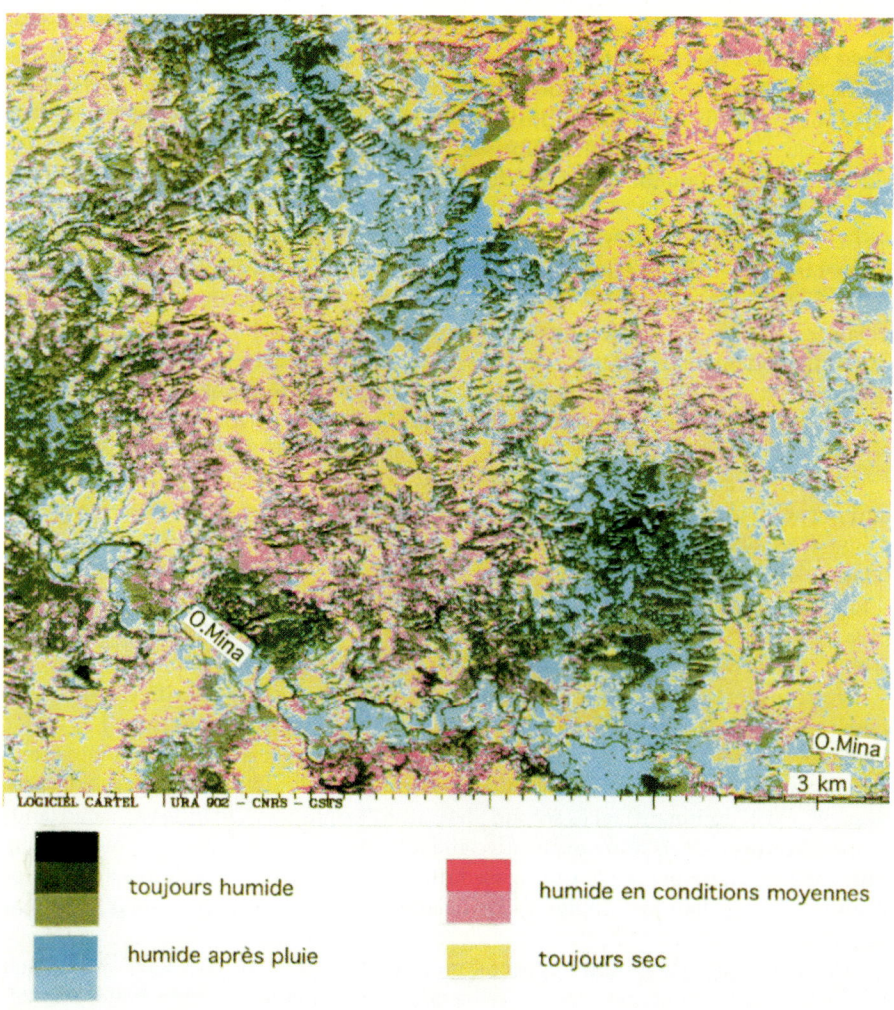

Fig. 4.4. Diachronic comparison of "wetness" factor scores (hyperboxes classification)

est to the main drainage axes being wetter. This corresponds to field observations on saturated areas. On the other hand, comparing this image to the soil classification it appears that the concentration of wetness is independent of soil types and better related to topography: in average humidity conditions water concentrates in the lower-lying areas.

In wet conditions: This first result had to be checked by applying the same method to different wetness conditions. A quarter scene Landsat TM of 19 March 1991 was chosen, which followed a rainy period (1–15 March: 74 to 102 mm in the different sub-basins). PCA reduces all variance in decorrelated axes and scores ranged between −1 and +1, i.e. the same space, whichever the original data is. Jaju (1988) demonstrated that PCA usefully replaces data correction and calibration, which is especially suitable for multitemporal data processing. This legitimises the procedure we employed.

The image of factor-scores in March differs from the first one, as humidity distribution is tributary of orographic situation: the wettest areas are west-facing slopes and thalwegs. Comparing the two images shows four types of behaviour (Fig. 4.4): (1) areas wet in every condition (north-facing slopes and valley bottoms); (2) areas wet immediately after a rainy period (highest hills, west-facing slopes, soils quickly soaked, but with little retention capacity); (3) areas wet in average conditions (topographically depressed areas near the drainage axes, where water draws in some days); (4) always dry areas (south-facing slopes and soils which became impermeable due to compaction by over-grazing or rain-crusting).

To verify the reliability of these results, two subfiles corresponding to the Telfifit catchment were extracted and the factor-scores were compared to hydrologic humidity parameters obtained by field measures, with a satisfying correlation (Gomer 1994).

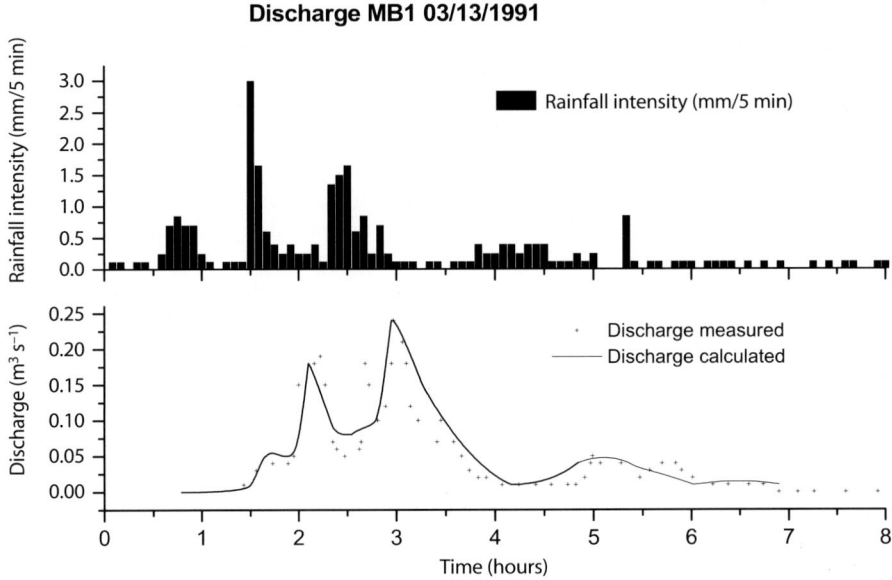

Fig. 4.5. Discharge measured/calculated

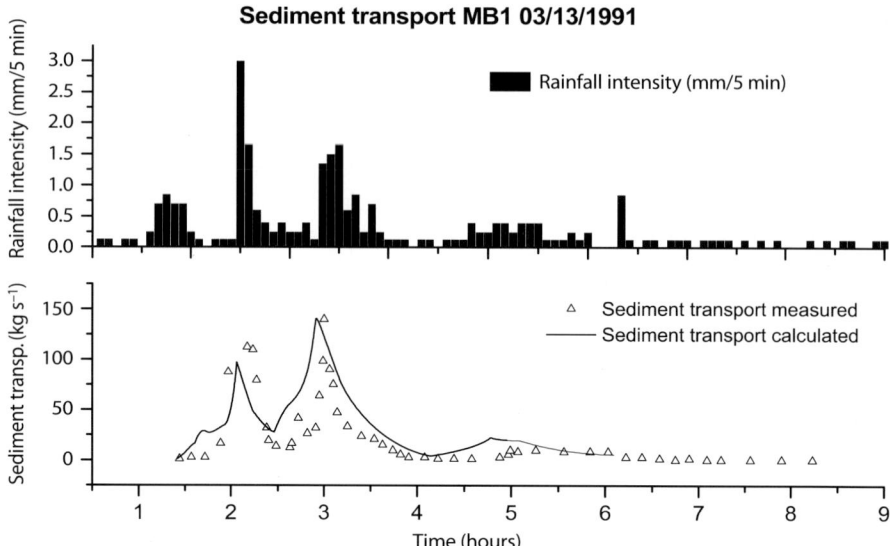

Fig. 4.6. Sediment transport measured/calculated

4.3.8
Model Results

The results of the model and the methodology for the determination of its parameters as described above matched quite well with field data, as shown by the example of an event from 13 March 1991. The calculated hydrograph fits with its peaks the measured values quite well (Fig. 4.5). Similar observations but with less precision can be made on the sediment graph, but sediment field data are less precise then discharge measurements (Fig. 4.6). As sediment transport occurred mainly as suspended load, the peak of the sediment graph is preceding the hydrograph.

As advantage of the distributed physically based model, erosion and deposition might be regarded within a watershed and not only the basin outlet. More important might be, that within that approach we have been able the simulate different scenario of erosion control measurements. With this tool we could confirm the positive impact of agriculture under the conditions of semi arid mediterranean climate and soils on marls.

4.4
Conclusions

The advantage of the described approach is, that it need only few physically based parameters. The most important tool to obtain this parameters is the described methodology to determine there spatial localisation by remote sensing. It has been shown that this approach, once checked by field work and measures, is reliable and give an appropriate, physically based tool for planning purposes. Further validations of this approach in Algeria have been hindered by the actual conditions in the project area.

References

Blum WEH, Gomer D (1996) Runoff from soils on marls under semi-arid mediterranean conditions. Int Agrophysics 10:1–10

Brandt CJ (1989) Simulation of kinetic energy of rainfall under vegetation. Paper pres To Brit Geom Res Group Symp On Veget and Geom 1989, Bristol

Chang HH (1988) Fluvial processes in river engineering. John Wiley, New York

Crist E, Cicone R (1984) Application of the Tasseled Cap concept to simulated TM data. Photogramm. Engineering and Remote Sensing 50:343–352

Engelund F, Hansen E (1967) A monograph on sediment transport in alluvial streams. Tekniks Verlag, Copenhagen

Estes JE, Ehrlich D, Scepan J (1991) Extracting agricultural information from satellite imagery for mapping purposes. Symp. Intern. Cartographie Thématique dérivée des images satellitaires, Saint-Mandé, 2-4 octobre 1990. Bull Comité Français Cartographie 127–128:68–74

Gomer D (1994) Oberflächenabfluß und Bodenerosion in Kleineinzugsgebieten mit Mergelböden unter einem semiariden mediterranen Klima. Mitt Inst f Wasserbau und Kulturtechnik, 191, Karlsruhe, 296 pp

Gunnink JL, de Jong SM, Riezebos HTh (1993) The use of a digital elevation model and experimentally derived infiltration of sealed soils to model surface runoff in a small mediterranean catchment. Workshop in Semi-Arid Mediterranean Areas, Taormina, Italy, 28–30 October 1993:123–136

Hirsch J, Schneider C (1983) CARTEL. Manuel d'utilisation du logiciel de traitement d'images et de cartographie de données de télédétection. Strasbourg: LCT/ULP

Horton RE (1939) Analysis of runoff plot experiments with varying infiltration capacity. Trans Am Geophys Union, Part IV

Imhoff K (1979) Taschenbuch der Stadtentwässerung. 25. Aufl., R. Oldenburg, München

Jaju L (1988) Development of principal component analysis applied to multitemporal Landsat TM data. Int J of Remote Sensing 7:1895–1907

Jensen JR (1986) Introductory digital image processing. A remote sensing perspective. Prentice Hall, Englewood Cliffs, NJ, 379 pp

Moore ID, Foster GR (1990) Hydraulics and overland flow. In: Anderson MG, Burt TP (eds) Process studies in hillslope hydrology. J. Wiley, Chichester.

Musick H, Pelletier R (1986) Response of some thematic mapper band ratios to variation in soil water content. Photogramm.Engineering & Remote Sensing 5:1661–1668

Rimbert S, Vogt T (1991) Données satellitaires et paysages factoriels. In: Pumain D (ed) Spatial analysis and population, INED (J.Libbey-Eurotext, Paris) 321–332

Roth CH (1992) Die Bedeutung der Oberflächenverschlämmung für die Auslösung von Abfluß und Abtrag. Bodenökologie und Bodengenese, H. 6

Schmid BH (1986) Zur mathematischen Modellierung der Abflußentstehung an Hängen. Wiener Mitt. Wasser, Abwasser, Gewässerkunde, Bd. 68

Schmidt J (1991) Entwicklung und Anwendung eines physikalisch begründeten Simulationsmodells für die Erosion geneigter landwirtschaftlicher Nutzflächen. Forschungsbericht BMFT 0339233A

Schramm M (1994) Ein Erosionsmodell mit räumlich und zeitlich veränderlicher Rillenmorphologie. Mitt Inst f Wasserbau und Kulturtechnik, 190, Karlsruhe, 220 pp

Schweickle V (1993) Genese und Standorteingeschaften von Böden auf Alluvionen Nordalgeriens. Mitt Dt Bodenkdl Gesellschaft 72:1051–1054

Semmel A, Nierste G (1987) Carte des associations de sols à 1 : 100 000. In: Projet d'aménagement intégré du bassin-versant de l'Oued Mina. IFG, Offenbach (RFA)

Sloff CJ (1993) Analyses of basic equations for sediment-loaden flows. Report 93-8, Delft

Vogt H, Vogt T (1996) Neotektonische Bedingtheit geoökologischer, durch Fernerkundung erkannter Verhältnisse im Nordelsass und Bienwald (südliche Rheinpfalz) (causes néotectoniques de la diversité spatiale géoécologique mises en évidence par télédétection). In: Mäusbacher R, Schulte A (eds) Beitr. z. Physiogeographie. Festschrift für Dietrich Barsch. Heidelberger Geogr Arbeiten 104:82–88

Vogt T (1987) Classification de terres limoneuses et repérage de sols humides dans l'Outre-Forêt (Alsace, France) à l'aide de données Landsat Thematic Mapper. Rech géogr Strasbourg 27:59–65

Vogt T (1991) Télédétection et risques de désertification: inventaire par télédétection des types de surfaces (données Landsat TM), oasis de Mareth, Sud Tunisien. "Symp Intern Cartogr Thématique dérivée des images satellitaires. Saint-Mandé, 2-4 octobre 1990". Bull Comité Français Cartographie 127–128:178–181

Vogt T, Lenco M (1995) Wetland mapping and monitoring in the Rhine Alluvial Plain (Alsace, France). Sistema Terra IV:75–78

Vogt T, Vogt H (1991) Utilisation de la télédétection pour la cartographie des zones humides. Symp. Intern. Cartographie Thématique dérivée des images satellitaires, Saint-Mandé, 2-4 octobre 1990. Bull Comité Français Cartographie 127-128:146-153

Woolhiser DA, Smith RE, Goodrich DC (1990) Kineros, a kinematic runoff and erosion model. USDA, ARS-77

Zanke U (1982) Grundlagen der Sedimentbewegung. Springer, Berlin

Chapter 5

Assessing the Impact of Lake Shore Zones on Erosional Sediment Input Using the EROSION-2D Erosion Model

S. Jelinek

5.1 Introduction

Non-point matter input in semiterrestrial- and aquatic ecosystems by groundwater flow and soil erosion becomes more and more important as the input by waste water is reduced by exhaustive and effective sewage plants. In this context the impact of shore zones – or general buffer zones – on erosional element input has been pointed out in several studies. All these investigations show in high filter-effects, even though the individual results differ greatly from each other. Mander (1989) reported phosphorus retention at different sites from 0 up to 100%; Fabis et al. (1993) reported an average phosphorus retention in shore-zones of 80%. However, as the authors pointed out these high filter effects are due to sheet flow, which occurs in only 30% of the investigated area.

All this experiment-oriented research is costly and time consuming. The application on other sites and natural areas is very difficult. The quantitative impact of such buffer stripes is controversially discussed in literature; experimental studies on this topic are rare.

This study examines the phosphorus input by soil erosion into Lake Belau in Schleswig-Holstein (Fig. 5.1) with special emphasis on the filter and buffer function of the surrounding shore zone using the EROSION-2D erosion model by Schmidt (1991). The main aim of this paper is not to quantify in detail the erosional sediment input into Lake Belau but to discuss whether EROSION-2D is an adequate tool in this sense.

In general, lake shore zones can be understood as ecotones, which are boundaries between adjacent ecosystems with different structures. The spatial gradients as well as the transfer resistances of these patches control the direction and intensity of biotic and abiotic interaction between the systems (Fränzle et al. 1996).

In this context, lake shore-ecotones (Fig. 5.2) as boundaries between lakes and adjacent terrestrial patches own a special position within the ecotone-types.

Characteristics for this special position are (Steinmann 1991; Naiman and Decamps 1990; Kluge and Fränzle 1992; Kluge and Jelinek 1995; Fränzle et al. 1996):

- high biotic productivity
- complex hydrology and hydrochemistry
- accumulation of organic matter and nutrients
- numerous ecological functions, like
  - habitat
  - connecting biotops
  - migration corridors
  - compensation of men activity
- and lake shore zones control non-point nutrient fluxes

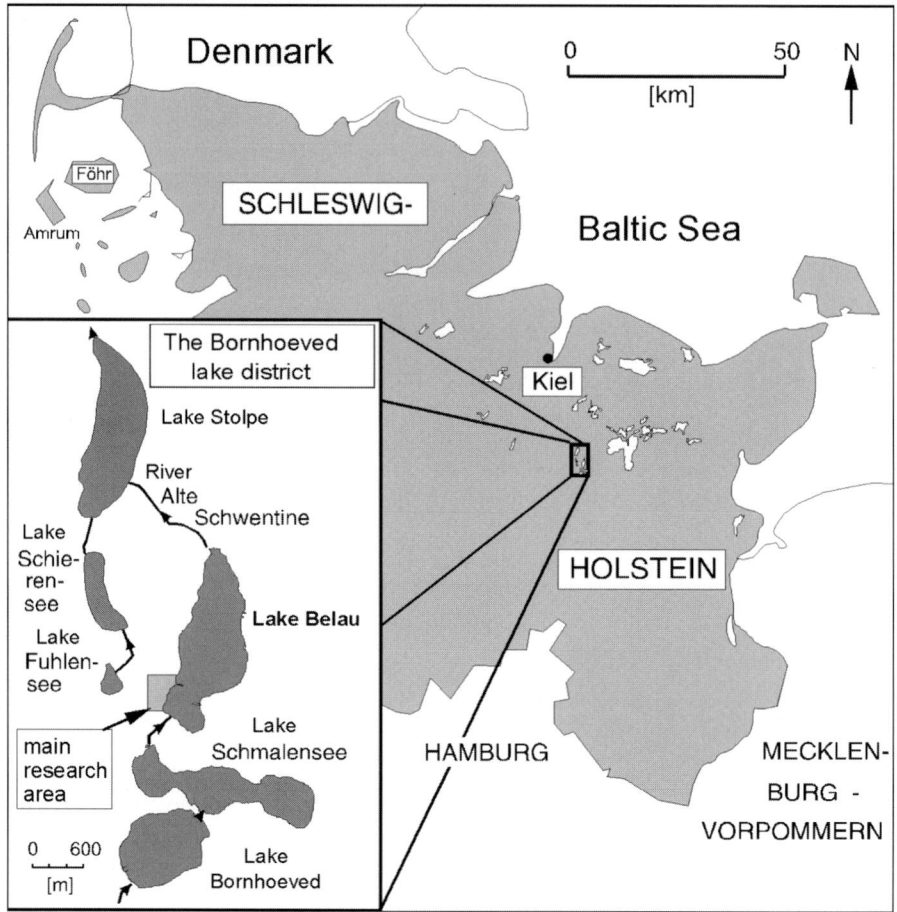

Fig. 5.1. Location of the Bornhoeved Lakes region in Northern Germany

5.2
Lake Belau and the Typifiying of its Shore Zone

Lake Belau is located in the border region of the Weichsalian glaciation, 30 kilometers south of Kiel, the capital of Schleswig-Holstein. It belongs to the hilly district of east Schleswig-Holstein and is part of the Bornhoeved lakes chain (Fig. 5.2). As a result of the anthropogenous lowering of the water level by 1 meter in 1934, a terrace now surrounds the lake (Blume et al. 1992) and defines the character of the shore zone and affects the erosive sediment input.

The relief is dominated by moraines, with heights of 20 m above lake level. Lake Belau is a flow-through lake, regulated by a small weir. As a result of the manifold glacial sediments, the morphology and the historic landuse, one find high variability in soils around the lake (Garniel 1988). Cambisols developed on glacial and fluvioglacial

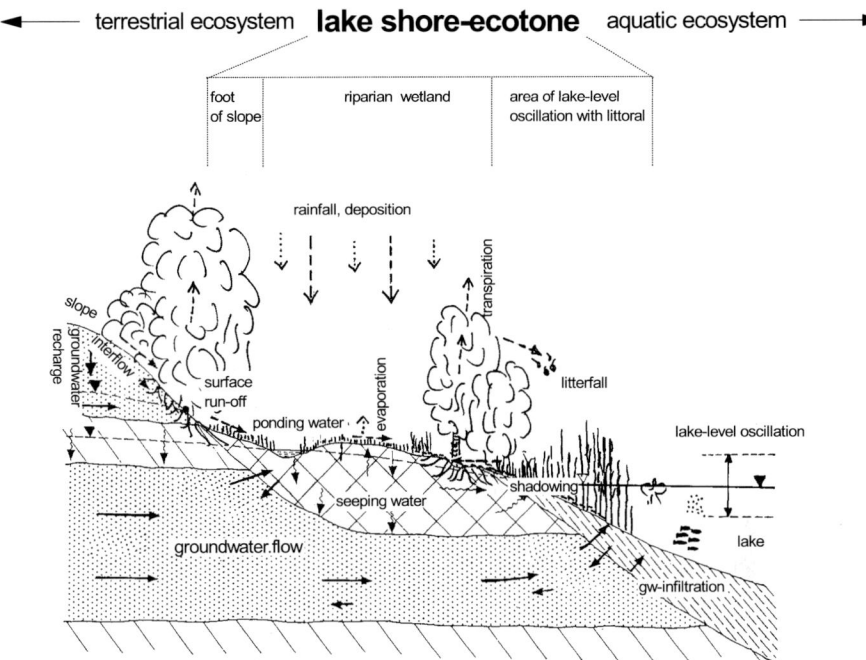

Fig. 5.2. Morphological and functional structure of a lake shore-ecotone

sands. Gleysols, colluvic gleysols and histosols occur on lowlands, colluvisols are found on the lower slope (Schleuß 1992).

The agricultural landuse in the catchment area is dominated by grass and arable land. In the shore zone some sites are used for grazing cattle, some parts are developed, large areas are not under any use. Most parts of the shore zone are forested.

Looking at erosional sediment input caused by surface runoff, element retention in the shore zone is only possible when the flow velocity of the sediment-laden flow decreases, so that adsorption processes in the upper soil region can occur. Typifying of the shore zone has to be related to the processes and parameters which affect the flow velocity of the surface runoff or which totally prevent erosive runoff. Mappings in the shore zone therefore stress the following parameters:

1. the oscillation range of the lake level
2. the width of the shore zone, the length and the gradient of the slope
3. the form of the shore zone (plain surface, concave or convex)
4. the micro- and nano relief (barriers, erosion runnels, surface damages caused by cattle, etc.)
5. the vegetation and ground cover
6. the hydrology of the shore zone like ponding water, springs, seepage water or ditches
7. the landuse

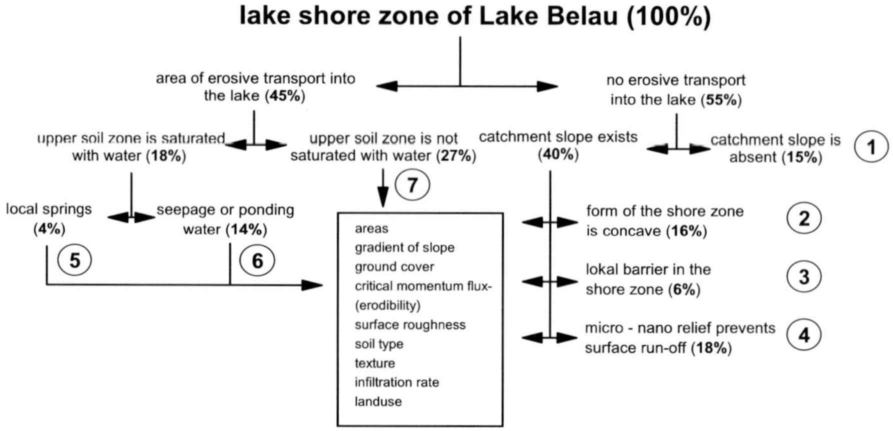

Fig. 5.3. Division of the shore zone of Lake Belau in representative types

Based on this mapping, the shore zone was divided into basic types of shore patches. The analysis thus leads to the definition of seven spatially and functionally distinct shore zone units (Fig. 5.3). No sediment input whatsoever reaches Lake Belau in 55% of the catchment area. The percentage data relates to the circumference of Lake Belau with a length of about 5 400 metres (100%). In 15% of the circumference of lake Belau there is no slope, thus no erosion can take place (1). The remaining 40% is divided into areas where the shore zone is either concave (2), local barriers occur (3), or where the relief with its ground cover prevents surface runoff (4).

Dividing the erosive catchment area (Fig. 5.3) it was important to include the special hydrology of the shore zone. The erosive catchment area is that part of the surface runoff catchment area from which sediment input can reach the lake – if only during extreme conditions, like rainfall on frozen soil, snow melting, or in case of heavy rainfall events. In other words, it is the potential erosive catchment area. In 4% of the catchment area springs emerge in the shore zone (5), and in 14% we find seepage or ponding water (6). So these features affect the sediment input, especially referring to the permanent water flux into the lake. The last type represents that area in which erosive transport into the lake occurs while the upper soil zone is not saturated with water most of the year (7).

Usually one would have to model all these areas, but too many input parameters are needed to do that. And of course EROSION-2D is a two dimensional model.

Because it is not possible to model the whole shore zone of Lake Belau with EROSION-2D, simulating was restricted to some representative shore profiles (Fig. 5.4).

5.3
Modeling

Several models for predicting rainfall erosion have been described recently, but most of them are complex and very difficult to apply in practice (Schmidt 1991). Modeling the erosion and deposition in shore zones is a complex task because of the great

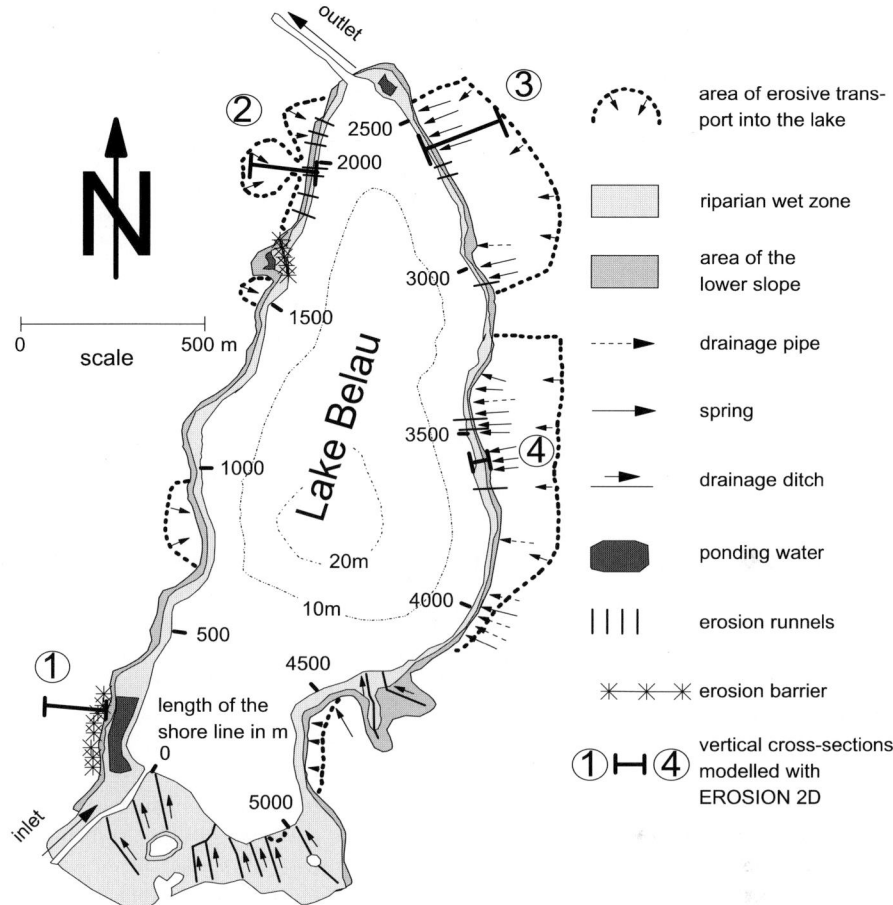

Fig. 5.4. Hydrological and erosional features in local scale in the shore zone of Lake Belau with the location of the modelled cross-sections

heterogeniety inside these patches (Chapter 5.1). EROSION-2D was used because it takes into account the deposition of eroded sediment. Input parameters can be taken from literature or be estimated rather easily in comparison to other models. Figure 5.5 shows the principles of the EROSION-2D model with its required input parameters.

Erosion and deposition in the shore zone of Lake Belau were simulated in four cross-sections (Fig. 5.4). All these cross-sections were modelled with rainfall of a duration of 1 h with varying intensities. The average input parameters for all following profiles are shown in Table 5.1. Geometric parameters, soil types and – textures and the ground cover by vegetation are known based on mappings in the shore zone, or were derived from topographical maps (1 : 5 000). The soil water content at the beginning of the storm event was estimated to an average parameter (Table 5.1). The

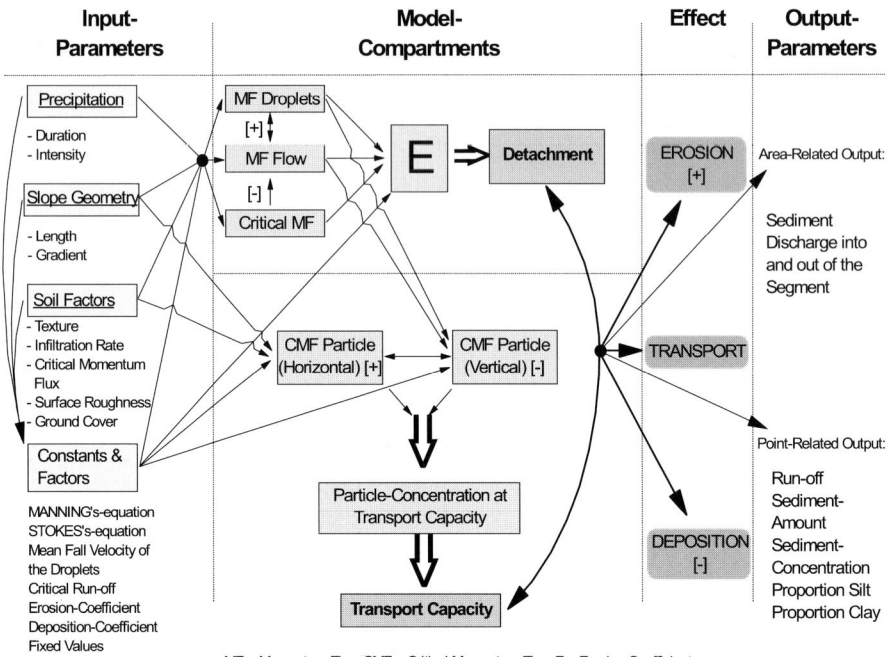

Fig. 5.5. Schematical description of the EROSION-2D model (following Schmidt 1992)

model shows high sensitivity in respect to the erosion-coefficient and surface-roughness, parameters describing the erosion risk. These parameters were determined by using modified standard parameters from Schmidt et al. (1996). They refer to arable land and meadows in general and do not yet fit the special demand of such complex patches like lake shore ecotones with its great heterogeniety and therefore were estimated to reasonable ranges, which are shown in Table 5.1.

The first cross-section (Fig. 5.6) shows the impact of a barrier which retents the whole sediment load. Note that this cross-section is not located inside the erosive catchment area. The reason for modeling this cross-section was to test the plausibility of the model results. The colluvium at the hedgerow with its sediment build-up (a Cumuli-Aric Anthrosol with a high of 1 m, Fig. 5.6) is known. An other aim was to demonstrate the principles of assessing the impact of buffer stripes on erosional sediment flow using the EROSION-2D model. A validation of the model, using the colluvium is rather difficult. The hedgerow is several hundred years old and the landuse has changed many times within this period. Without detailed historic information, a validation of the model is not possible, but to asses the order of magnitude of recent sediment build-up nowadays. Further research has to be done on this topic. Remember, model validation is not the aim of this paper, but to give new aspects for using erosion models in landscape ecology.

Recent landuse of the upper region of cross-section 1 is maize monoculture on a cambic arenosol, the gradient of the slope is only about one degree. The main slope is pastured with a ground cover of 100% (Table 5.1). The critical rainfall inten-

Table 5.1. Mean input parameters for all cross-sections; Schmidt (1990); Schleuß (1992); Schmidt et al. (1996); Schmidt and Schleuß (verbal information)

Cross-section/site	(1) cs	(1) ts	(1) sz	(2) cs	(2) ts	(2) sz	(3) cs	(3) ts	(3) sz	(4a) cs	(4a) ts	(4a) sz
Length (m)	30	53	5	90	7	3	150	3	12	5	4	9
Clay (%)	10	6	4	9	15	15	10	5	20	15	10	15
Silt (%)	24	15	33	26	30	30	20	20	25	30	25	30
Sand (%)	66	79	63	65	55	55	70	75	55	55	65	55
Upper soil density (kg m^{-3})	1450	1300	1100	1400	1450	1350	1400	1500	1250	1300	1400	1300
Org matter content (%)	2.0	3.0	4.0	1.5	1.0	2.0	1.0	0.5	6.0	3.0	3.0	5.0
Water content (%)	30	40	45	30	30	50	30	60	100	30	70	100
Erosion-coefficient (N m^{-2})	0.002	0.010	0.010	0.010	0.001	0.001	0.010	0.005	0.003	0.010	0.001	0.010
Surface roughness (s m$^{-1/3}$)	0.050	0.150	0.150	0.200	0.040	0.050	0.100	0.050	0.080	0.300	0.010	0.035
Ground-cover (%)	0	100	100	100	50	70	100	80	85	100	70	90

cs Catchment slope;
ts Terrace slope;
sz Shore zone.

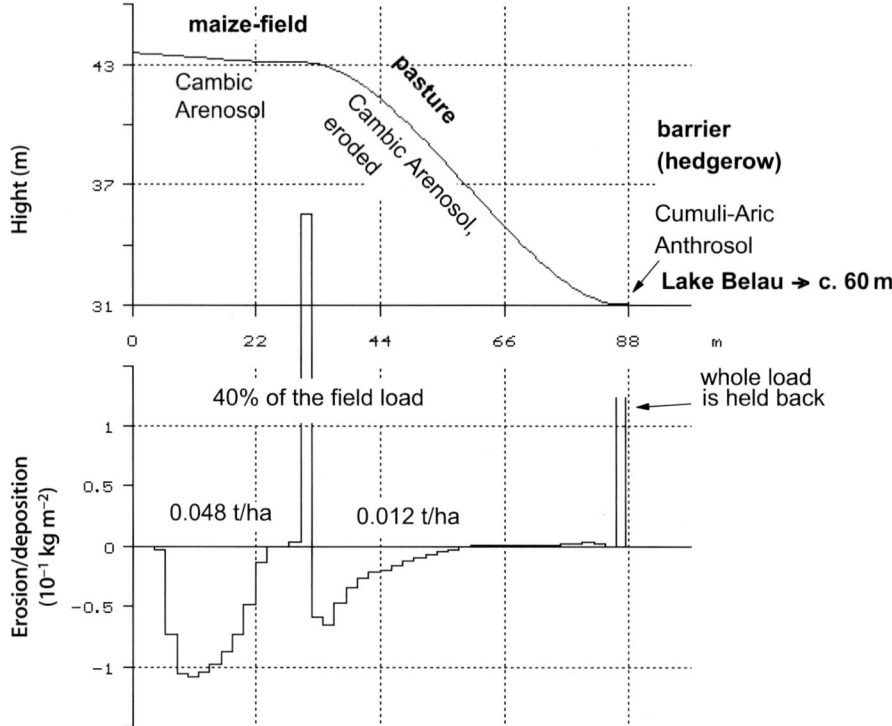

Fig. 5.6. Cross-section 1 (EROSION-2D simulation, input parameters in Table 5.1)

sity reaches 40 mm h^{-1}, which means that the erosion starts at rainfall intensities above this value. The average erosion of 0.048 t ha^{-1} on the field and 0.012 t ha^{-1} on the pasture is very low. 40% of the field-load is held back when it reaches the pasture. Certainly the hedgerow holds back the whole load. The impact of buffer stripes can be demonstrated quite well with this cross-section and that it is possible to quantify these effects using the EROSION-2D erosion model.

Cross-section 2 is located in the north-western region of Lake Belau inside the direct erosive catchment area and shows a very steep gradient of the slope of the first lake-terrace (Fig. 5.7, Table 5.1). The gradient reaches 40 degrees and the ero-dibility (erosion-coefficient), the surface roughness, and the ground-cover is strongly affected by cattle (Table 5.1). The shore zone is narrow (3 m) and the mollic gleysol is not saturated with water in the upper soil region. The critical rainfall intensity was calculated to 10 mm h^{-1}. The average erosion occurring at the slope is 0.16 t ha^{-1}, the retention in the shore zone was calculated at 12.5%. Simulating a rainfall intensity of 30 mm h^{-1}, the average erosion rises up to 26 t ha^{-1}, the retention is then irrelevant. This is the typical phenomenon found in every cross-section. With increasing rainfall intensity, the percentual retention decreases disproportionately (Table 5.2).

Chapter 5 · Assessing the Impact of Lake Shore Zones on Erosional Sediment Input

Table 5.2. Calculated erosion E (kg yr^{-1}) and deposition D (kg yr^{-1}) amounts in the shore zone of Lake Belau (1989–1994)

Cross-section	Length (m) of the shoreline (5 400 m)	Critical rain-fall intensity h_N (mm h^{-1})	1989		1990		1991		1992		1993		1994	
			E	D	E	D	E	D	E	D	E	D	E	D
I	300	40	–	–	–	–	–	–	–	–	–	–	–	–
II	300	10	320	40	130	19.3	71	8	–	–	130	19.3	4000	199
III	200	20	10	–	10	–	–	–	–	–	10	–	30	–
IVa	100	10	243	–	68.5	–	46	–	–	–	68.5	–	159	–
IVb	200	10	390	125	141	28.7	93	36	–	–	141	28.7	331	48.7
IVc	600	30	–	–	–	–	–	–	–	–	–	–	120	–
IVd	1100	30	–	–	–	–	–	–	–	–	–	–	200	–
Sum (kg yr^{-1})			963	165	350	48	210	44	0	0	350	48	4840	258
Sediment-input (kg yr^{-1})			798		302		166		0		302		4582	
Phosphorus-input (kg yr^{-1})			0.4		0.15		0.08		0		0.15		2.3	

IVa   Slope under pasture, wetzone saturated with water (springs);
IVb   Slope under pasture, wetzone is not saturated with water;
IVc   Slope is of use, wetzone saturated with water (springs);
IVd   Slope is of use, wetzone is not saturated with water;
E   Erosion;
D   Deposition.

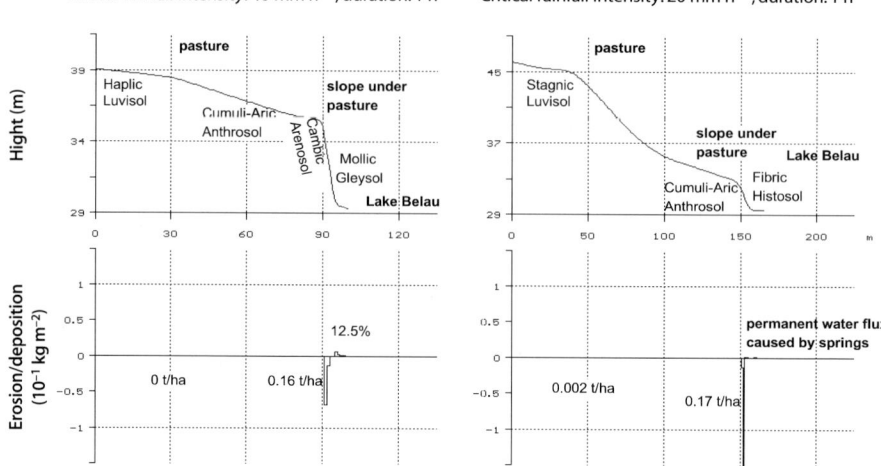

Fig. 5.7. Cross-section 2 (left) and 3 (right; EROSION-2D simulation, input parameters in Table 5.1)

Cross-section 3 is very similar to the previous one, but the ground-cover of the slope of the first lake terrace is not as strongly affected by cattle like in profile 2 (Fig. 5.6). The hydrology of this shore zone is different from the one described before. Springs emerge at the foot of the slope, which leads to a permanent water flux into Lake Belau. Thus, the whole sediment load reaching the shore zone will reach Lake Belau. ERO-SION-2D is able to handle springs by using negative infiltration rates.

Cross-section 4 reveals a slope, typical for the lake terrace (Fig. 5.8). Simulation demonstrates the effect that grazing and springs, ponding water or seepage water in the shore zone may have.

The degree of retention in the shore zone depends on certain hydrological and morphological features. All sites in cross-section 4, described in Fig. 5.8 and Table 5.2, were modelled with a rainfall-intensity of 10 mm h^{-1}. The first example (Fig. 5.8b) simulates a slope under pasture with a natural shore zone. The whole sediment eroded at the slope is held back in the shore zone.

The following site (Fig. 5.8a) demonstrates the impact of springs, emerging in the shore zone or at the foot of the slope.

The last site (Fig. 5.8c and d) shows a natural slope, erosion processes occur only at rainfall intensities higher than 30 mm h^{-1}.

5.4
Upscaling

All cross-sections (1–4) were modelled with different rainfall intensities, the last one with different inputs as well. As a next step the cross-sections have been assigned to the mapped erosive catchment area (Table 5.2).

Then the erosion and the deposition in the shore zone were calculated for every cross-section, referring to the number of rainfall events occurring in the different years (Table 5.3).

Fig. 5.8. Cross-section 4 (ERO-SION-2D simulation, input parameters in Table 5.1)

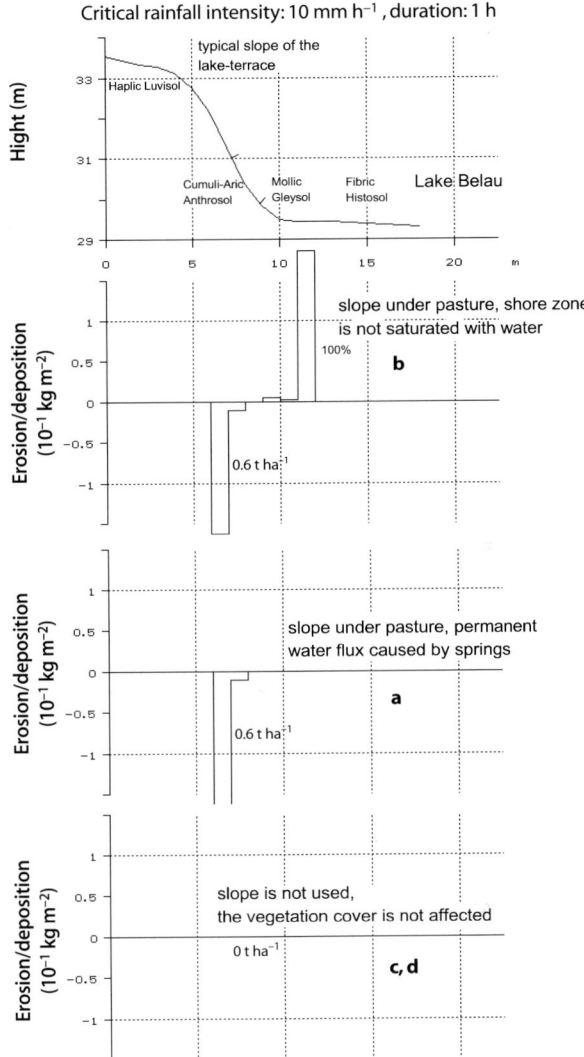

The first column in Table 5.2 shows the modelled cross-section. Cross-section IV was modelled with different input parameters referring to grazing and different hydrological features.

The second column shows the corresponding shore length, the third the critical rainfall-intensity which starts the erosion, and the following columns show the calculated annual erosion and deposition amounts in kg yr^{-1}.

The phosphorus concentration was estimated 500 mg kg^{-1} sediment (Scheffer and Schachtschabel 1992). The annual phosphorus input rates reaches from 0 up to 2.3 kg yr^{-1} phosphorus input, corresponding to the number of heavy rainfall events in the years 1989 to 1994 (Table 5.3).

Table 5.3. Number of heavy rainfall-events at Lake Belau (1989–1994)

Year	Number of rainfall events reaching an intensity of		
	10 mm h^{-1}	20 mm h^{-1}	30 mm h^{-1}
1989	9	1	–
1990	1	1	–
1991	3	–	–
1992	–	–	–
1993	1	1	–
1994	1	1	1

Some remarks on the uncertainties of this method:

1. The input parameters for the different cross-sections were taken from literature or were estimated because they do not yet fit the special demands of lake shore zones and their experimental determination was not possible in this study.
2. The assessment of the sedimental input referring to only 3 geometric cross-sections has to be seen critically (cross-section 1 is not situated inside the erosive catchment area and cross-section 4 was calculated with only 1 geometric profile but different soil parameters).
3. Taking into account only rainfall-intensities with a duration of 1 hour may lead to an underestimation of sediment transport.
4. The phosphorus concentration in the sediment was estimated at 500 mg kg^{-1}, but it may be rather higher in soils enriched with organic matter which often occur in lake-shore zones.

In spite of all these problems, the calculated sediment and phosphorus amounts seem to be in a reasonable range. In comparison with another study (Meyer 1996) using the modified united soil loss equation USLE-DABAG (differenzierte allgemeine Bodenabtragsgleichung) in combination with a Geographic Information System and complex submodels, the average annual phosphorus input into Lake Belau was calculated to 0.15 kg yr^{-1}, rated as too low by the author himself.

The average phosphorus input calculated with EROSION-2D is 0.5 kg yr^{-1} (1989–1994). The reason for this (minimal) gap could be the imprecise resolution of the GIS in the shore zone. This is a common problem of working with a GIS. Certainly the resolution depends on the digitized data source, but nowadays in most cases the resolution of the GIS is not detailed enough to give good results for hydrological questions, especially in the flatlands of northern Germany.

5.5
Conclusions

Looking at other phosphorus input paths into Lake Belau, it becomes clear that the phosphorus input caused by surface runoff plays a minor role for the ecological balance of the lake (Naujokat 1996). Figure 5.9 demonstrates the dominance of the inlet (83%).

The river Alte Schwentine is responsible for the largest amount of phosphorus input. The eutrophication of the Bornhoeved Lake is the reason for this high phosphorus load of the river Alte Schwentine connecting all the lakes of the Bornhoeved lakes chain (Fig. 5.2). Within the past hundred years, the eutrophication of the Bornhoeved Lake has considerably increased till 1975. Since then the concentration of total phosphorus has decreased in the main inlet of the Bornhoeved Lake due to reduced phosphorus export of municipal wastewater corresponding with the erection of a effective sewage plant (Naujokat 1996). The phosphorus accumulation in the sediment in the past leads to high resuspension rates from sediment into the free water body nowadays, which illustrates the dominant roll river Alte Schwentine plays in the ecological balance of Lake Belau.

It has to be pointed out that the very low phosphorus input rate into Lake Belau caused by soil erosion is certainly not representative for other lakes or rivers. The special character of the shore zone of Lake Belau referring to the lake lowering in 1934 is the reason for this, and there is no arable land adjacent to the lake. Slopes of shore zones and of course the shore zone itself have to be taken out of pasture use, because not only the direct input of liquid manure by cattle leads to high phosphorus input rates (Peter and Wohlrab 1990), the erosion risk referring to surface damages caused by cattle can also lead to increasing eutrophication.

In spite of all the problems like estimations and upscaling it could be shown that it is possible to use EROSION-2D to calculate the impact of buffer zones on erosional

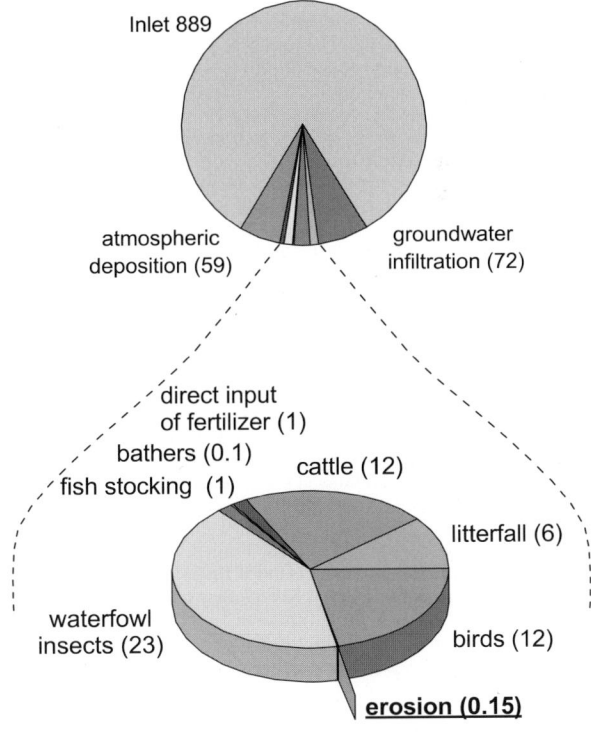

Fig. 5.9. Total phosphorus input rates into Lake Belau in 1993 (*TP* in kg a^{-1}); data from Naujokat 1996; Jelinek 1995

sediment input. Certainly, one has to generate new input parameters for this special use focussing a little more on the deposition effects rather than on the erosion itself.

Calculating the erosional phosphorus input into lakes as demonstrated here has some advantages against other methods like the USLE-equation or other, more complex models:

1. It is an easy and relatively fast method to get a statement on magnitude of erosional sediment input into aquatic- or semiaquatic ecosystems.
2. one may calculate the retention-capability of buffer zones.
3. one gets results, taking into account single rainfall-events, not only annual average measurements.
4. it is an adequate tool for landscape planners to get the figures needed in discussions about the important protection-efforts of buffer zones like lake shore ecotones.

References

Blume HP, Fränzle O, Kappen L, Nellen W, Widmoser P, Heydemann B (1992) Ökosystemforschung im Bereich der Bornhöveder Seenkette. EcoSys 1, Kiel
Fabis J, Bach M, Frede HG (1993) Stoffretention im Uferstreifen. Mitt Dt Bdkdl Gesellschaft 72:1153–1156
Fränzle O, Kluge W, Jelinek S (1996) Die Bedeutung von Uferökotonen für den Wasser- und Nährstoffhaushalt von Ökosystemen. In: Mäusbacher R, Schulte A (eds) Beiträge zur Physiogeographie. Festschrift für Dietrich Barsch, 450–459 pp
Garniel A (1988) Geomorphologische Detailaufnahme des Blattes L 1926 Bordesholm. Staatsexamensarbeit, Kiel
Jelinek S (1995) Einsatz hydrologischer Modelle zur Bewertung des Einflusses von Seeuferzonen auf diffuse Stoffeinträge. Dipl.-Arbeit, Geogr Institut Kiel
Kluge W, Fränzle O (1992) Einfluß von terrestrisch-aquatischen Ökotonen auf den Wasser und Stoffaustausch zwischen Umland und See. Verh Ges Ökol 21:401–407
Kluge W, Jelinek S (1995) Hydrological processes in lake shore zones control the nonpoint nutrient exchange between a lake and its catchment. XX General Assembly Europ Geophys Soc in Hamburg. Ann Geophys 13, Suppl II, C447
Mander Ü (1989) Kompensationsstreifen entlang der Ufer und Gewässerschutz. Landesamt für Wasserhaushalt und Küsten Schleswig-Holstein, Kiel
Meyer M (1996) Erprobung und Anwendung von Methoden zur einzugsgebietsbezogenen Modellierung der Phosphatdynamik terrestrischer Ökosysteme. Dipl.-Arbeit, Geogr Institut Kiel
Naiman RJ, Décamps H (eds) (1990) The ecology and management of aquatic-terrestrial ecotones. Casterton Hall
Naujokat D (1996) Nährstoffbelastung und Eutrophierung stehender Gewässer. Möglichkeiten und Grenzen ökosystemarer Entlastungsstrategien am Beispiel der Bornhöveder Seenkette. Dissertation, Universität Kiel
Peter M, Wohlrab B (1990) Auswirkung landwirtschaftlicher Bodennutzung und kulturtechnischer Maßnahmen. Uferstreifen an Fließgewässern. DVWK 90:55–133
Scheffer F, Schachtschabel P (1992) Lehrbuch der Bodenkunde, 13. Aufl, Stuttgart
Schleuß U (1992) Böden und Bodenschaften einer Norddeutschen Moränenlandschaft – Ökologische Eigenschaften, Vergesellschaftung und Funktionen der Böden im Bereich der Bornhöveder Seenkette. EcoSys Suppl 2
Schmidt J (1991a) A mathemaical model to simulate rainfall erosion. Catena Suppl. 19:145–165
Schmidt J (1991b) The impact of rainfall on sediment transport by sheet flow. Catena Suppl 19: 9–17
Schmidt J (1991c) Anwendung eines theoretischen Modells zur Langfristsimulation von Erosions- und Akkumulationsprozessen an Hängen. Freiburger Geographische Hefte 33:145–165
Schmidt J, Werner M, Michael A (1996) EROSION-2D/3D. Ein Computermodell zur Simulation der Bodenerosion durch Wasser. Sächsisches Landesamt für Landwirtschaft, Dresden
Steinmann F (1991) Die Bedeutung von Gewässerrandstreifen als Kompensationszonen im Grenzbereich zwischen landwirtschaftlichen Nutzflächen und Gewässern für die Immobilisierung der löslichen Fraktionen von Stickstoff und Phosphor aus der gesättigten Phase. Dissertation, Universität Kiel

Chapter 6

Modeling the Sediment and Heavy Metal Yields of Drinking Water Reservoirs in the *Osterzgebirge* Region of Saxony (Germany)

J. Schmidt · M. von Werner

6.1
Introduction

The long-term, continuous delivery of sediments into the surface water system leads to the aggradation and silting-up of stillwater reaches and reservoirs, and gradually to the loss of their ecological and economic functions. Maintenance and restoration of these functions requires extensive technical measures such as impoundment structures and/or continuous dredging of the accumulated sediments. The removal and disposal of the trapped sediments often cause high expenses, in particular if the sediments are contaminated so that special treatment or deposition is required.

Problems of this kind were the reason for conducting the study discussed in this paper. The study aims to estimate the yields of sediment and sediment-bound heavy metals for three drinking water reservoirs in the *Osterzgebirge* Region of Saxony, Germany. Based on this estimation it is examined, how sediment production and sediment transport could be controlled more efficiently within the respective watersheds.

The study makes use of the EROSION-3D simulation model, which allows

- to identify the main areas and causes of sediment production
- to locate the points at which mobilized sediments enter the surface water system, and
- to estimate the yields of sediments and sediment-bound contaminants

6.2
Modeling Principles and Methods

EROSION-3D is a physically-based computer model for predicting soil erosion by water on agricultural land. The model simulates the detachment of soil, the transport of detached soil by overland flow, and the delivery of suspended soil to the surface water system (von Werner 1995; von Werner and Schmidt 1996).

The event-based model consists of two modules. The so-called "pre-processor" calculates the amount and the direction of overland flow by taking account of the slope and the exposition of the considered land surface, and the infiltration rate which is estimated by an infiltration subroutine based on the approach of Green and Ampt (1911). The algorithm for calculating the spatial distribution of flow paths uses a raster-based digital elevation model.

The main module of EROSION-3D simulates

- the detachment of soil particles due to overland flow and raindrop impact
- the hydraulic transport of detached particles by overland flow, and
- the deposition of suspended particles and/or their delivery into the surface water system

The fundamental erosion equations of the main module are based on the momentum flux approach of Schmidt (1996). In this approach, the sum of all mobilizing forces (i.e. of the overland flow) acting on the soil particles are compared to the sum of those forces which prevent the particles from being detached and transported (i.e. cohesion, gravity). Erosion occurs if the sum of the mobilizing forces is greater than that one of the resisting forces. In all other cases, no particles are eroded from the soil surface. Deposition occurs if the balance between mobilizing and resisting forces changes in favor of the latter.

The application of EROSION-3D requires information on site-specific relief, soil and rainfall conditions. This information is supplied to the model using the following parameters:

- *Relief parameters:* x, y, z coordinates (digital elevation model)
- *Rainfall parameters:* Date of rainfall event (dd.mm), rainfall duration, rainfall intensity
- *Soil parameters:* Texture, bulk density, content of organic matter, initial soil moisture, erosional resistance, hydraulic roughness of the soil surface and percentage ground cover

The effects of different types of land use and agricultural management practices on erosion are accounted for by varying the values for erosional resistance, hydraulic roughness and percentage soil cover. Suggested values for these input parameters can be estimated from tabular data (parameter catalogue) which are available for various soils, surface conditions, and management options.

EROSION-3D uses a grid-cell data representation of the watershed. The values of all input parameters are assumed to be spatially uniform below the scale of grid resolution.

The model produces raster-based, quantitative estimates of soil loss, soil deposition, and the sediment delivery into the surface water system. The following data are provided for each grid cell:

- *Parameters related to area:*
    - Erosion and deposition for a chosen grid cell (mass/unit area)
    - Erosion, deposition and net erosion for the watershed draining into a chosen grid cell (mass/unit area)
- *Parameters related to cross-section of flow:*
    - Runoff (volume/unit width)
    - Sediment delivery (mass/unit width)
    - Sediment concentration (mass/unit flow volume)
    - Particle-size distribution of the transported sediment (percentages of clay, silt and sand by mass)

The predicted spatial distribution of erosion and deposition can be plotted as a colored map (see Fig. 6.3) or a three-dimensional block diagram.

6.3
Watersheds

The model was applied to the following reservoir watersheds (Fig. 6.1): Malter/Rote Weißeritz (A_o = 81 km^2), Klingenberg-Lehnmühle (90 km^2) and Saidenbach (61 km^2).

Table 6.1 shows that the sediments accumulated in the reservoirs are highly contaminated with heavy metals. This high metal content makes the dredging and secure disposal of the sediments extremely expensive and may also cause technical problems and further environmental risks.

Since the sediment production cannot be related to any point sources (i.e. mine tailings), it is most likely that the sediments are caused by soil erosion on agricultural land which covers about 30% (Malter, Lehnmühle) to 70% (Saidenbach) of the respective watersheds. However, metal concentrations in the soils are considerably lower than those in the reservoir sediments (Table 6.2). Therefore, there must be an enrichment process which results in higher metal concentrations in the sediments.

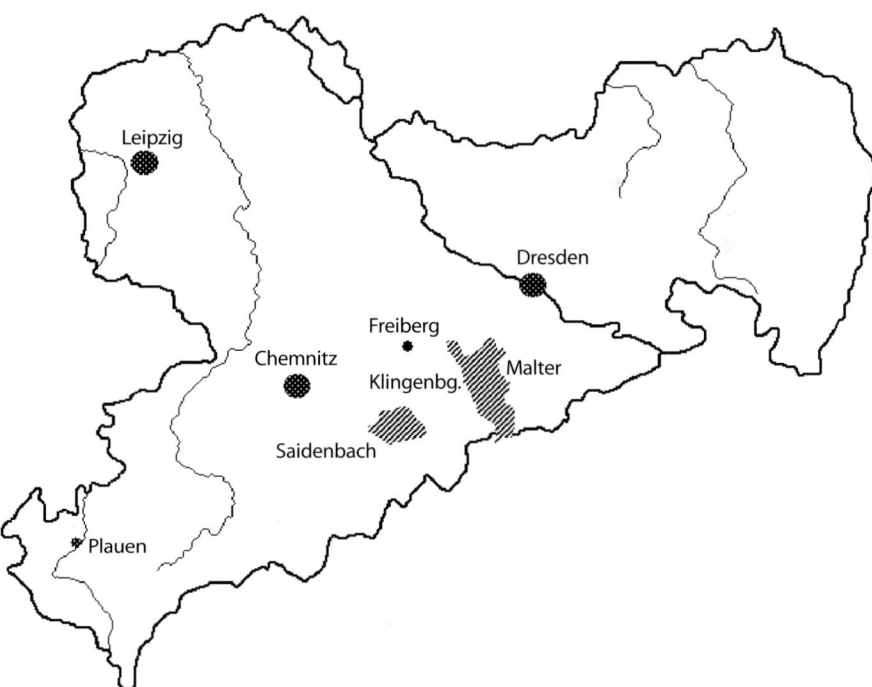

Fig. 6.1. Sketch map of Saxony (Germany) showing the reservoir watersheds which were studied

Table 6.1. Concentrations of heavy metals in the sediment of the Klingenberg-Lehnmühle reservoirs (data from Terra Nova Engineering Ltd. 1993)

Element	Reservoir	
	Klingenberg (mg kg^{-1})	Lehnmühle (mg kg^{-1})
Pb	200 – 300	200 – 400
Zn	600 – 1 500	300 – 800
Cd	10 – 30	4 – 12
As	50 – 70	40 – 60
Hg	0.3 – 0.7	0.1 – 0.3

Table 6.2. Mean concentrations of heavy metals in the soils of the Malter and Saidenbach watersheds (data for Malter reservoir from Nitsche et al. 1993; data for Saidenbach reservoir from Engelhardt 1996; Schmidt 1996)

Element	Reservoir	
	Malter (mg kg^{-1})	Saidenbach (mg kg^{-1})
Pb	–	88
Zn	–	114
Cd	1.1	1.0
As	55	21
Hg	0.13	0.16

6.4
Results

6.4.1
Erosion Simulation

Since EROSION-3D can only be applied to watersheds with a maximum size of 1 000 ha (= 10 km^2), all reservoir watersheds had to be divided into partial watersheds (subwatersheds). The watershed of the Klingenberg-Lehnmühle Reservoirs, for example, was divided into 11 subwatersheds (Fig. 6.2).

Input parameter values were gathered from the following data or maps:

1. *Relief parameters (digital terrain model)*
   - 1 : 10 000 topographic map (scanned sheets)
2. *Rainfall parameters*
   - Digital rainfall data of the German Weather Service (*Deutscher Wetterdienst*)
3. *Soil parameters/land use*
   - 1 : 10 000 topographic map, aerial photos, mapping, soil sample data, EROSION-2D/3D-parameter catalogue

Figure 6.3 shows the spatial distribution of soil erosion and soil deposition as predicted for a single rainfall event on subwatershed 1 of the Klingenberg-Lehnmühle

Fig. 6.2. Division of the Klingenberg-Lehnmühle watershed into subwatersheds

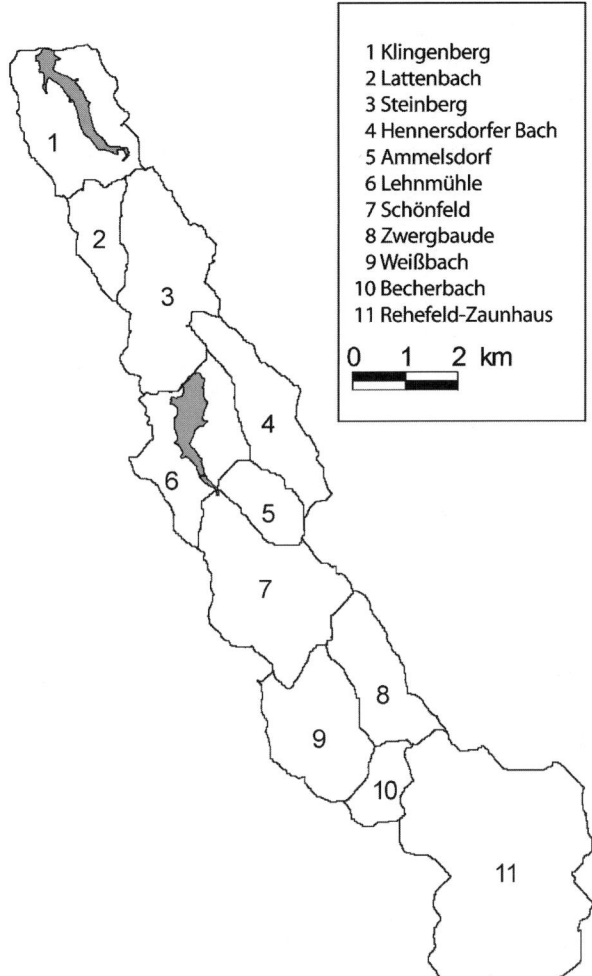

reservoirs. Areas of erosion are indicated by yellow to red, and green to blue indicates areas of deposition. The map shows that, in some locations, deposition fans reach the reservoir despite the forest strip which surrounds it. Areas of maximum erosion are usually found in slope depressions which are drained by concentrated overland flow.

Sediment delivery to any of the reservoirs was predicted from annual estimates of soil loss. These estimates are based on a baseline or reference year scenario which reflects the average changes in the climatic and farming conditions during one year. The simulation predicts an annual sediment delivery into the Klingenberg reservoir of 5.6 kt with 62% passed to the main reservoir by smaller streams, and 38% entering the reservoir directly. Further results from the simulations are listed in Table 6.3.

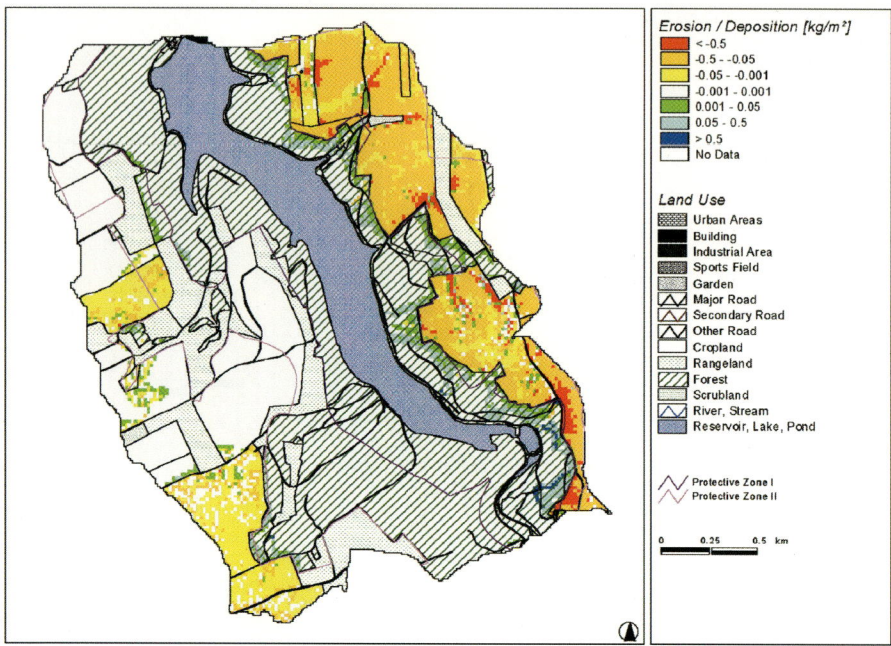

Fig. 6.3. Erosion map for subwatershed 1 of the Klingenberg-Lehnmühle watershed (predicted for the rainfall event on 7 July of the reference-year rainfall scenario; Schmidt and von Werner 1998)

Table 6.3. Predicted mean annual sediment delivery to the Klingenberg-Lehnmühle, Malter and Saidenbach reservoirs (Schmidt and von Werner 1998)

Reservoir	Watershed area (km^2)	Predicted sediment delivery (kt yr^{-1})
Lehnmühle	61	2.6
Klingenberg	29	5.6
Malter	81	5.6
Saidenbach	61	7.9[a]

[a] Preliminary value.

6.5
Comparison of Predicted and Measured Sediment Yields

Measured data on the sediment yield of the investigated reservoirs are either based on

- information on the dredged volume of trapped sediments, or
- the surveying and volumetric estimation of the trapped sediment volume

Mean annual sediment yield was then estimated from the time interval between two dredgings, the volume, and the specific weight of the accumulated sediments. Continuous data on sediment yield are not available.

Table 6.4. Measured and predicted particle-size distribution of the sediment dredged out of the retention basin of the Malter reservoir

Particle class	Measured sediment inflow (kt yr^{-1})	Predicted sediment inflow (kt yr^{-1})
Sand	1.38	1.61
Silt	1.11	3.28
Clay	0.45	0.67

Table 6.5. Particle-size distribution of the sediment dredged out of the retention basin, and out of the main basin of the Malter reservoir

Particle class	Retention basin	Main reservoir
Sand (%)	46	1
Silt (%)	37	67
Clay (%)	15	32

6.5.1
Malter Reservoir

The *Rote Weißeritz* drains approximately two thirds of the total watershed area of the Malter reservoir. Most of the suspended sediment transported by the *Rote Weißeritz* is trapped in a retention basin located at the head of the Malter lake. The sediment in this basin is removed approximately every ten years. The dredging for the period from 1984 to 1994 yielded a mean annual sediment delivery of about 3 kt.

Model calculations estimate the annual sediment delivery from the *Rote Weißeritz* to be approximately 5.6 kt. This is a surplus of 2.6 kt yr^{-1} compared to the mass of sediment trapped in the retention basin. Although no quantitative evidence can be given this surplus appears plausible, because the trapping efficiency of the retention basin may be less than 100%.

Comparing of predicted and measured particle fractions shows that there is a much better agreement for calculated and measured yields of sand than for calculated and measured yields of silt and clay (Table 6.4). This result suggests that the trapping efficiency of the retention basin is much higher for sand than for silt and clay. The data in Table 6.5 also show that virtually no sand has accumulated in the main reservoir. On the other hand a much higher silt and clay content was found in the main reservoir compared to the retention basin.

6.5.2
Saidenbach Reservoir

Like the Malter reservoir, the Saidenbach reservoir is fed by several streams. Due to the high sediment load, retention basins have been built at the mouths of all major streams. For one of the tributary streams, the Hölzelbach, sediment delivery into the retention basin and the main reservoir could be estimated by measuring the volume of the accumulated sediments (Engelhardt 1996, pp. 33–40).

The last dredging operation in the Hölzelbach retention basin dates back to 1960. From measurements of the thickness of the sediment layer in 1995, the mean annual rate of sediment delivery into the retention basin was calculated to be approximately 24 t. Since probably not all the sediments are retained by the retention basin, an attempt was made to quantify the amount of sediment which passes through the retention basin into the main reservoir. This estimation was helped by the location of the retention basin: this drains into a shallow inlet of the main reservoir in which the sediments are deposited as a long alluvial fan. Since the volume of this fan could be accurately surveyed, it was possible to estimate the average annual rate of sediment accumulation to be approximately 96 t since the flooding of the reservoir in 1935. Thus, the Hölzelbach watershed (0.73 km^2) delivers a total sediment mass of approximately 120 t per year to the Saidenbach reservoir and the retention basin respectively.

From simulation runs performed by Engelhardt (1996, pp. 60–65), a sediment delivery of 143 t yr^{-1} was predicted; this value overestimates the measured delivery by about 20%. Considering the uncertainty associated with the surveying of the sediment fan in the Hölzelbach inlet, this level of agreement between the observed and predicted annual delivery rates is satisfactory.

Comparison of the particle fractions (Engelhardt 1996, p. 64) shows that more clay and silt is observed for the measured sediment (i.e. 5% sand, 67% silt, 28% clay) than for the predicted one (i.e. 21% sand, 62% silt, 17% clay). The higher predicted sand content may result from the general overprediction of the sediment delivery from the Hölzelbach watershed.

Unfortunately, the simulations for the watershed of the Klingenberg-Lehnmühle reservoirs cannot be validated in this way because of the lack of measured sediment yield data.

6.6
Estimating the Delivery of Particle-Attached Heavy Metals

Due to the high contamination with heavy metals (Table 6.1), the dredged reservoir sediments usually have to be treated by special decontamination procedures before re-use, or have to be disposed in secured landfills. High costs associated with these procedures must be considered in addition to the dredging costs.

Since the contamination of the reservoir sediments is not the result of point sources, and since the heavy metal contents of the soils are much lower than those of the sediments, some reasons for the observed heavy metal enrichment must exist.

A more detailed analysis of the processes of erosion and sediment transport which are involved here leads to the following hypotheses:

- Heavy metals are adsorptively bound to the soil particles. Particularly clayey and silty particles are loaded with contaminants due to their high specific surface area.
- Once detached by erosion, the smaller soil particles can be carried over long distances by flow, whereas the coarser particles are deposited after short distances preferentially at or near the bottom of the slopes. As a result of this, the smaller, heavily-loaded particles become more and more dominant relative to the coarser, less heavily-contaminated ones as travel distance increases.

The main objective of the simulations presented in the following section is to provide an answer to the question of whether the observed contamination of the reservoir sediments can be explained by the above mentioned processes.

6.6.1
Methodology

The sediment delivery predicted by EROSION-3D provides the basis for the estimates of sediment-bound heavy metal delivery. The calculation includes the following steps:

1. EROSION-3D estimates the sediment yield totals from a given watershed (in mass/unit width) and the percentages of clay, silt and sand. In the first step, the predicted percentages are converted to absolute amounts, such as mass totals for each particle fraction (conversion from percent by mass to mass/unit width).
2. These mass fractions of sediment delivery are multiplied by the respective heavy metal concentrations for each textural fraction of the parent soil (in mass/mass). The resulting product is the absolute heavy metal delivery from the watershed, given for each particle fraction of the delivered sediment (in mass/unit width).
3. These fractional contributions are added in order to yield the total mass of heavy metals contained in the delivered sediment (in mass/unit width).
4. The absolute amount of heavy metals which is delivered from the watershed into the reservoir is obtained by multiplying the total output of heavy metals (in mass/unit width) with the raster width (or spatial resolution) of the grid-cell data representation.

6.6.2
Plot Rainfall Simulations

The described methodology has been experimentally validated by rainfall simulations which were carried out on four test plots (see Fig. 6.4). Main data for these experiments are summarized in Table 6.6.

With the except of cadmium, the enrichment of heavy metals in the delivered sediments could be observed during all rainfall experiments. This result is particularly surprising, since the flow length was limited to 22 m or 14 m, depending on the experimental layout of the plot simulations. The enrichment of finer particles in the sediments is also indicated by the differences between the particle-size distributions shown in Fig. 6.5.

Table 6.6. Main data of the plot rainfall simulations

Plot	1	2	3	4
Length (m)	22	22	22	14
Slope (%)	10	9	10	9
Intensity (mm min^{-1})	0.55	0.62	0.62	0.55
Soil cover (%)	0	10	5	20

Fig. 6.4. Plot rainfall simulation used for experimental investigation of soil erosion and sediment-bound heavy metal transport

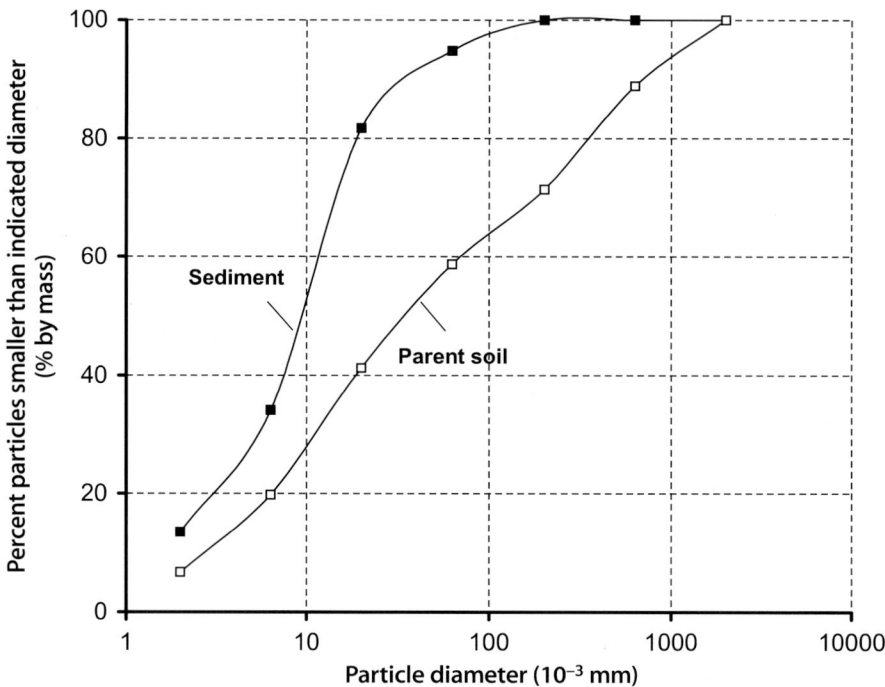

Fig. 6.5. Comparison between the particle-size distribution of the parent soil and the delivered sediment (plot No. 3; Schmidt 1996)

Table 6.7. Heavy metal concentrations (mg kg^{-1}) in the parent soil and the delivered sediment (Schmidt 1996)

Plot	Element	Parent soil				Sediment			
		Total	Clay	Silt	Sand	Total	Clay	Silt	Sand
1	Pb	74	154	68	70	100	157	87	71
	Zn	145	410	143	109	266	461	219	246
	Cd	3.1	4.6	3.2	2.8	1.9	2.1	1.8	1.8
	Hg	0.20	0.44	0.13	0.23	0.37	0.78	0.17	0.73
2	Pb	87	178	94	68	117	191	98	60
	Zn	153	434	171	100	290	523	230	110
	Cd	2.9	3.6	2.7	3.0	1.8	2.3	1.9	0.9
	Hg	0.17	0.17	0.25	0.07	0.41	0.40	0.40	0.54
3	Pb	92	216	98	66	117	183	109	63
	Zn	135	330	156	80	215	271	207	166
	Cd	2.6	3.6	2.1	3.1	2.4	3.0	2.3	2.0
	Hg	0.20	0.32	0.23	0.14	0.31	0.34	0.29	0.48
4	Pb	95	192	94	82	174	178	105	70
	Zn	133	354	154	87	448	441	249	210
	Cd	1.0	1.5	1.0	1.0	2.4	1.8	1.6	1.1
	Hg	0.29	0.43	0.25	0.30	–	0.68	0.26	0.30

Table 6.8. Measured and predicted percentages of clay, silt and sand in the sediment and the parent soil

Plot	Textural class	Parent soil	Sediment	
			Measured	Predicted
1	Clay	6	20	10
	Silt	51	78	89
	Sand	43	2	1
2	Clay	6	23	12
	Silt	45	71	87
	Sand	49	6	1
3	Clay	6	15	10
	Silt	51	80	88
	Sand	43	5	2
4	Clay	6	28	13
	Silt	40	71	86
	Sand	54	1	1

In a comparison of the heavy metal concentrations in the textural fractions of the parent soils with those of the sediments, only slight differences were observed (Table 6.7). The specific load of contaminants clearly increases with decreasing particle diameter. Compared with the parent soil, the higher contamination of the sediments seems to be a result of the enrichment with more heavily contaminated clay and silt particles, due to selective transport and deposition. These experimental results appear to confirm the hypotheses mentioned previously.

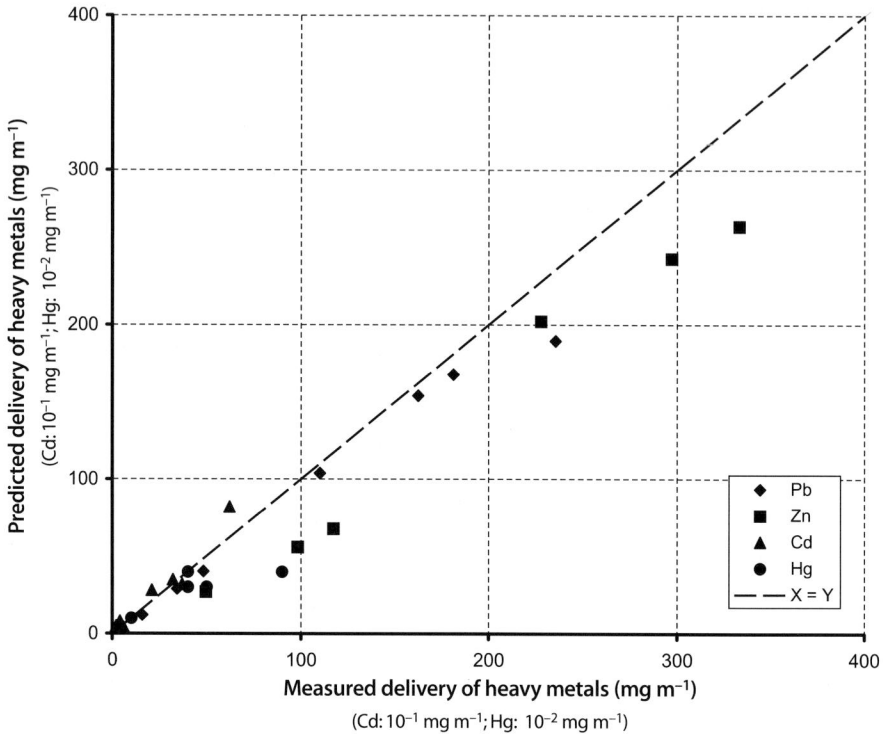

Fig. 6.6. Comparison of measured and predicted deliveries of heavy metals

The main processes of the selective, particle size-dependent transport of soil particles are represented in the modeling approach. However, the predicted deliveries of sediment-bound contaminants are very dependent on the accuracy with which these processes are simulated. An assessment of this accuracy can be made from the comparison of the measured and predicted sediment textures as shown in Table 6.8. The table also includes the respective textures of the parent soils. The data show that the model tends to underestimate the content of clay and, at the same time, to overestimate the silt fraction. The mean deviations amount to −9% for the clay fraction, +12% for the silt fraction, and −3% for the sand fraction.

Measured and predicted deliveries of heavy metals are plotted in Fig. 6.6 for comparison. The diagram shows that – in most cases – the total delivery is notably underpredicted. The mean deviation amounts to −13.5% for lead, −28.4% for zinc and −21.8 for mercury. A positive deviation of +43.6% is observed for cadmium only.

6.6.3
Saidenbach Reservoir (Hölzelbach Subwatershed)

The experimentally-tested estimation procedure was applied to the Hölzelbach watershed in order to estimate its metal delivery to the Saidenbach reservoir. Heavy metal

Element	Element concentration in soil (mg kg^{-1})			Element concentration in sediment (mg kg^{-1})		
	Clay	Silt	Sand	Clay	Silt	Sand
Pb	179	85	65	150	89	68
Zn	317	119	55	366	230	163
Cd	2.2	1.0	1.0	1.6	1.5	3.0
As	49	21	11	45	27	15
Hg	0.4	0.2	0.1	0.3	0.2	0.0

Table 6.9. Mean heavy metal concentrations in the main textural classes of the soils of the Hölzelbach watershed, and in the sediments of the retention basin and the Hölzelbach inlet of the Saidenbach reservoir (Schmidt 1996; Engelhardt 1996)

Table 6.10. Comparison of measured and predicted annual heavy-metal deliveries from the Hölzelbach watershed (Saidenbach reservoir)

Element	Total annual delivery (kg yr^{-1})	
	Measured	Predicted
Pb	13	14
Zn	15	20
Cd	0.2	0.2
As	3.7	3.4
Hg	0.03	0.03

concentrations in the textural fractions of the parent soils were determined from 16 representative samples. Values for the metal concentrations in the accumulated sediments were derived from 11 samples. Mean values for the concentration in the soils and the sediments are compiled in Table 6.9.

Values for the measured and predicted annual delivery of heavy metals from the Hölzelbach watershed are summarized in Table 6.10. The table shows that the model reproduces the measured metal-deliveries with an acceptable accuracy. A larger variation is observed for the delivery of zinc, which is considerably overestimated.

6.7
Strategies for Minimizing the Sediment Delivery

The model predictions can be used to locate those areas within the watersheds which are subject to serious erosion and which thus make a major contribution to the total delivery of sediments and heavy metals into the reservoirs. The erosion map produced by EROSION-3D also helps to point out those locations at which eroded soil enters streams or lakes. Possible measures to reduce erosion on these particular areas and to minimize the sediment transport into the reservoirs are presented and discussed in the following section.

In general, active and passive measures can be differentiated:

- *Active measures* aim to reduce or even prevent the mobilization of soil particles. On agricultural sites, this objective can be attained either by leaving residues of the main crop on the soil surface, or by cultivating a cover crop. In both cases the following crop is planted into the residues of the previous crop (mulch seeding). The main effect of these measures is the protection of the soil surface from splash impact. The vegetative cover also reduces the eroding impact of surface runoff due to the increased roughness of the soil surface.

    A further option of active erosion control is the use of plowless or non inverting tools for primary tillage. The most important effect of these tillage techniques is the maintenance of stable soil aggregates and fast-draining macropores (e.g. earthworm burrows, root channels). This usually results in an improved structural stability of the soil, higher infiltration capacity, and higher resistance to erosion. The combination of cover cropping, mulch seeding and plowless tillage is also called "conservation tillage".
- *Passive measures* aim to intercept and retain the detached soil particles before they enter the surface water system. Such measures may consist of riparian buffers, grassed waterways, diversion ditches or retention basins.

Most of the reservoirs investigated in this study and their tributary streams are protected by riparian forests (see Fig. 6.3 for example) and retention basins. As the simulation runs show these buffers might fail to retain all the transported sediment, just as the retention basins. Nevertheless, the improvement of the trapping efficiency of existing forest buffers and retention basins could be a possible option for minimizing the sediment yield of the reservoirs.

The second option involves the application of active protection measures which, however, can be implemented only in cooperation with the local farmers. A major problem might be the investment in new tillage machinery which is required when changing to conservation management practices. The reliability of crop yields may also decrease, and sometimes crop rotations have to be changed. For that reasons a successful implementation of active soil erosion control measures usually requires intensive agricultural counseling and financial compensation for the farmers involved.

Beyond that it is particularly important that farmers and reservoir authorities should be provided with site specific informations on the effectiveness of active as well as passive soil conservation measures. Related to the subwatersheds of the Klingenberg-Lehnmühle reservoirs some EROSION-3D simulations were performed for this purpose. At first a complete conversion to conservation management practices was assumed ("best-case" scenario) for the entire watershed. The results of these simulations are summarized in Fig. 6.7.

The results of a "worst-case" scenario, for which conventional management practices (i.e. seedbed, no soil cover) were assumed, are plotted for comparison. The effectiveness of the active soil erosion control measures is clearly demonstrated. The sediment delivery from some subwatersheds can even be reduced to zero if all agricultural land is managed using conservation tillage.

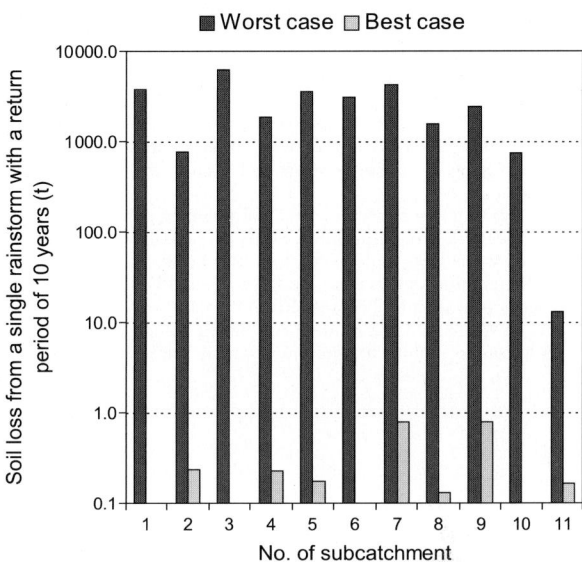

Fig. 6.7. Sediment deliveries from the subwatersheds of the Klingenberg-Lehnmühle reservoirs, predicted for a rainfall event with a return period of 10 years. Seedbed conditions have been assumed for both the conventional ("worst case") and conservation ("best case") management scenario

6.8
Conclusions

The EROSION-3D model has been successfully applied for estimating the yields of sediments and sediment-bound heavy metals of drinking water reservoirs in the *Osterzgebirge* Region of Saxony, Germany. The model predictions enable the local authorities to identify the main areas of accelerated erosion and to locate the points at which eroded sediments enter tributary streams or directly the reservoirs. With this the model supply important informations in order to implement soil and water conservation measures within these watersheds.

Model simulations performed for the watershed of the Klingenberg-Lehnmühle reservoirs show that the sediment yield can be considerably reduced if all agricultural land is managed using conservation tillage. Since conservation tillage not only minimizes the sediment and contaminant delivery, but also prevents soil erosion, it should be clearly preferred. Passive measures of soil erosion control, such as riparian buffers, grassed waterways, diversion ditches, or retention basins should be considered only for selected locations where the risk of direct sediment delivery into the surface water system is extremely high.

References

Engelhardt S (1996) Partikelgebundener Schwermetalltransport aus einem kleinen Einzugsgebiet – Hölzelbergbach, östliches Erzgebirge. Unpubl M Sc thesis. Depart Geogr, Göttingen Univ

Green WH, Ampt GA (1911) Studies on soil physics. I: The flow of air and water through soils. Journal of Agricultural Sciences 4:1–24

Nitsche C, Schönherr A, Rößner U, Kemmesies O, Ehrlich B (1993) Ergebnisbericht über die Durchführung laborativer Untersuchungen zur Deponierung der Sedimente aus der Vorsperre Malter. Groundwater Research Center, Dresden

Schmidt J (1996) Entwicklung und Anwendung eines physikalisch begründeten Simulationsmodells für die Erosion geneigter landwirtschaftlicher Nutzflächen. Berliner Geographische Abhandlungen 61

Schmidt J, von Werner M (1998) Einsatz hochauflösender Erosionsprognosekarten zur Verbesserung des vorsorgenden Schutzes von Böden und Gewässern. Talsperren Klingenberg/Lehnmühle, Niederstriegis. Research report on behalf of the Saxonian Agency for the Environment and Geology

Terra Nova Engineering Ltd (1993) Qualität und Quantität des Sediments in den Trinkwassertalsperren Klingenberg und Lehnmühle. Study on behalf of the Saxonian Reservoir Authority

Werner M von (1995) GIS-orientierte Methoden der digitalen Reliefanalyse zur Modellierung von Bodenerosion in kleinen Einzugsgebieten. Ph D thesis, Depart Geogr, Berlin Free University

Werner M von, Schmidt J (1996) EROSION-3D – Ein Computermodell zur Simulation der Bodenerosion durch Wasser. Bd. III: Modellgrundlagen – Bedienungsanleitung. Saxonian Agency of Agriculture & Saxonian Agency for the Environment and Geology (eds), Leipzig Dresden-Radebeul

Chapter 7

A Multiscale Approach to Predicting Soil Erosion on Cropland Using Empirical and Physically Based Soil Erosion Models in a Geographic Information System

V. Wickenkamp · R. Duttmann · T. Mosimann

7.1 Introduction

In view of the ecological and economic damage caused by soil erosion, investigating soil erosion processes in order to develop measures to control erosion is of great practical relevance. Erosion processes are often not limited to the field plot on which they originate. In many cases harmful effects can also be observed in neighboring fields. When the erosion systems connect up with running water, the fertilizer and pesticides transported by the surface runoff and the sediment load also enter this running water, leading to the well-known pollution of creeks, rivers and more distant seas. Not least for this reason the identification and analysis of soil erosion processes require an approach that overcomes spatial or dimensional boundaries.

A number of soil erosion models are available for estimating and quantifying soil erosion and for evaluating the risk of sediments being transported into surface water (Bork 1991; De Roo 1993; Bork and Schröder 1996). These can be integrated into Geographic Information Systems (GIS) and applied in the planning of soil and water conservation measures. Depending on the questions that are to be studied and the level of observation (e.g. size of the catchment area, size of the region under study), the models must be able to satisfy different requirements. This in term affects the extent, quality and resolution of the input data that have to be provided. For instance, when evaluating the erosion risk in large catchment areas it makes sense to identify potentially erosion prone areas by means of estimative procedures that require only a few input data, but data that are usually available for the entire area. At this stage the erosion rates are not yet quantified exactly. Only a preselection takes place. Areas selected because they are more susceptible to erosion or because of their greater role as sources of sediment yield can be investigated subsequently in more detail. For the actual quantification of the erosion and deposition processes occurring within a delimited area and for a more exact estimation of the amounts of sediment entering the surface water more highly differentiated erosion models need to be employed. The altered level of observation requires a database with a higher degree of spatial and temporal resolution.

On a macroscale scale level we identify erosion prone areas by means of simple estimative procedures. Then we move down to a microscale level, at which we attempt to quantify the erosion and deposition processes with the physically based model EROSION-3D (von Werner 1995; J. Schmidt 1996). This process of moving "downscale" will be described in the following on the basis of the data and methods contained in a Geoecological Information System (Duttmann and Mosimann 1995).

7.2
Multiscale Investigation and Modeling of Soil Erosion Processes

In accordance with Neef's (1963) "theory of geographical dimensions" developed for research in landscape ecology, soil erosion processes can be viewed at various scale levels. The following viewing dimensions or scale ranges can be distinguished:

- The subtope dimension (*nanoscale*): erosion processes and the dynamics of these processes are analyzed experimentally at very brief time intervals on areas measuring only a few square meters (e.g. test plots, miniature fields, small plots for rainfall simulators; R.-G. Schmidt 1992). Investigations at this scale level help to recognize regularities and relationships between the individual factors that influence the process of soil erosion (e.g. the influence of the initial water content on the erosion process, the erosion resistance of soils, infiltration rates). Examples of the experimental investigation of the processes occurring at the level of individual particles with the help of rainfall simulators were described by Auerswald (1993) and J. Schmidt (1996). Such field and laboratory experiments with sprinklers are essential for the understanding of erosion processes. They also help to identify model parameters and to calibrate the model. The knowledge gained in the process generally cannot be applied to larger areas, however, or if so only to a limited degree.
- The dimension of the tope (*microscale*): at this level erosion processes and systems are analyzed and modeled on site, i.e., on individual slopes, individual fields or in low-order catchment areas (Frielinghaus et al. 1994). In a given area erosion can be either areal or linear with concentrated runoff. The predominating type depends on various things, such as the natural conditions in the region, spatial elements resulting from land use, or crop management. Different processes are generally taking place in these two types of erosion and they vary in their spatial effectivity. To study them under field conditions it is accordingly necessary to choose different methodological approaches (e.g. quasi-areal field measurements by means of sediment deposition boxes or point measurements by means of sediment and runoff collectors in main runoff channels). For these reasons, strictly speaking one should distinguish between a *microscale I* (linear erosion forms) and a *microscale II* (areal erosion forms).
- The dimension of the chore (*mesoscale*): erosion processes are identified and analyzed at the level of landscape units. These units are distinguished by a spatial mosaic of interconnected elementary areas with different natural conditions and different types of management. They have similar mesoclimatic conditions and regional water budgets. Examples of spatial units at this level are the catchment areas of small streams or entire catchment areas of rivers (Leser 1997).
- The regional dimension (*macroscale*): soil erosion phenomena are investigated at the level of large regions, which are distinguished from neighboring landscape regions because of their similar natural and cultivation conditions.

The transitions between the different levels are always fluid, especially if the erosion models employed have a high degree of spatial resolution and use physical models to describe processes. If a database with an accordingly high degree of detail is available, soil erosion can be calculated for a large number of very small catchment

Chapter 7 · A Multiscale Approach to Predicting Soil Erosion on Cropland

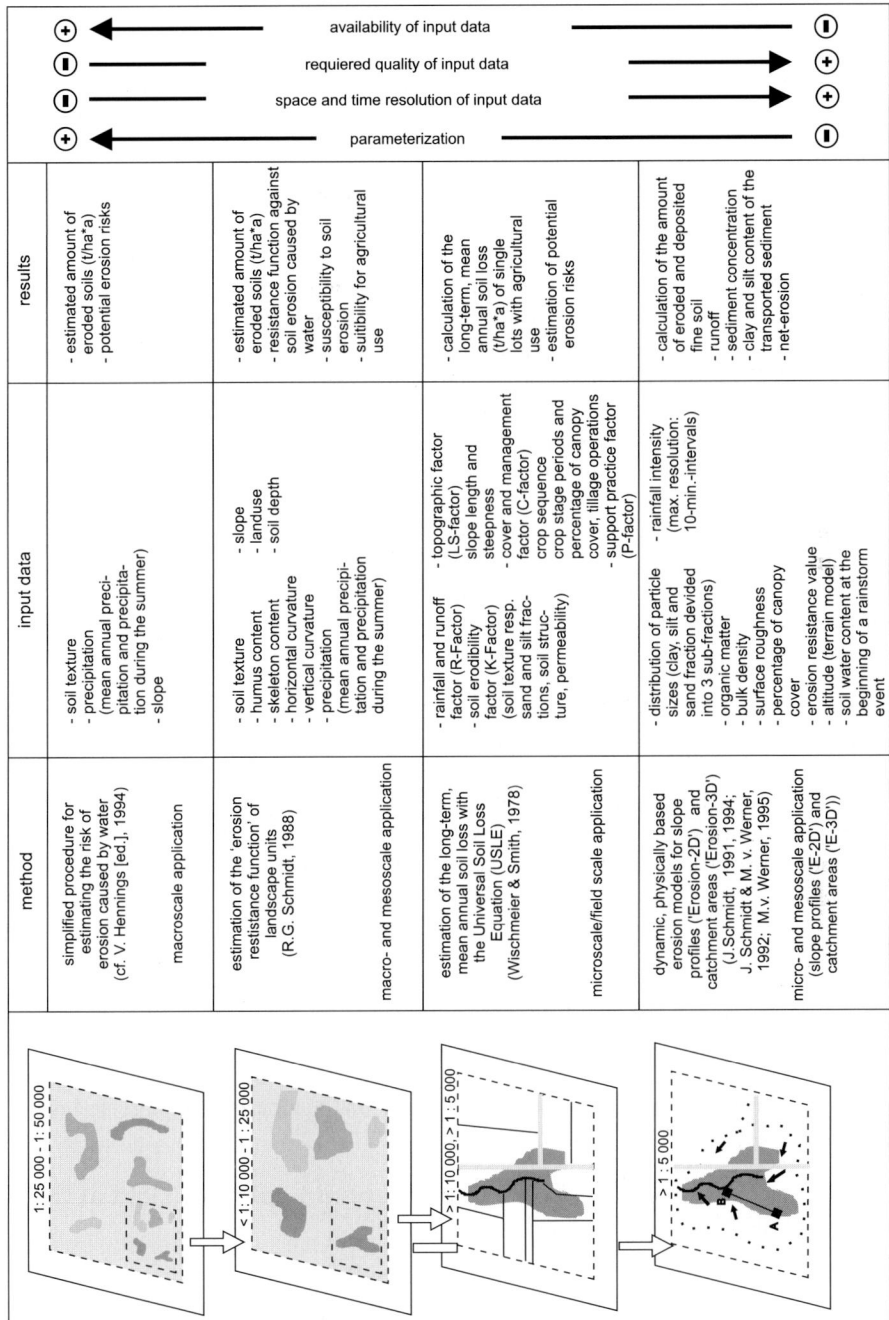

Fig. 7.1. GIS-based "downscaling"-procedure for estimating soil losses in different geographic dimensions

areas or individual plots (*microscale*). The spatially differentiated results can then be depicted for areas of the chore dimension (*mesocale*). The modeling of the elementary processes, such as, e.g., particle detachment and particle transport, is always done at the nanoscale level, i.e. on the scale of individual particles. Examples of other scale graduations used in the investigation and simulation of erosion processes are described, e.g. by Kirkby et al. (1996) and Poesen et al. (1996).

Employing physically based models within the framework of practical applications only makes sense, however, if

- the required model parameters are available with the necessary spatial and temporal resolution or can be made available without undue effort,
- the spatial variability of the model's input values, which are usually sampled at points, is known. This is essential before the data can be extrapolated to a spatial unit. It is indispensable for an estimation of the amount of error in the model computations,
- the changes in the model's input values over time (e.g. water content of the soil at the beginning of an erosion event) can be calculated or estimated with sufficient accuracy and
- the employed models are still operable despite the volume of input parameters.

In contrast to empirical models, physically based models allow the amount of eroded and deposited fine soil to be calculated in relation to a single rainfall event. In the majority of cases it is also possible to estimate the sediment yield. They also provide the option of experimenting with various scenarios by, e.g., varying the rainfall intensity or the management conditions. Consequently they are very important for the planning of erosion control measures (J. Schmidt 1996). In practice, however, the required model input data are not available for large regions, so that the spatial application of the models is highly limited, i.e., to areas that are particularly erosion prone. These areas are selected within the framework of the "downscaling" process (Fig. 7.1). Contrary to the definition of the term "downscaling" as given, e.g., in hydrology where it is used to mean "areal" disaggregation or differentiation of information about larger areas (for example, grid areas, elementary areas etc.) (Becker 1992), the term here is used in its literal sense of transition from a higher to a lower scale or dimension. The procedure followed thus corresponds to the in landscape ecology so-called "approach from above", a deductive spatial analysis. This approach to modern digital landscape analyses has proved to be appropriate. In this way, the depiction of landscape household processes based on data available on a small scale serves the "orientation in space" (Leser 1997) which is the precondition for research in the tope, i.e. the large-scale level.

7.3
Assessment of Erosion Hazard and Potential Sediment Yield at the Macro- and Mesoscale Level

7.3.1
Procedures for Areal Estimation of the Potential Erosion Hazard

Complex numerical models cannot be applied to quantify soil erosion in entire regions or to assess the amounts of particle-bound nutrients and pollutants entering surface waters with large catchment areas. Model input data are lacking for this scale level and it would not be worthwhile to try to generate them. It can nevertheless be desirable to be able to estimate the potential soil erosion or sediment yield. For this purpose simple, usually empirical models are applied. They include only a few parameters, but the ones they do include are those with a particularly great influence on soil erosion. Examples of such easy-to-handle empirical estimation procedures are the methods described by Hennings (1994) for estimating "erosion sensitivity."

The areal database required for these procedures can be compiled from soil maps or from topographical maps and transferred into Geographic Information Systems. Some of these data are already available in digital form (e.g. Digital Soil Map 1 : 25 000 [BK25], General Soil Map 1 : 50 000 [BÜK50], Digital Terrain Model [DGM50] and the Digital Landscape Model [ATKIS-DLM25/1]) and can be obtained from the appropriate state agencies and transferred into a GIS. With these procedures the erosion hazard can only be roughly estimated. The GIS based application of these methods can, however, be useful for rapidly identifying and preselecting areas that should be given priority in soil and water conservation work. To help in planning such measures these areas can then be investigated more intensively at the microscale level.

R.-G. Schmidt (1988) developed a procedure for calculating the so-called resistance function of spatial units or sites vis-à-vis erosion on the basis of the Universal Soil Loss Equation (USLE) (Wischmeier and Smith 1978). This procedure yields much more

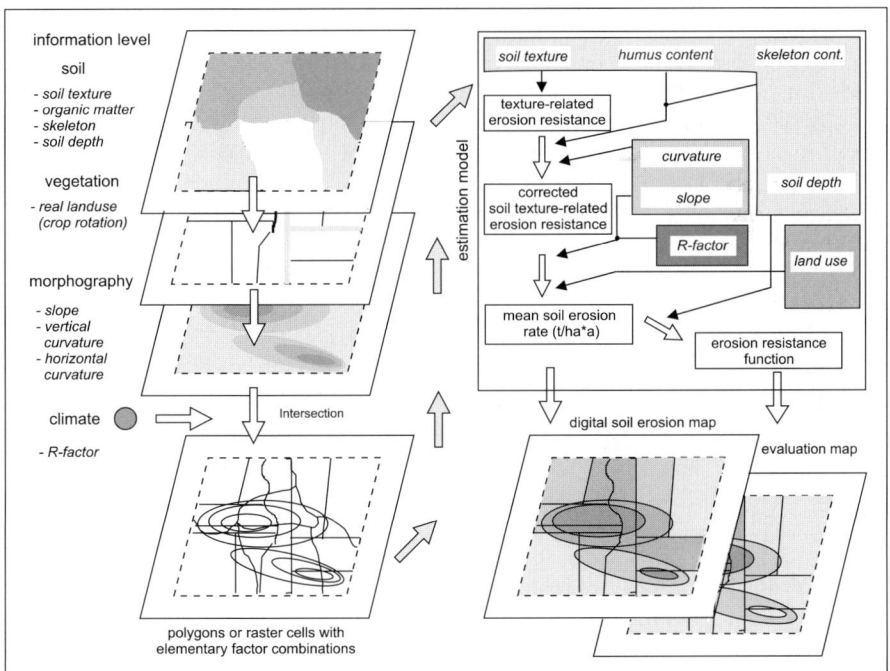

Fig. 7.2. Method for the spatial derivation of erosion resistance functions of landscape units with the help of a Geographic Information System

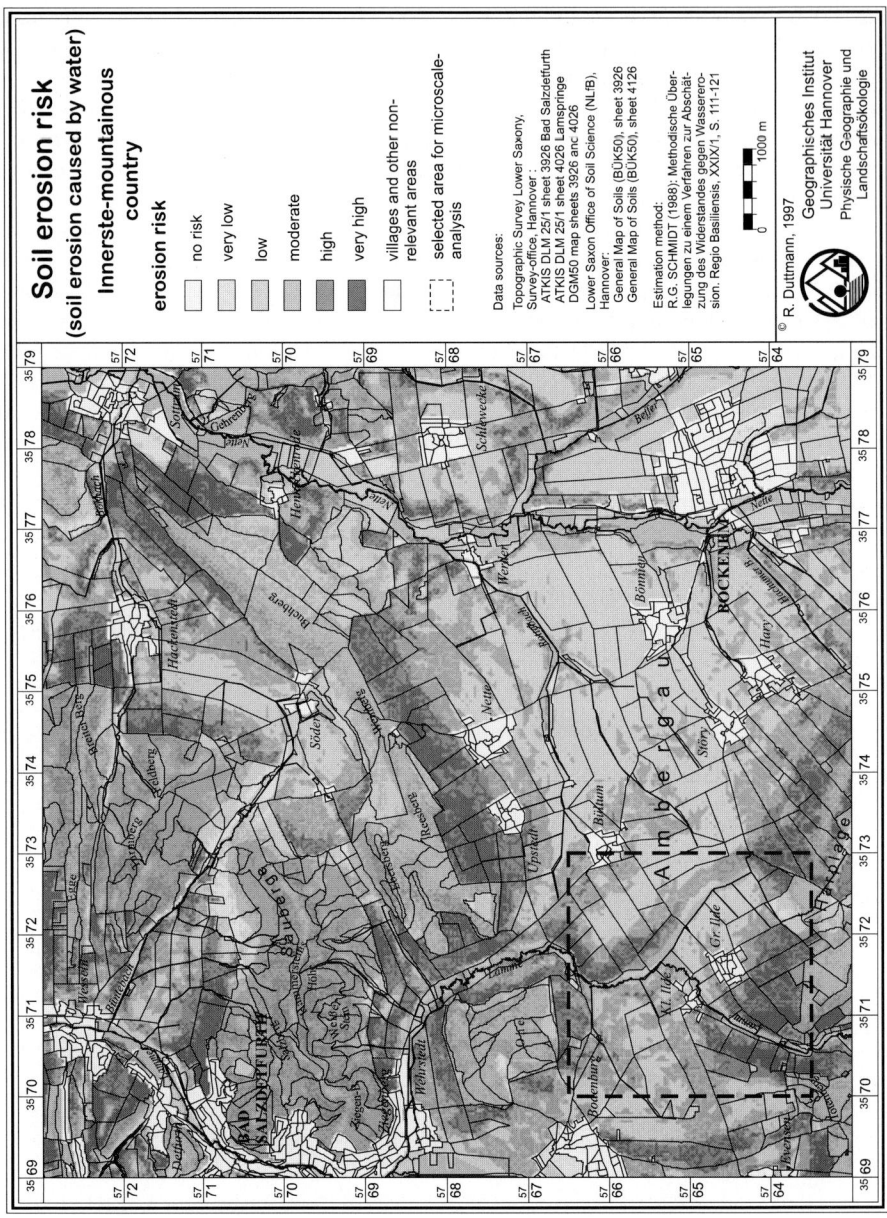

Fig. 7.3. Modeled susceptibility of landscape units to soil erosion in a section of the lower Saxon mountainous country

differentiated results. This erosion resistance function is dependent primarily on soil and relief properties and is subject to modification by land use. It is not to be confused with the physically defined term "erosion resistance" described by J. Schmidt (1991). The required areal data comprise data on soil properties (soil type, humus content, content of skeletal material and root development depth), land use (types of management) and relief (slope gradient, horizontal and vertical concavities and convexities), which can be used as information layers in Geographic Information Systems. Figure 7.2 shows how the individual parameters are linked within the model. It also illustrates a procedure for assessing the erosion hazard, which employs a GIS that was developed as a Geoecological Information System (Sect. 7.4.2.1). The erosion rates computed with the aid of this model are to be understood as relative values (R.-G. Schmidt 1988). They can be used as a measure of the susceptibility of a given section of the landscape to soil erosion (Fig. 7.3).

7.3.2
Identifying the Potential Sites at which Fine Sediments Enter the Surface Water and Estimating the Risk of Sediment Delivery

One of the ecologically most important off-site effects of soil erosion is that eroded fine soil matter may get into the surface water, taking with it adsorbed nutrients and pesticides from cropland. This leads to a number of detrimental effects (e.g. siltation of channels or reservoirs, lowering of the storage capacity of reservoirs, eutrophication, toxification, damage to waterworks). Especially if we are dealing with the prevention of water pollution we therefore need procedures at the macroscale level for not only estimating the danger of soil erosion from farmed land but additionally for

- identifying potential channels where surface runoff and the eroded fine soil it carries are likely to be concentrated,
- predicting points where surface runoff and the fine soil it carries will enter surface water and
- making a preliminary estimate of the risk of sediments being discharged into the surface water.

Neufang et al. (1989) already described a GIS based approach toward predicting water pollution caused by erosion, an approach that included compiling large-scale maps of predicted erosion and water pollution. They used the procedure known in German as "differenzierende Allgemeine Bodenabtragsgleichung" (dABAG) (differentiating universal soil loss equation) (Auerswald et al. 1988; Neufang et al. 1989; Flacke et al. 1989). This approach links the Universal Soil Loss Equation (USLE) (Wischmeier and Smith 1978) with a digital terrain model. Individual areas can be subdivided according to their location in the relief and the type of relief and linked with each other on the basis of the catchment area to which they belong. Based on the surface runoff conditions, the catchment area and the size of the catchment area are determined for each part of the area and for each segment of the watercourse. The soil losses for the catchment area in question are assessed in terms of balances.

In many cases, however, sediments do not enter running water areally or diffusely, but at certain points at the end of main channels of surface runoff that have developed either due to the relief or to tillage. Spatz et al. (1996) pointed out that runoff

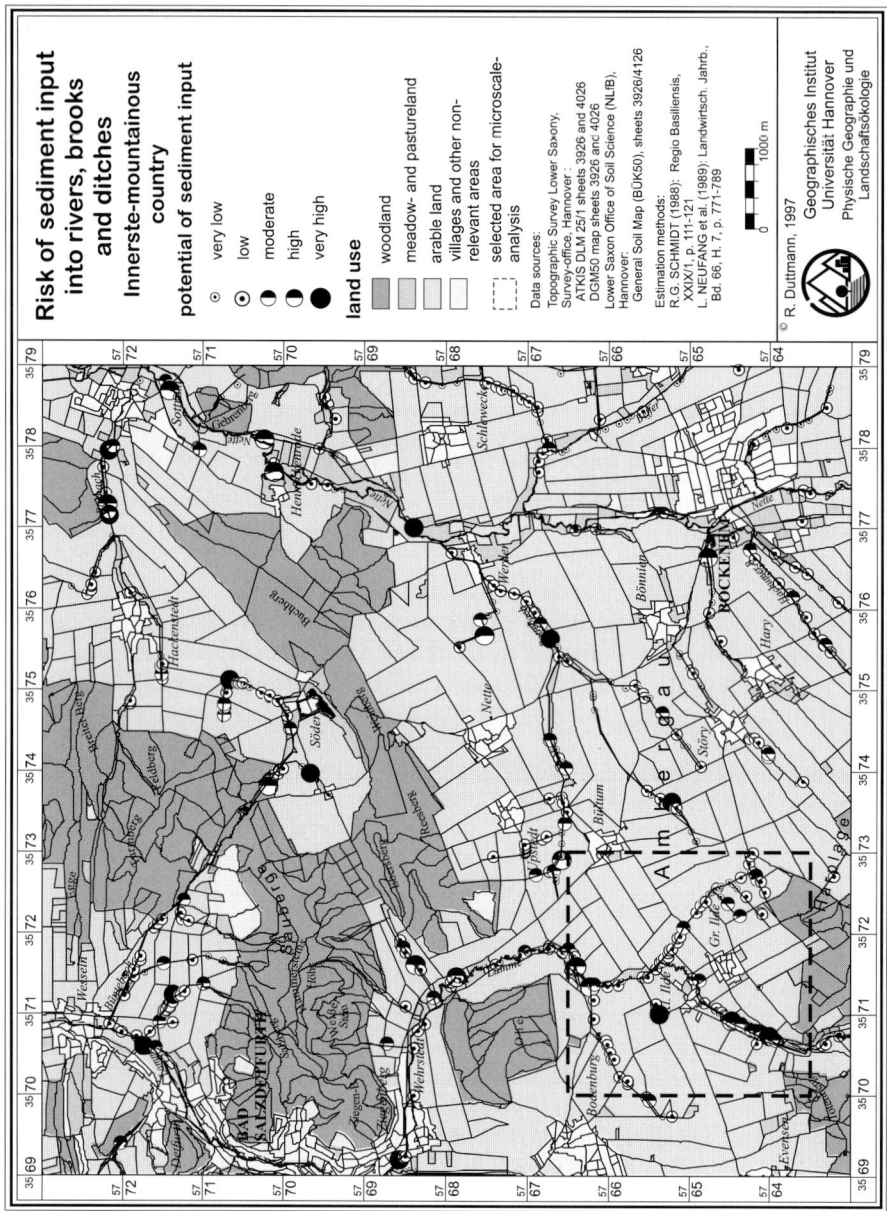

Fig. 7.4. Prediction of the risk of sediment inputs into rivers and ditches

becomes concentrated in the erosion rills developing on cultivated fields, resulting in concentrated discharge at certain points. To be able to predict the risk of sediment delivery, it is important to look particularly at those segments of watercourses that have not only a (direct) connection to main erosion channels but also to areas yielding surface runoff and sediments. In contrast to the procedure described by Neufang et al. (1989), we did not try to predict the risk of erosion caused sediment delivery for all sections of the watercourses shown in Fig. 7.4. Rather, we looked at the "points of intersection" of rivers, creeks and ditches containing running water with the channels dictated by the relief. They were treated as potential transfer points for fine soil matter. Every predicted transfer point simultaneously forms a "pour point" for a catchment area behind it. For each such point the parameters required for determining sediment yield, such as "size of catchment area" and "amount of detachment in the catchment area" can be determined with the aid of the appropriate analytic functions of a GIS. The sediment delivery can be estimated according to the above described procedure.

7.3.2.1
Model-based Identification of Transfer Points and Estimation of Potential Sediment Delivery

The identification of transfer points where fine soil transported from cropland enters surface water and of the potential yields at the transfer points by means of a GIS uses a grid based or raster procedure. This type of procedure is particularly effective for extensive spatial analyses like those described here. The data structure is easy to handle; direct geometric access to the data is possible, and it is comparatively simple to link data and to calculate neighborhood relationships (Göpfert 1991). To identify transfer points and estimate their potential sediment yields the required areal data were transferred into a 25×25 m grid, so that each basic data set consists of 160 000 grid cells. The following data should be provided for the calculation of the potential sediment yield:

- Morphometric data (database: DGM50 from the geodetic survey for Lower Saxony): *slope gradient, vertical and horizontal concavity and convexity, catchment area, size of catchment area, direction of runoff, linear runoff channels*
- The network of watercourses (database: ATKIS DLM25/1 from the geodetic survey for Lower Saxony, with the addition of all ditches containing running water)
- Soil data (datebase: BÜK50 from the Lower Saxony Bureau of Soil Research): *soil type, humus content, content of skeletal material, bulk density, root development depth*)
- Land use data (database: ATKIS DLM25/1 from the geodetic survey for Lower Saxony and LANDSAT-TM images for studying the current land use and changes in land use)
- Data on linear and areal elements of the spatial structure that influence erosion (e.g. roads, grassland, rows of shrubs, woods, towns, industrial and commercial areas, parks and recreational and sports facilities; database: ATKIS DLM25/1 from the geodetic survey for Lower Saxony)

The risk of sediment delivery is estimated by means of a computer assisted procedure using a GIS. This requires several steps:

1. Generating a hydrologically correct terrain model. In this model the terrain elevation data are adjusted to fit the actual surface watercourses (creeks, rivers and ditches containing running water). The direction of runoff is calculated on the basis of this elevation model.
2. Calculating the routes taken by surface runoff. Structural elements that obstruct runoff and erosion and areas that do not play an important role in soil erosion must also be included.
3. Estimating the amount of soil loss areally according to the procedure for determining erosion resistance described by R.-G. Schmidt (1988).
4. Identifying the points where runoff channels on the soil surface intersect with rivers, creeks and ditches. The intersections correspond to potential transfer points where the concentrated surface runoff enters flowing water.
5. Distinguishing the catchment areas belonging to the various intersection points and calculating the size of the catchment area from which the surface runoff is derived.
6. Calculating the total soil loss from cultivated fields for each catchment area.
7. Computing the sediment delivery ratio for each catchment area of a potential transfer point and estimation of the sediment delivery by means of the above mentioned equations.
8. Classifying the calculated sediment yields at the transfer points according to the degree of danger of sediment delivery.

7.3.2.2
Potential Transfer Points and Risk of Sediment Delivery Computed by the Model

The results of the GIS based prediction of potential transfer points at which fine sediments will enter surface water are shown in Fig. 7.4. For purposes of clarity the map shows only those transfer points for which sediment yields of more than 1 t yr^{-1} and catchment areas of more than 0.5 ha were computed. A total of 785 potential points of sediment delivery were localized in the section of the Leine-Innerste hill country examined here (Table 7.1).

Figure 7.5 shows the percentage of the total sediment yield computed by the model for each of the different types of watercourses. In comparison with Table 7.1 it is evident that in our study area

- the largest number of potential transfer points are located in areas with *ditches* containing running water. The reason for this is that there is a dense network of ditches in areas with a large amount of soil loss. Although each transfer point characteristically has only a small potential sediment yield (Table 7.1), there are so many transfer points that the ditches provide 42% of the total sediment yield calculated for all water bodies in the area. Transfer points with potential yields of 10 to 50 t yr^{-1} account for approximately 18% of the total yield (Fig. 7.5). With the database used here it is impossible to assess how much of the sediment yield is accumulated in the ditches and hence not transported any further. Some of the ditches have no drain or carry water only periodically. The further transport into larger bodies of water is disregarded here.
- approximately 40% of the total sediment yield in the area of our map sheet takes place via the runoff channels directly connected to the *rivers* (Lamme and Nette).

Table 7.1. Mean potential sediment yields at predicted transfer points entering rivers, creeks and ditches in parts of the catchment areas of the rivers Lamme and Nette (Leine-Innerste hill country)

Sediment delivery into	Number of predicted transfer points	Mean size of catchment area of transfer points (ha)	Mean soil loss in catchment area of transfer points (t ha^{-1} yr)	Mean potential sediment yield of the transfer points (t yr^{-1})	Percentage of total yield in sampled area (%)
Rivers	177	4.9	23	49	39.4
Creeks	175	4.0	14	23	18.6
Ditches	433	3.5	11	21	42.0

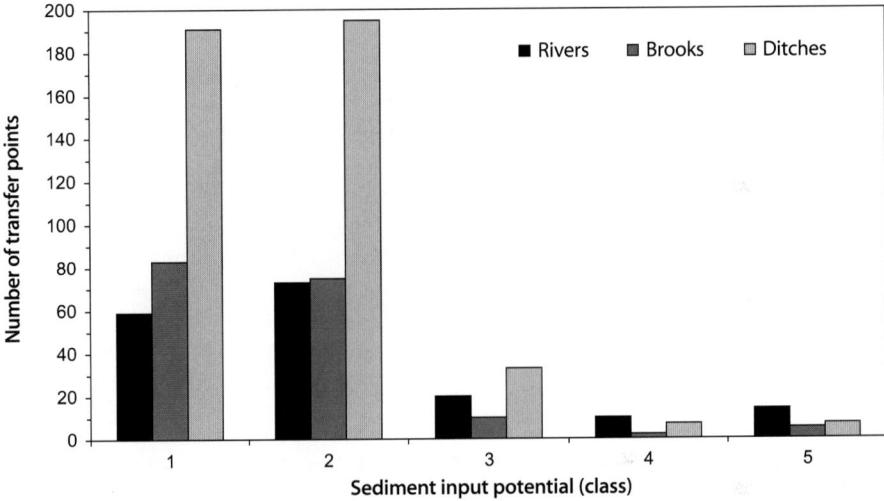

Fig. 7.5. Predicted sediment yields at transfer points in parts of the catchment area of the rivers Lamme and Nette, differentiated according to type of watercourse and potential yield of the transfer points

On average the catchment area is larger and the soil loss is greater in the catchment areas of the rivers Lamme and Nette, although the number of transfer points is relatively small. As a result their share of the total sediment yield is comparable to that of ditches (Table 7.1). The 14 transfer points with potential yields of more than 150 t yr^{-1} alone in the area of the rivers make up approximately one fifth of the estimated sediment yield for the entire region. Since these transfer areas are localizable single points, the sediment yield could be reduced considerably by expedient erosion control measures.
- *small creeks* make up approximately 20% of the total yield calculated for the region. This low value can be explained by the peculiarities of the region in question. The network of creeks tends to have long flowing stretches, and it is denser in areas without much tendency toward soil erosion. This holds for large parts of the Ambergau basin in the southern and southeastern part of the map sheet and likewise for the flat valleys along the Büntebach in the northwest.

In the area of Bodenburg-Bültum and Groß-Ilde erosion damage was mapped parallel to the work with the model. The mapping of erosion damage, runoff channels and sites at which fine sediments enter the surface water were based on the procedure described by the DVWK (1996). For estimating the sediment yield the types of erosion were measured according to Rohr et al. (1990). Comparisons revealed a high degree of correspondence between the location of transfer points predicted by the model and the actual ones especially for those transfer points with high amount of predicted sediment yield. Approximately 60% of sites with a potential for sediment deposition above 20 t yr^{-1} were explained in this way. The aim was a qualitative analysis of the occurrence of such sites on the edges of water bodies. The model can only predict a certain proportion of all transfer points existing in reality, however, because many factors that affect the linear runoff and soil detachment occurring at the soil surface in small areas (e.g. concentrated water influx from roads, artificial drainage channels, small water outlets) are out of the reach of this scale level.

7.4
Areal Prediction of Soil Loss at the Microscale Level

Now that we have used the above described procedures to select the areas with a greater risk of soil loss, we can investigate them in more detail at the microscale level. For this purpose we can again employ empirical-statistical procedures to estimate soil loss and the threat to soil fertility on individual farm fields. There are also, however, a number of physically based soil erosion models available for the microscale level. With such models the actual soil erosion events can be predicted and quantified with a much greater degree of precision.

7.4.1
Estimating the Threat to Soil Fertility for individual Plots Employing the Allgemeine Bodenabtragsgleichung (ABAG)

Figure 7.6 shows the soil losses calculated with the help of the ABAG (Allgemeine Bodenabtragsgleichung = universal soil loss equation) (Schwertmann et al. 1990) for the individual fields in an area selected in the process of "downscaling". It shows the erosion situation, taking into account the actual management conditions (crop rotation 1993 to 1996). The highest soil losses occur, as expected, on fields with steep slopes and crop rotations involving sugar beets or corn (though these are negligible in area) on which conventional tillage methods are employed. On some slopes minimum tillage techniques are employed. Because of this plus the use of intercropping and direct tilling (e.g. direct tilling of beets in mustard seed after it has been killed by frost), the factor for the influence of slope length and steepness (LS factor) accounts for only 50% of the calculated soil loss (for $n = 10\,211$). There is, however, no definite correlation between the other ABAG factors and the amount of soil loss. If we assume a uniform crop combination for all fields in the section (here: winter wheat-winter barley-sugar beets) the LS factor accounts for around 75%.

In addition to estimating the danger of soil loss and possible sediment yields, another aspect that is of particularly great interest from the viewpoint of soil conservation is the threat to soil fertility due to soil erosion. This can be computed, e.g. according

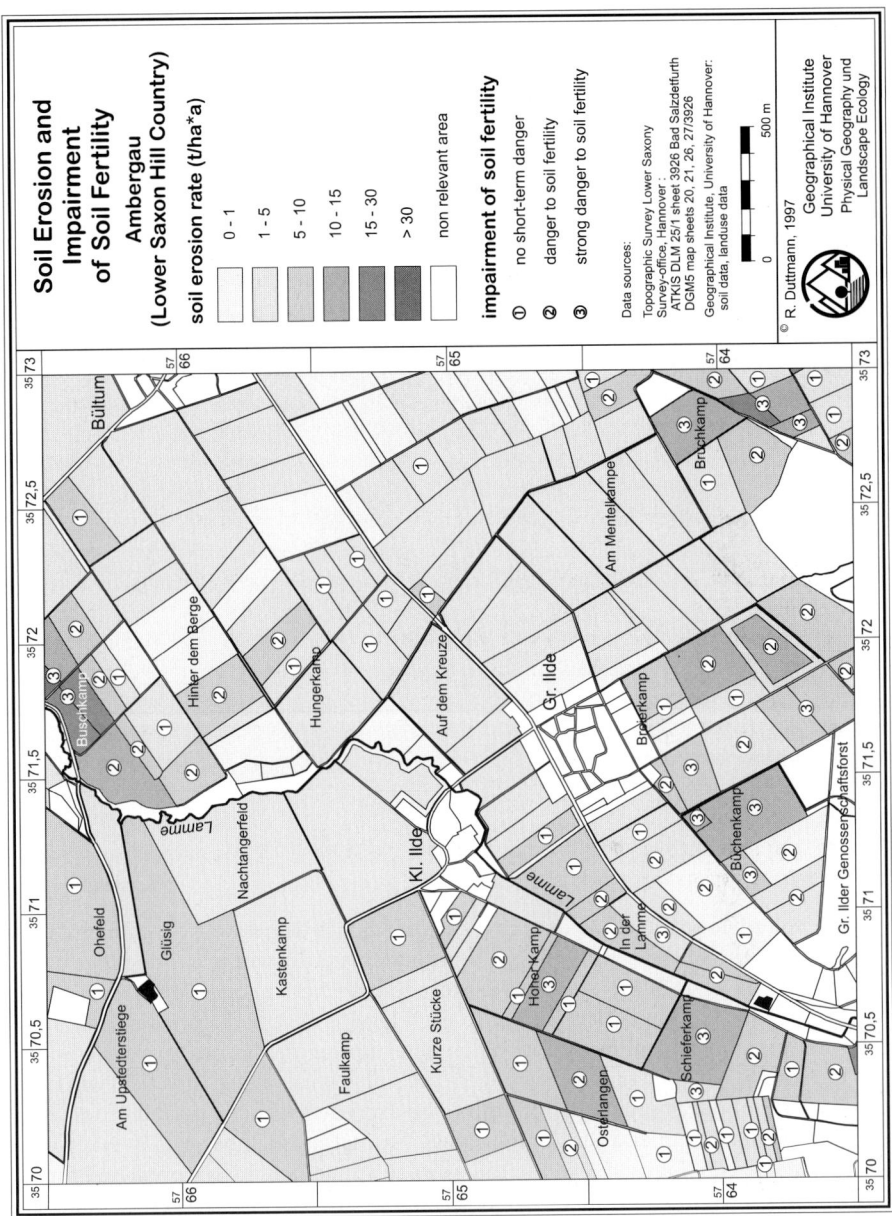

Fig. 7.6. Estimation of soil erosion risks on agricultural lots by using the Universal Soil Loss Equation and evaluating the impairment of soil fertility

to a concept developed by Mosimann and Rüttimann (1996). Depending on the root development depth of the soil, this concept contains guidelines stipulating the "tem-

Table 7.2. Soil loss (t ha^{-1} yr^{-1}) and hazard levels for selected farm fields in the region of Ilde (Leine-Innerste hill country) according to various land use scenarios

Field	Actual crop rotation	Scenario 1		Scenario 2		Scenario 3		Scenario 4	
		Soil loss	Hazard level	Soil loss	Hazard level	Soil loss	Hazard level	Soil loss	Hazard level
17	Fallow – sugar beets – fallow	25	3	34	3	8	1	8	1
91	Winter wheat – winter wheat – winter barley	14	3	21	3	2	0	2	0
94	Corn – sugar beets – spring wheat	15	3	10	2	2	0	3	0
180	Winter wheat – winter wheat – sugar beets	10	3	9	3	2	1	2	1
213	Winter wheat – sugar beets – winter wheat	13	3	12	3	4	0	4	0
215	Winter wheat – winter wheat – winter wheat	19	3	22	3	5	2	5	2

Hazard levels:
0 = soil fertility not threatened,
1 = soil fertility not threatened on the short term (conservation measures recommended),
2 = soil fertility threatened (conservation measures necessary),
3 = soil fertility highly threatened (conservation measures very urgent).

porarily tolerable soil loss." As long as the actual values lie below these standards it is assumed that the soil fertility will not suffer fundamentally within a time range of 300 to 500 years (level 0). If this standard is exceeded, the field in question is classified in one of three hazard levels (levels 1–3). To each level is assigned a certain level of urgency for erosion control measures. When determining the hazard levels the important question is whether and to what degree the soil loss calculated for a field will cause a specified root development depth limit to be exceeded within a period of 100 years (for details Mosimann and Rüttimann 1996). Estimating such hazards is easily done by means of a GIS. Figure 7.6 shows the average annual soil losses predicted for the Ilde region on the basis of the ABAG. It also shows the calculated levels of danger to soil fertility for specified farm fields. The highest hazard levels occur, as described above, primarily on the steeper slopes.

Additionally we now see some areas in which the soil fertility is threatened, but for which smaller amounts of soil loss were calculated. These are primarily fields on not overly steep hilltops and upper slopes with an root development depth of less than 50 cm (rendzina, brown earth-rendzina, rendzina-brown earth). Mosimann and Rüttimann (1996) generally accord to such sites a "short-term acceptable soil loss" of 0 t ha^{-1} yr.

For the purposes of GIS based land management, it is possible to identify fields whose soil fertility is threatened and and to initiate the appropriate conservation measures. Moreover, it is of course also possible to experiment with various land use scenarios. In this manner management strategies can be developed for such fields to ensure the long-term conservation of the soil substance required for profitable cultivation. The results of four such scenarios for selected fields in the area shown in Fig. 7.6 are illustrated in Table 7.2.

With otherwise constant input variables the amount of soil loss and the degree of danger to soil fertility were determined for the following cultivation variants:

- Scenario 1: actual crop rotation and types of tillage in the period between 1993 and 1996
- Scenario 2: crop rotation: winter wheat, winter barley, sugar beets, conventional tillage
- Scenario 3: crop rotation: winter wheat, winter barley, sugar beets, minimal tillage and direct tilling of sugar beets in mustard seed
- Scenario 4: actual crop rotation, minimal tillage, mulching of fields planted to corn, direct tilling of beets in mustard seed

As Table 7.2 shows, on individual fields (Nos. 17, 180 and 215) the hazard level "0" cannot be achieved, even if we assume the most favorable tillage and management types. These are areas with shallow soil and steep slopes. For these areas further agricultural use should be reconsidered from the standpoint of conservation.

7.4.2
Areally Differentiated Erosion-Deposition Modeling with Physically Based Soil Erosion Models: Application of the Model EROSION-3D within a Geographic Information System

In contrast to most empirical-statistical procedures, the application of dynamic, physically based soil erosion models at the microscale level allows an event-related, process-oriented and areally differentiated analysis of the actual process of soil erosion. The processes that are taking place can be analyzed with a high degree of resolution, and the effect of individual countermeasures can be evaluated. Data collection and modeling are generally done at the *microscale* level, whereas the results are presented at the *mesoscale* level.

The model used here, EROSION-3D (von Werner 1995), is a physically based soil erosion model for small catchment areas. It employs the basic physical model of the slope profile model EROSION-2D (J. Schmidt 1991, 1996; J. Schmidt and von Werner 1992).

EROSION-3D has the following features:

- event-related modeling with variable time steps ranging from 1 to 15 minutes
- a distributed watershed model with a high degree of spatial resolution (it is possible to process more than 500 000 grid elements)
- a relatively small number of input parameters, though some of these are relatively difficult to obtain

Table 7.3. How the input parameters required by EROSION-3D are assigned to the information layers in the Geoecological Information System

EROSION-3D input	Time-related variable/invariable	Assigned information layer	Allocated attributes stored in the
Elevation	Invariable	Morphology	Space-oriented database
Texture	Invariable	Soil	Space-oriented database
Organic carbon			
Bulk density	Variable	Soil	Site-oriented database
Erodibility			
Soil moisture			
Plant cover	Variable	Vegetation	Site-oriented database
Roughness coefficient (Manning)			
Precipitation	Variable	Climate	Site-oriented database

The main area of application of the model is in environmental and agricultural planning.

With EROSION-3D the redistribution of soil occurring in small catchment areas can be simulated, i.e. the amount of soil eroded and deposited per spatial unit, on the basis of a uniform square grid. The required morphographic and hydrographic parameters (slope gradient, runoff distribution, grid cell related size of the catchment area, runoff concentration, stream network) are computed from a digital terrain model by means of a digital relief analysis. The parameters influencing the processes that we wish to model are considered to be homogeneous within both a grid cell and a selected time step. The model system EROSION-3D is divided into two component models adopted from EROSION-2D:

- Infiltration model: computation of matrix potential, hydraulic conductivity and difference in water content. The results enter directly into the erosion model.
- Erosion model: calculation of detachment, transport and deposition

The physical approach of the erosion model is based on calculating the detachment of soil particles from the soil surface, dependent on the momentum fluxes exerted by overland flow or by falling drops of rain. They are compared with a "critical" momentum flux (J. Schmidt 1991), which characterizes the specific erosional characteristics of the soil in question. With the resulting dimensionless coefficient, which is entered into an empirical equation, the detachment of solid matter can be computed. On the basis of individual rainfall events the model computes detachment, deposition and net erosion for each grid cell of a catchment area. Additionally the amount of runoff, sediment concentration and the grain size distribution (% of clay and silt) are computed. The simple physical description of processes (von Werner 1995) employed by EROSION-3D to model soil erosion leads to the following limitations:

- Precipitation that does not fall directly onto the soil, but first hits the plant cover is not considered.
- The input parameters resistance to erosion and surface roughness, which vary over the course of a rainfall event, remain constant in the calculations.
- Rill erosion processes are not calculated.
- The kinematic wave equation is not employed to describe the discharge out of the catchment area.

Whereas the number of input parameters that must be provided is relatively small (Table 7.3), these parameters must be of high quality and very precise. With the exception of precipitation data, the input data can be provided in the form of raster files. Since the results of the simulation are also outputted in the form of raster files, EROSION-3D can be integrated into an appropriately structured GIS.

7.4.2.1
Integration of EROSION-3D into the Geoecological Information System GOEKIS

In the following we will describe how EROSION-3D can be integrated into a GIS that was developed as a Geoecological Information System. For this purpose, based on standard GIS software a geodatabase was created whose structure depicts the regular systems of circulation and process correlation (Mosimann 1997). In the process we will explain the structures required to ensure that data can be exchanged between the model and the GIS (Wickenkamp et al. 1996).

As Fig. 7.7 shows, the Geoecological Information System consists of a methodological sector and a data sector (Duttmann 1993; Duttmann and Mosimann 1995). The data sector can be divided into information layers that correspond to the components of the landscape ecosystem (climate, land use, soil, groundwater, morphography and geology). In addition to the geometry files containing the boundaries of the features stored as vector or raster data (e.g. soil units, land use), the space-oriented database contains the allocated attribute data, which do not change with time. The site-oriented database contains data gathered as points (e.g. soil profile data) or data from serial measurements (e.g. precipitation). Along with the more or less static data (e.g. soil type), variable data are stored. These can be values that change with the annual rhythm, such as plant cover, or data that are dependent on the period of observation, such as soil water.

The methodological sector of the GIS comprises the documentation of the methods, data and sampled sites and the procedure itself. This includes methods for deriving structural and process variables (e.g. process variables relating to climate ecology, soil properties), for the digital relief analysis, for evaluating component functions of the landscape budget (e.g. estimating the danger of soil erosion) and for modeling dynamic processes (e.g. soil moisture budget modeling, soil erosion modeling with EROSION-3D) (Duttmann and Mosimann 1995).

The data required by EROSION-3D can be assigned to certain information layers in the Geoecological Information System, as shown in Table 7.3. The time-variable data are linked with the geometry files of the assigned information layers stored in the space-oriented database via a common code and can thus be represented spatially.

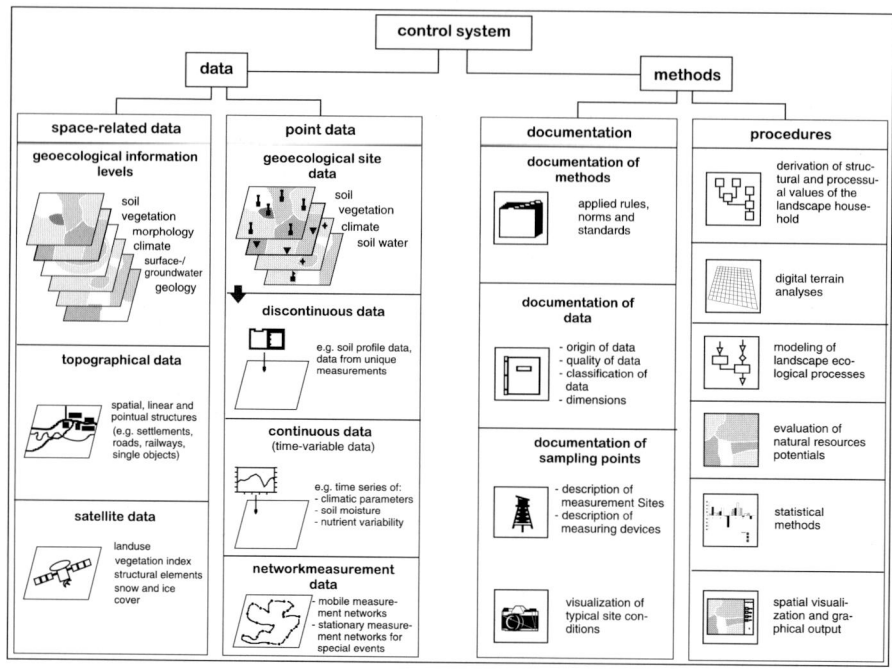

Fig. 7.7. Structure of the Geoecological Information System

Depending on their temporal resolution one can think of the result as a set of time maps. The plant cover, e.g. is stored at a resolution of 1 month, so that 12 maps of the spatial distribution of the plant cover dependent on land use would be available in the GIS. This would increase the required storage capacity unnecessarily, however. Therefore the data files are only linked when they are needed for the time point that is being considered. The precipitation values are a special case. They hold for the entire study area and are therefore not linked with the space-oriented database.

Figure 7.8 shows the integration of the model system EROSION-3D into the database of a Geoecological Information System and the organization of the data flows. The required input data must be entered into EROSION-3D in raster format. Therefore the spatial information existing in vector format is first converted into thematic grids. Difficulties arise in depicting narrow linear structures, such as creeks, ditches, hedges or roads. When vectorial data are converted into raster format the resulting cells generally are assigned the value of the object that takes up the largest proportion of the area of the cell in question. Consequently only fragments of linear elements are included if the grid squares are more than 5 m in width. Since such linear elements have considerable influence on erosion, e.g. because they interrupt a slope or serve as deposition strips, it is necessary to depict these elements of the spatial structure in their entirety and to incorporate them into the modeling. For this purpose weighting factors were introduced that guarantee that relevant linear elements will be treated with priority in the compiling of the grid. A grid cell is thus given the value "ditch"

Chapter 7 · A Multiscale Approach to Predicting Soil Erosion on Cropland

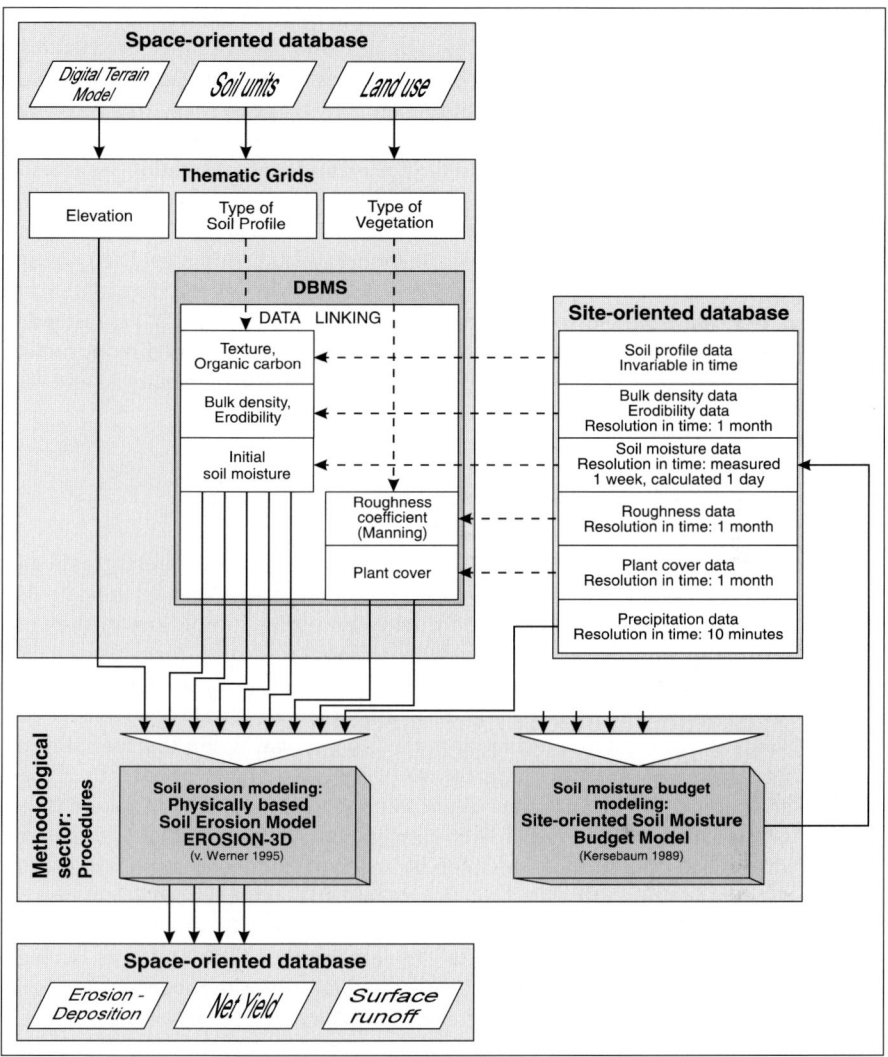

Fig. 7.8. Integration of the model EROSION-3D into the Geoecological Information System

even if the ditch only takes up a small proportion of the area of the cell. In this manner the network of ditches and roads can be transferred almost completely into the grid structure and accordingly be included in the process of modeling. However, the linear structure elements will unavoidably be overemphasized, since they too are now depicted in the preset grid cell width (here 12.5 m).

During the conversion step when vectorial data are transferred to a grid, not only the area that is to be modeled is determined, the size of the grid cells and the necessary weighting of the spatial structure elements are selected. The resulting digital terrain model can be fed directly into EROSION-3D and used. In contrast, the model input data assigned to the information layers soil and land use are first stored in coded form, i.e. the grid cells contain as values only the codes for the corresponding soil and land use units. In a second step the required site and sequence data are linked with the grid cells from the site database. In the process the time period to be modeled is selected, in that only the values relevant to the period under consideration are accessed and linked.

The so developed raster data files are read into the model system EROSION-3D so that the computations can be performed. The model yields for each grid cell not only the amount of erosion and deposition but also data on grain size distribution, surface runoff, etc. (above). Since these again can be accessed from EROSION-3D as raster data files and are thus available in spatial form, it is possible to input them directly into the spatial database of the GIS. There they are available for both further analyses and visualization (Wickenkamp 1995).

7.4.2.2
Modeling Soil Erosion in the Region of Mehle with EROSION-3D

The sample region selected here was Mehle in the loess covered transitional area between the fertile plain and the hill country of Lower Saxony. Most of the data necessary for the modeling were gathered during mapping campaigns in 1993 and 1994. Additionally we were able to use data available in digital form from ATKIS DLM 25/1 and the digital terrain model DGM 5 (geodetic survey for Lower Saxony). The model computations used a grid cell width of 12.5 m for all erosive excessive rainfall events during the measurement period 1993–1994. When the spatial structure elements were weighted, the watercourses were given higher priority than roads and buffer strips, to make sure that ditches and creeks would be included in any case. Thus their influence on the erosion process can be investigated and amounts of sediment entering the surface water can be analyzed. Figure 7.9 shows the erosion and deposition amounts calculated for the storm of 26–27 May 1993 (total precipitation 41.9 mm, maximum intensity I_{30max} = 33.8 mm h^{-1}). The western part of the investigated catchment area is completely covered with woods and thus does not contribute to erosion. In our modeling this area serves only as the source of (an insignificant amount of) surface runoff. Therefore only the eastern part of the catchment area, the relevant area for the erosion process, is depicted here. In most of the study area only minor to moderate erosion takes place. Only for one single field which was planted to corn a higher amount of erosion was calculated. The ditches, which generally run alongside roads, stand out clearly as areas in which transported material is deposited. Areas with woods and grassland and paved areas (roads, pathways, settlements) are depicted as being erosion neutral. A remarkable finding is the high erosion rates in the lee of the current, i.e., on slopes beneath roads. The explanation for this is that the model shows the surface runoff containing a relatively lower sediment load after flowing over paved surfaces. This increases the absorptive capacity of the water, leading to increasing erosion. The same effect explains the high soil losses in the cells adjoining the woods to the east. Here the surface runoff from the woods is modeled still devoid of sediments and therefore possesses a high erosion and transport capacity.

Because the values for all modeled factors are available for each grid cell in the space-oriented database of the GIS, it is possible to perform analyses that go beyond the mere identification of areas and amounts of erosion and deposition. For instance, with the GIS it is possible to isolate grid cells that represent watercourses and to determine the amount of sediment carried into them and the simulated particle size distribution. On the basis of these data the amount of phosphates and other nutrients and pollutants entering the water via the transported sediment can be estimated. If we were to include the values for the amount of surface runoff also outputted by the model, we could also calculate the substances dissolved in it.

Figure 7.10 shows the sum of the calculated net yields for all erosive events modeled during the period of observation. The results for the individual events were summed up with GIS assistance. Here a further effect of the ditches becomes evident. Because they interrupt the slope, they function not only as areas of deposition. There is an obvious decrease in erosion in the adjacent areas below them. Areas with the highest net erosion losses can be distinguished well, so that areas where erosion control measures are particularly urgent can be identified. With EROSION-3D the influence of changes in land use on the erosion process can also be modeled. Figure 7.11, e.g., shows the situation calculated for the storm of 26–27 May 1993 assuming that the entire arable land of the study area was planted to corn. Small changes in cultivation methods (e.g. the effect of buffer strips of grass left in large fields) can be simulated. Scenarios for changes in the networks of roads and ditches can also be experimented with, for instance as part of programs to reallocate and consolidate arable land. In this manner the most acceptable variants from the viewpoint of soil and water conservation can be selected.

The model yields plausible results, that is, the structures of the erosion and deposition areas they identify in the sample area are credible. The influence of the networks of watercourses and roads can be envisaged well with EROSION-3D. However, the model does not yet make allowances for the direction of tillage on cultivated fields, and that is an important factor in the processes occurring during erosion. Thus in some cases there are obvious differences between the model results and actual nature.

Satisfactory results can be obtained with a model only if the input parameters are sufficiently precise. Particularly problematic here is the availability of data on erodibility (Schramm 1994; von Werner 1995; J. Schmidt 1996) and on soil moisture.

7.5
Linking a Soil Moisture Budget Model with EROSION-3D for the Model Assisted Estimation of the Parameter Soil Moisture

Due to the underlying process dynamics soil moisture is one of the most difficult input parameters to estimate. For a model run with EROSION-3D spatially differentiated data on the soil moisture content prevailing at the beginning of a given rainfall event must be provided. Since these data cannot be recorded with gauges or inferred from other parameters, they have to be obtained from field measurements involving a great deal of personal or instrumental effort. This can be reduced, however, by applying appropriately validated soil water budget models (Wickenkamp et al. 1996).

The dynamic soil water budget model employed here is based on the mathematical relationships described by Kersebaum (1989). It is site-oriented model that divides

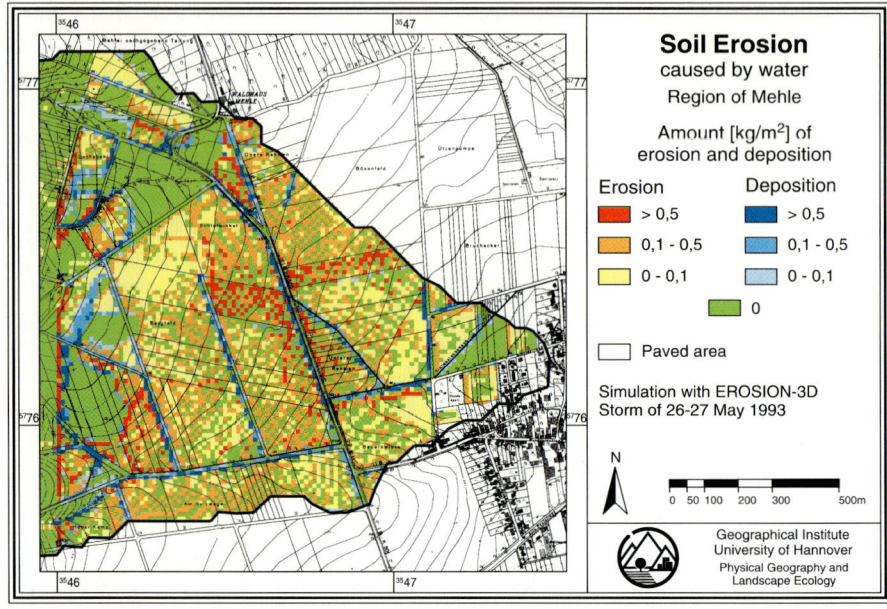

Fig. 7.9. EROSION-3D Simulation: Erosion and deposition amounts calculated for the single storm 26–27 May 1993

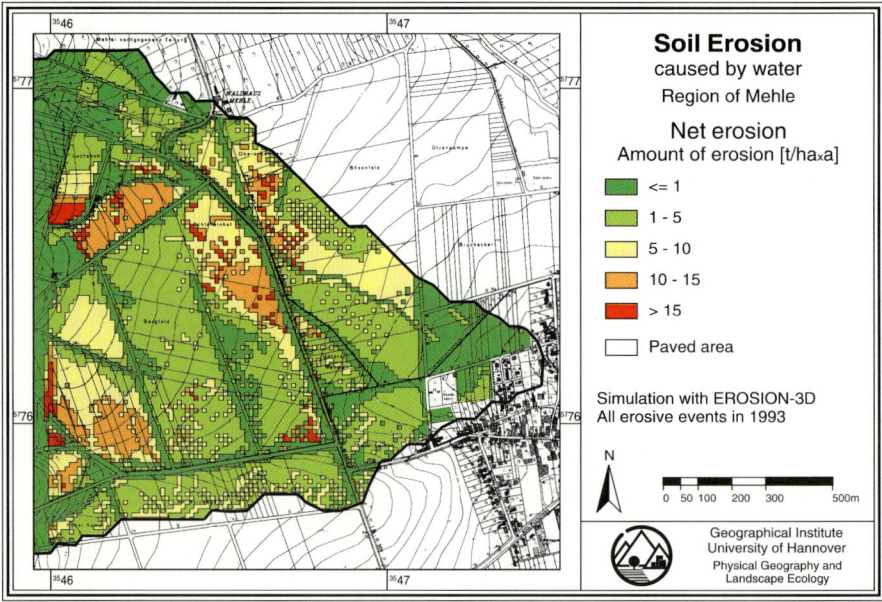

Fig. 7.10. EROSION-3D Simulation: Sum of the calculated net yields for all erosive events modeled during the period of observation

CHAPTER 7 · **A Multiscale Approach to Predicting Soil Erosion on Cropland**   131

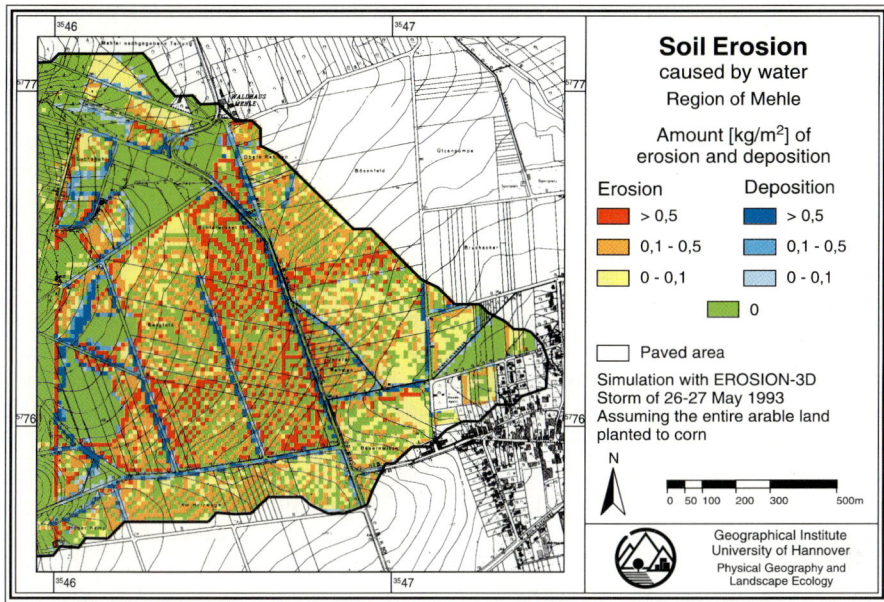

Fig. 7.11. EROSION-3D Simulation: Erosion and deposition amounts calculated for the single storm 26–27 May 1993 assuming that the entire arable land was planted to corn

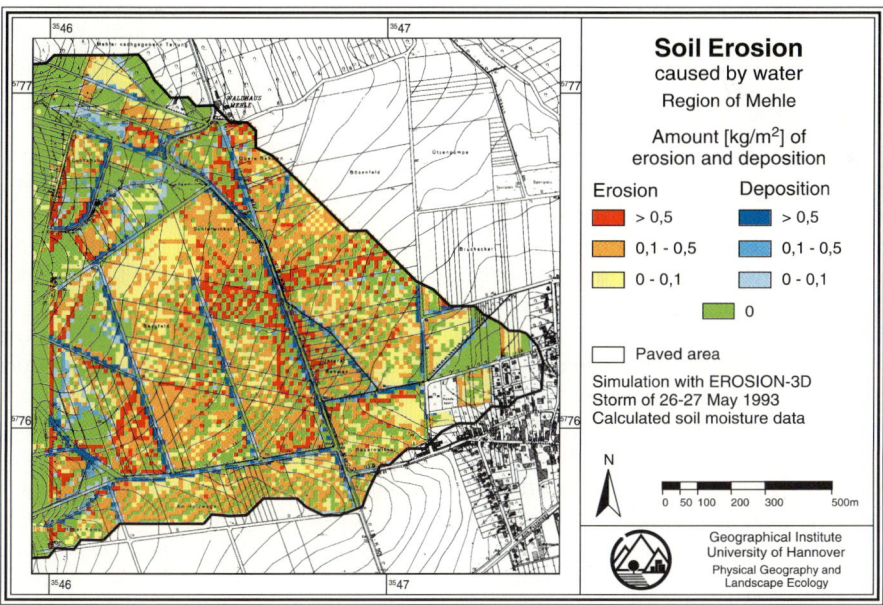

Fig. 7.12. EROSION-3D Simulation: Erosion and deposition amounts calculated for the single storm 26-27 May 1993 on the basis of the calculated soil moisture data

the soil body into individual layers connected with each other by flows. It also was integrated into the Geoecological Information System and linked with EROSION-3D, as shown in Fig. 7.8. For the combinations of soil, land use and groundwater conditions identified with GIS assistance, the soil moisture must be measured once in the field at the beginning of the modeling period to serve as starting value for the water budget model. The model then supplies soil moisture data at time steps of one day, guaranteeing that the required soil moisture data will be available on an individual event basis.

Figure 7.12 shows the erosion and deposition values for the storm of 26–27 May 1993 simulated on the basis of the calculated soil moisture data. A comparison with Fig. 7.9 shows considerable deviations in some areas from the measured values. The deviation between measured and computed soil moisture contents is less than 15% in most cases. Test runs showed that deviations of 5% already cause considerable differences in the model results. The reason for this is that EROSION-3D is highly sensitive to the input parameter soil moisture (J. Schmidt 1996). In calculating the required input data it is therefore recommendable to apply only sufficiently validated models. Otherwise the danger is that although greater efforts have to be put into the computations, the advantages of the physical description of processes will be lost.

7.6
Prospects

By means of the process of "downscaling" it is possible to rapidly identify erosion prone areas or areas in which there is a risk of sediment delivery by means of empirical methods. In these areas physically based soil erosion models requiring large amounts of data and higher-powered computers can be applied. With these models the amounts of eroded and deposited material can be quantified, and the areas in which erosion is occurring can be localized. They allow various scenarios to be tried out, in which e.g., the influence of changes in land use and tillage methods, the use of buffer strips or changes in the routing of roads can be investigated. In this manner the most suitable erosion controls, and with them soil and water conservation measures, can be planned and carried out specifically, e.g., within the framework of measures to reallocate and consolidate arable land or in connection with reservoir maintenance. In the future we can expect laws on soil conservation to target the avoidance of soil erosion. By integrating models into Geographic Information Systems, we can guarantee that the data required to analyze various scenarios are rapidly available, because they can, e.g., be linked with further models. Moreover, the results can be depicted spatially with GIS assistance, and they can be linked with other data contained in GIS and used for more detailed evaluations. The main obstacle to effective application of the models in practice is that in some cases complex space-oriented data are not available in the requisite high spatial resolution. Along with the development of simple estimation procedures or dynamic models for estimating the required parameters, publications of data collections are needed. The prerequisite is, however, that the data be transferable to other regions and to the natural conditions prevailing there. The development of new methods and models for simulating ecological processes will play an increasingly important role in the future. By integrating such methods into GIS we acquire instruments that aim at providing the high level, rapidly available information needed by decision makers in conservation and regional planning.

References

Auerswald K (1993) Bodeneigenschaften und Bodenerosion, Relief, Boden, Paläoklima, Bd 8. Gebrüder Bornträger, Berlin Stuttgart

Auerswald K, Flacke W, Neufang L (1988) Räumlich differenzierende Berechnung großmaßstäblicher Erosionsprognosekarten – Modellgrundlagen der dABAG. Z Pflanzenern Bodenkd 151:369–373

Becker A (1992) Methodische Aspekte der Regionalisierung. In: Kleeberg H-B (ed) Regionalisierung in der Hydrologie. Ergebnisse von Rundgesprächen der Deutschen Forschungsgemeinschaft. Mitt XI Senatskomission Wasserforschung, VCH, Weinheim, 16–32

Bork H-R (1991) Bodenerosionsmodelle – Forschungsstand und Forschungsbedarf. In: Schwertmann U, Auerswald K (eds) Bodennutzung und Bodenfruchtbarkeit, 3: Bodenerosion. Berichte über Landwirtschaft, Sonderheft 205:51–67

Bork H-R, Schröder A (1996) Quantifizierung des Bodenabtrages anhand von Modellen. In: Blume H-P, Felix-Henningsen P, Fischer WR, Frede H-G, Horn R, Stahr K (eds) Handbuch der Bodenkunde, 1. Erg Lfg 12/96:1–43

De Roo APJ (1993) Modeling surface runoff and soil erosion in catchments using geographical information systems: validity and applicability of the ANSWERS model in two catchments in the loess area of South Limburg (The Netherlands) and one in Devon (UK). Nederlandse geografische studies 157, Utrecht

DVWK (Deutscher Verband für Wasserwirtschaft und Kulturbau) (1996) Bodenerosion durch Wasser – Kartieranleitung zur Erfassung aktueller Erosionsformen. Merkblätter zur Wasserwirtschaft 239/1996. Wirtschafts- und Verlagsgesellschaft Gas und Wasser, Bonn

Duttmann R (1993) Prozessorientierte Landschaftsanalyse mit dem geoökologischen Informationssystem GOEKIS. Geosynthesis H 4, Veröff Abt Physische Geogr u Landschaftsökologie. Dissertation, Geogr Inst Univ Hannover

Duttmann R, Mosimann T (1995) Der Einsatz Geographischer Informationssysteme in der Landschaftsökologie – Konzeption und Anwendungen eines Geoökologischen Informations-systems. In: Buziek G (ed) GIS in Forschung und Praxis. Wittwer, Stuttgart, 43–59

Flacke W, Auerswald K, Neufang L (1989) Combining a modified universal soil loss equation with a digital terrain model for computing high resolution maps of soil loss resulting from rain wash. Catena 17:383–397

Frielinghaus M, Helming K, Deumlich D, Funk R (1994) Gegenstand und Defizite in der regionalen Bodenerosionsforschung. Mitt Dt Bodenkdl Gesellschaft 74:71–74

Göpfert W (1991) Raumbezogene Informationssysteme. Grundlagen der integrierten Verarbeitung von Punkt-, Vektor- und Rasterdaten, Anwendungen in Kartographie, Fernerkundung und Umweltplanung. 2. Aufl. Wichmann, Karlsruhe

Hennings V (1994; ed) Methodendokumentation Bodenkunde. Auswertungsmethoden zur Beurteilung der Empfindlichkeit und Belastbarkeit von Böden. Geol Jb F 31

Kersebaum KC (1989) Die Simulation der Sticktoff-Dynamik von Ackerböden. Dissertation, Univ Hannover

Kirkby MJ, Imeson AC, Bergkamp G, Cammeraat LH (1996) Scaling up processes and models from the field plot to the watershed and regional areas. J Soil Water Cons 51(5): 391–396

Leser H (1997) Landschaftsökologie. 4. Aufl. Ulmer, Stuttgart

Morgan RPC, Quinton JN, Rickson RJ (1993) EUROSEM – A User Guide, EUROSEM User Guide. Version 2. Silsoe

Mosimann T (1997) Prozess-Korrelations-System des elementaren Geoökosystems. In: Leser H, Landschaftsökologie. 4. Aufl. Ulmer, Stuttgart, 262–270

Mosimann T, Rüttimann M (1996) Abschätzung der Bodenerosion und Beurteilung der Gefährdung der Bodenfruchtbarkeit. Grundlagen zum Schlüssel für Betriebsleiter und Berater mit den Schätztabellen für Südniedersachsen. Geosynthesis H 9

Neef E (1963) Dimensionen geographischer Betrachtungen. Forsch u Fortschr 37:361–363

Neufang L, Auerswald K, Flacke W (1989) Automatisierte Erosionsprognose- und Gewässerverschmutzungskarten mit Hilfe der dABAG – ein Beitrag zur standortgerechten Bodennutzung. Bayer Landwirtsch Jb 66(7):771–789

Poesen JW, Boardman J, Wilcox B, Valentin C (1996) Water erosion monitoring and experimentation for global change studies. J Soil Water Cons 51(5): 386–390

Rohr W, Mosimann T, Bono R, Rüttimann M, Prasuhn V (1990) Kartieranleitung zur Aufnahme von Bodenerosionsformen und -schäden auf Ackerflächen. Materialien Physiogeogr 14, Basel

Schmidt J (1991) A mathematical model to simulate rainfall erosion. In: Bork H-R (ed) Erosion and transport processes: Theories and models. Heinrich Rohdenburg Memorial Symposium. Catena Supplement 19:101–109

Schmidt J (1996) Entwicklung und Anwendung eines physikalisch begründeten Simulationsmodells für die Erosion geneigter landwirtschaftlicher Nutzflächen. Berliner Geogr Abh 6

Schmidt J, Werner M von (1992) EROSION-2D. Computergestützte Simulation der Bodenerosion auf landwirtschaftlichen Nutzflächen. Bedienungshandbuch (Vers. 1.5), FU Berlin

Schmidt R-G (1988) Methodische Überlegungen zu einem Verfahren zur Abschätzung des Widerstandes gegen Wassererosion. Regio Basiliensis XXIX(1 u 2):111–121

Schmidt R-G (1992) Methoden in der Bodenerosionsmessung – ein aktueller Überblick. Flensburger Regionale Studien Sonderh 2:172–194

Spatz R, Müller K, Hurle K (1996) Verringerung des Pflanzenschutzmitteleintrages in Oberflächengewässer durch die Anlage von grasbewachsenen Pufferstreifen. Mitt Biol Bundesanstalt 321:381

Schramm M (1994) Ein Erosionsmodell mit räumlich und zeitlich veränderlicher Rillenmorphologie. Mitt 190. Inst Wasserbau u Kulturtechnik, Dissertation, Univ Karlsruhe

Schwertmann U, Vogl W, Kainz M, unter Mitarb. von Auerswald K, Martin W (1990) Bodenerosion durch Wasser. Vorhersage des Abtrags und Bewertung von Gegenmaßnahmen. 2. Aufl. Eugen Ulmer, Stuttgart

Werner M von (1995) GIS-orientierte Methoden der digitalen Reliefanalyse zur Modellierung der Bodenerosion in kleinen Einzugsgebieten. Dissertation, Fachber Geowiss FU Berlin

Wickenkamp V (1995) Flächendifferenzierte Erosions-Akkumulationsmodellierung im Gebiet Mehle (Niedersachsen) unter Einsatz des Geoökologischen Informationssystems GOEKIS. Konzeption von Datenstrukturen und Datenausgabe, Anwendung des Modells EROSION-3D. Unveröff Dipl-Arbeit, Geogr Inst Univ Hannover

Wickenkamp V, Beins-Franke A, Duttmann R, Mosimann T (1996) Ansätze zur GIS-gestützten Modellierung dynamischer Systeme und Simulation ökologischer Prozesse. Salzburger Geogr Mat 24:51–60

Wischmeier W-H, Smith DD (1978) Predicting rainfall erosion losses – a guide to conservation planning. USDA Agricultural Handbook 537, Washington DC

Chapter 8

SMODERP – A Simulation Model of Overland Flow and Erosion Processes

T. Dostál · J. Váška · K. Vrána

8.1
Introduction

In the Czech Republic, about 50% of agricultural land is affected by water erosion that adversely influences soil productivity and the environment. SMODERP (A Simulation Model of Overland Flow and Erosion Processes) was developed at the Department of Irrigation, Drainage and Landscape Engineering of the Faculty of Civil Engineering, CTU Prague, in 1988 (Holý et al 1988). The model was intended for being used as an engineering tool for decision-making and design in soil and water conservation. The results of field and laboratory experiments carried out in the Czech Republic and other countries were used for developing the model (Holý et al. 1982).

SMODERP is a single-event physically-based model for simulating of overland flow and erosion at the field scale up to 1 km^2. The model simulates overland flow and erosion from heavy rainfall of variable intensity. Topography, soil, vegetation and land use can vary along the slope.

The model is used to estimate

- the volume, peak flow rate, velocity, and tangential stress of overland flow
- the critical (permissible) slope length (i.e. the distance between point of origination of overland flow and its concentration)
- the soil loss

8.2
Concept of the SMODERP Model

8.2.1
Concept of the Overland Flow Model

The model of overland flow is based upon the equation of continuity and the equation of motion. A kinematic wave approach was used to simplify the overland flow process. The investigated slope profile is divided into homogenous elements in terms of gradient, soil and land use. Interception, depression storage and infiltration of water into the soil are taken into account for overland flow simulation (Holý et al. 1982).

At the beginning of a rainfall event, interception occurs. Rainfall which is not intercepted reaches the soil surface (net rainfall) and starts to fill the depression storage of the soil surface. Water stored in depressions infiltrates into the soil depending on its infiltration capacity. Surface runoff starts when the depth of water on the soil surface exceeds the depression storage capacity. After the rainfall ends, overland flow continues until the level of the depression storage is reached, and the remaining water infiltrates into the soil.

The simulation of overland flow proceeds from the beginning till the end of the rainfall event in simulation steps. The time interval for a simulation step is 0.2 min. At the end of each simulation step the model balances the rainfall-runoff relations in all elements along the slope. The following relationships were used for water balance:

$$NS_{i,t} = BS_{i,t} - INT_{i,t} \tag{8.1}$$

$$ES_{i,t} = NS_{i,t} - RC_i - INF_{i,t} \tag{8.2}$$

$$O_{i,t} = f(H_{i,t}) \tag{8.3}$$

$$H_{i,t} = f(ES_{i,t}) \tag{8.4}$$

where:
- $NS_{i,t}$ = net rainfall (mm)
- $BS_{i,t}$ = gross rainfall (mm), i.e. total rainfall depth
- $INT_{i,t}$ = interception (mm) (see Eq. 8.14)
- $ES_{i,t}$ = effective rainfall (mm)
- RC_i = soil surface depression storage (mm)
- $INF_{i,t}$ = infiltration (mm)
- $O_{i,t}$ = overland flow rate from effective rainfall (mm)
- $H_{i,t}$ = overland flow depth (mm)
- i = number of an element of the investigated slope
- t = simulation step

For overland flow simulation, the governing balance equation is written:

$$H_{i,t} = H_{i,t-1} + ES_{i,t} + O_{i-1,t} - O_{i,t} \tag{8.5}$$

where:
- $H_{i,t}$ = depth of overland flow over the soil depression storage (mm)
- $ES_{i,t-1}$ = effective rainfall (mm) (see Eq. 8.2)
- $O_{i-1,t}$ = runoff depth (mm)
- i = number of the element
- $i-1$ = number of the preceding element
- t = time of the simulation step (from the start of simulation)
- $t-1$ = time of the preceding simulation step

The overland flow rate is based on a kinematic wave approach, i.e. it is calculated from the relation between *overland flow depth H* and *runoff rate q* in the form:

$$q = aH^b \tag{8.6}$$

where:
- q = overland flow rate (m³ s⁻¹ per unit width)
- H = overland flow depth (m)
- a, b = parameters

Parameters a, b were derived from laboratory and field experiments for three basic soil types. Their values are for

heavy soils	$a = 47.521G^{0.561}$	$b = 1.585$	(8.7)
medium heavy soils	$a = 29.462G^{0.552}$	$b = 1.736$	(8.8)
light soils	$a = 25.472G^{0.491}$	$b = 1.859$	(8.9)

where:
- G = slope gradient (%)

Parameters a, b apply for bare soils. For soils with vegetation cover, parameter a should be adjusted using Manning's roughness coefficient by an equation written

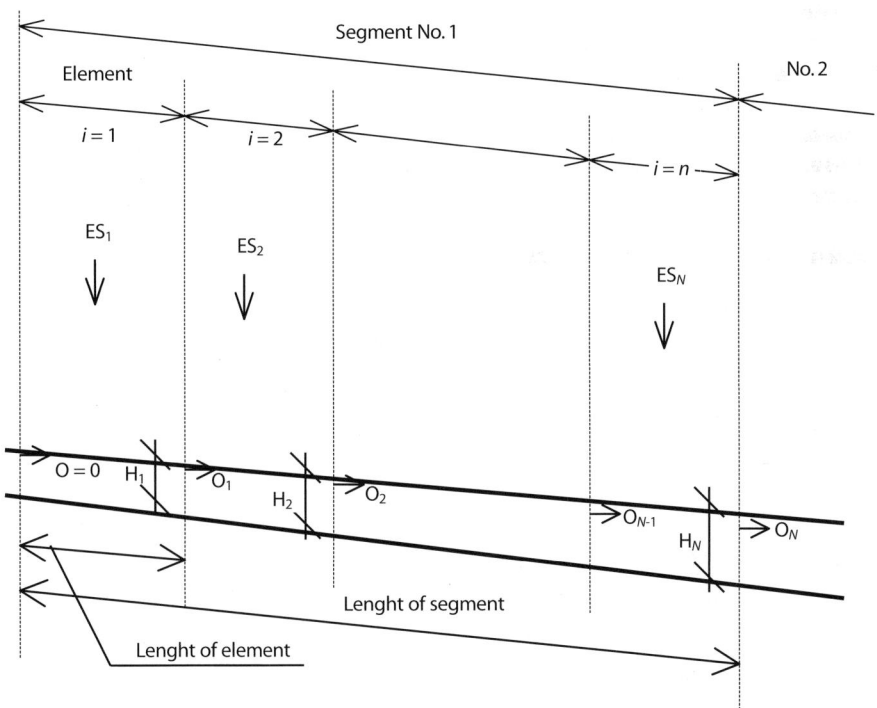

Fig. 8.1. The scheme for determining surface runoff

$$A = a(100N)^{-1} \tag{8.10}$$

where:
- A = parameter including the influence of vegetation cover
- N = Manning's roughness coefficient for overland flow

Parameter b remains unchanged for soils with vegetation cover. The overland flow rate O in Eq. 8.5 is given by the relation:

$$O_{i,t} = [(6 \times ADT)L_i^{-1}](0.1 \times H_{i,t-1})^b \tag{8.11}$$

where:
- $O_{i,t}$ = overland flow rate (mm)
- A = parameter (see Eq. 8.10)
- DT = simulation step (DT = 0.2 min in the model)
- L_i = length of element (m)
- $H_{i,t-1}$ = overland flow depth in element i in the previous simulation step (mm)
- b = parameter

The scheme for determining surface runoff is given in Fig. 8.1.

8.2.2
Concept of the Erosion Model

The erosion model is based on a dynamic concept of erosion; i.e. soil loss and sediment yield are determined from the amount of soil detached by the kinetic energy of rainfall and overland flow and the transport capacity of overland flow (Holý et al. 1982).

The same principle as was applied for simulation of overland flow, i.e. division of the slope into homogenous elements, was also used for erosion simulation. The amount of detached soil particles and the transport capacity of overland flow are determined for each element. Comparing the detached amount with the transport capacity simulates transport of soil particles along the slope and/or their sedimentation.

The detachment of soil particles by rainfall and overland flow is modelled by including the joint effect of both erosion agents, i.e. rainfall and overland flow

$$DP_{i,t} = a_0 E_{i,t}^{a1} O_{i,t}^{a2} TR^{a3} SE_i C_i \tag{8.12}$$

where:
- $DP_{i,t}$ = amount of detached soil particles (kg m^{-2} min^{-1})
- $E_{i,t}$ = kinetic energy of rainfall (J m^{-2} min^{-1})
- $O_{i,t}$ = overland flow rate (l m^{-2} min^{-1}) (output of the overland flow model)
- TR = total rainfall duration (min)
- SE_i = relative soil erodibility
- C_i = crop and management factor (USLE)
- i = element of the investigated slope

- t = simulation step
- a_0, a_1, a_2, a_3 = empirical parameters: $a_0 = 2.391\ E(-04)$
$$a_1 = 1.588$$
$$a_2 = 1.216$$
$$a_3 = 0.768$$

The transport capacity of overland flow is expressed as the function of the overland flow rate and the slope gradient

$$Tc_{i,t} = b_0 O_{i,t}^{b_1} G^{b_2} \qquad (8.13)$$

where:
- $Tc_{i,t}$ = transport capacity of overland flow (kg m^{-2} min^{-1})
- $O_{i,t}$ = overland flow rate (lm^{-2} min^{-1})
- G_i = gradient of the investigated element (%)
- b_0, b_1, b_2 = empirical parameters: $b_0 = 5.494\ E(-04)$
$$b_1 = 1.240$$
$$b_2 = 1.490$$

Parameters a_0 to a_3 in Eq. 8.12 and parameters b_0 to b_2 in Eq. 8.13 were derived from research results of long term field measurements of erosion, and from laboratory experiments (Holý et al. 1982).

8.3
Variable Description

8.3.1
Rainfall

Rainfall data are the basic input for the simulation of overland flow. Heavy rainfalls, occurring mainly in the vegetation period, are decisive for design of soil conservation measures in the Czech Republic. Design data or actual break-point rainfall data are required as an input.

8.3.2
Interception

The maximum amount of intercepted rainfall is given by the potential interception of the respective plant. Interception is dependent on its growth stage and estimated by the following equation:

$$INT_{i,t} = PLP_i BS_{i,t} \qquad (8.14)$$

where:
- $INT_{i,t}$ = interception (mm)
- PLP_i = relative leaf area
- $BS_{i,t}$ = gross rainfall (mm)

The remaining rainfall is treated as net rainfall that is determined from the equation:

$$NS_{i,t} = (1 - PLP_i)BS_{i,t} \tag{8.15}$$

where:
- $NS_{i,t}$ = net rainfall (mm)

After the potential interception has been subtracted, direct throughfall is calculated from:

$$NS_{i,t} = BS_{i,t} \tag{8.16}$$

8.3.3
Surface Depression Storage

The depression storage of the surface depends on the actual soil condition. The overland flow starts after the storage capacity has been filled up. Depth of 1–3 mm is recommended for the depression storage of a smooth soil surface.

8.3.4
Manning's Roughness Coefficient for Overland Flow

Manning's roughness coefficient for overland flow is used for incorporating the variable roughness of different types of surface cover and its reducing effect on overland flow velocity. Manning's coefficient is used in Eq. 8.10 for reducing the value of the parameter a.

8.3.5
Infiltration

The Philip equation is used for calculating the infiltration rate

$$INF_{i,t} = (0.5 S_i T^{-0.5} + K_i/\alpha) DT \times 10 \tag{8.17}$$

where:
- $INF_{i,t}$ = infiltration rate (mm) during each simulation step
- S_i = soil sorptivity (cm min$^{-0.5}$)
- K_i = saturated hydraulic conductivity (cm min^{-1})
- DT = simulation step (min); (in the model DT = 0.2 min)
- α = coefficient that is dependent on the soil type:
  sandy soils: $\alpha = 1.50$
  sandy loam, loamy and clay loam soils: $\alpha = 1.65$
  clay soils: $\alpha = 1.80$

Soil sorptivity depends on soil moisture. Sorptivity $S = 0$ for higher soil saturation (ratio of actual moisture to the moisture of saturation >0.85), which represents conditions for occurrence of gully erosion; Eq. 8.17 for infiltration rate takes now the form

$$INF_{i,t} = 10DTK_i/\alpha \tag{8.18}$$

If available, field values of K and S are recommended as input data for the runoff simulation.

8.3.6
Rainfall Kinetic Energy

The kinetic energy of rainfall is determined from the relationship

$$E_{i,t} = (206 + 87\log RI_{i,t})RD_{i,t} \tag{8.19}$$

where:
- $E_{i,t}$ = rainfall kinetic energy (J m^{-2})
- $RI_{i,t}$ = rainfall intensity (cm h^{-1})
- $RD_{i,t}$ = rainfall depth (cm)

For a rainfall intensity of $RI_{i,t}$ greater than 7.6 cm h^{-1}, $E_{i,t}$ is calculated as

$$E_{i,t} = 283 RD_{i,t} \tag{8.20}$$

8.3.7
Relative Soil Erodibility

The parameter SE_i in Eq. 8.12 that is used to determine the amount of detached soil particles were derived for medium heavy soils, for which the value of relative soil erodibility $SE = 1.0$. Values of relative soil erodibility for other soil types were estimated from experimental results as follows:

- sandy soils: $SE = 0.6$
- sandy loam soils: $SE = 0.7$
- loamy soils: $SE = 1.0$
- clay loam soils: $SE = 0.9$
- clay soils: $SE = 0.8$

The values of relative erodibility enter the simulation as specified foron the basis of the respective soil type.

8.3.8
Vegetation Cover and Management (C-Factor)

The influence of vegetation cover and management on erosion is accounted for by the C-factor of the Universal Soil Loss Equation (USLE).

8.4
Model Inputs and Outputs

8.4.1
Input Files

Two input files are required for the simulation of overland flow and erosion. The input data for the rainfall file are

- rainfall depth and duration (break-point rainfall data), or uniform design rainfall intensity

The input data for the slope file are:

- slope geometry (slope length, width and gradient)
- soil data (soil type, saturated hydraulic conductivity, sorptivity, depression storage, Manning's roughness coefficient)
- vegetation characteristics (vegetation or crop type, potential interception, relative leaf area, vegetation and management factor)

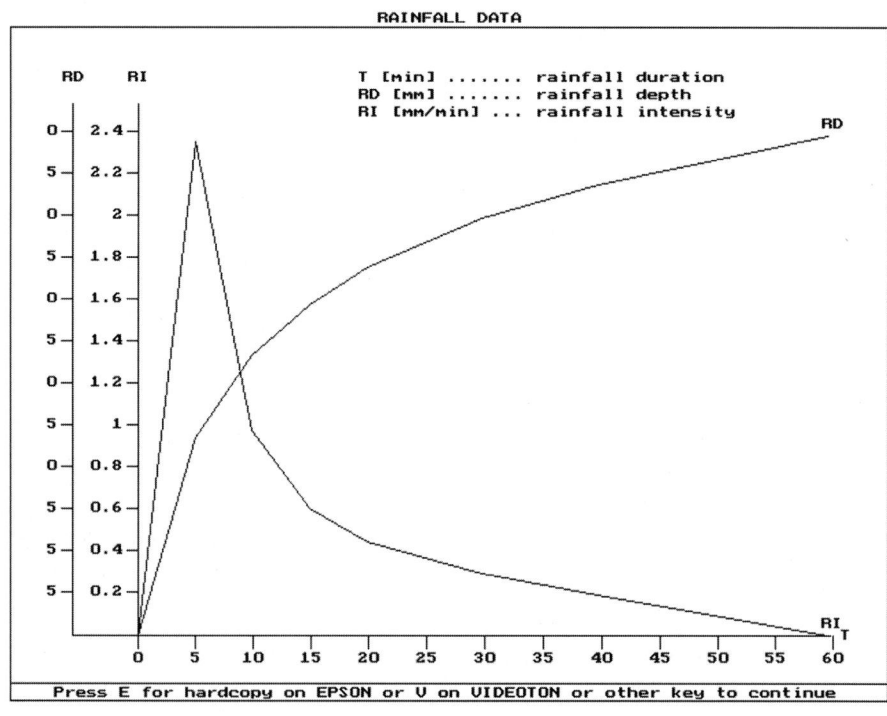

Fig. 8.2. Rainfall input data

Guide values for the input data are given in the SMODERP User's Guide for ones in the Czech Republic (Holý et al. 1989).

Examples of input files are given in Fig. 8.2, 8.3a, 8.3b.

CHARACTERISTICS OF THE INVESTIGATED SLOPE	
FARM NAME:	test
Field register number:	123
Number of segments:	2
a SEGMENT No. 1	
Morphological characteristics:	
– segment length:	125 m
– segment average width:	10 m
– segment gradient:	13.5 %
Soil characteristics:	
– soil type:	loam
– soil sorptivity:	0.1 cm/min 0.5
– coefficient of hydraulic conductivity:	0.01 cm/min
– soil surface retention:	3 mm
Vegetation and management characteristics:	
– vegetation cover:	bare soil
– Manning´s roughness coefficient:	0.030
– relative leaf area:	0
– potential interception:	0 mm
– vegetation and management factor:	1.0
b SEGMENT No. 2	
Morphological characteristics:	
– segment length:	90 m
– segment average width:	10 m
– segment gradient:	10.0 %
Soil characteristics:	
– soil type:	loam
– soil sorptivity:	0.115 cm/min 0.5
– coefficient of hydraulic conductivity:	0.014 cm/min
– soil surface retention:	0
Vegetation and management characteristics:	
– vegetation cover:	corn
– Manning´s roughness coefficient:	0.035
– relative leaf area:	0.16
– potential interception:	0.13 mm
– vegetation and management factor:	0.6

Fig. 8.3. a Slope characteristic input data-segment 1. **b** Slope characteristic input data-segment 2

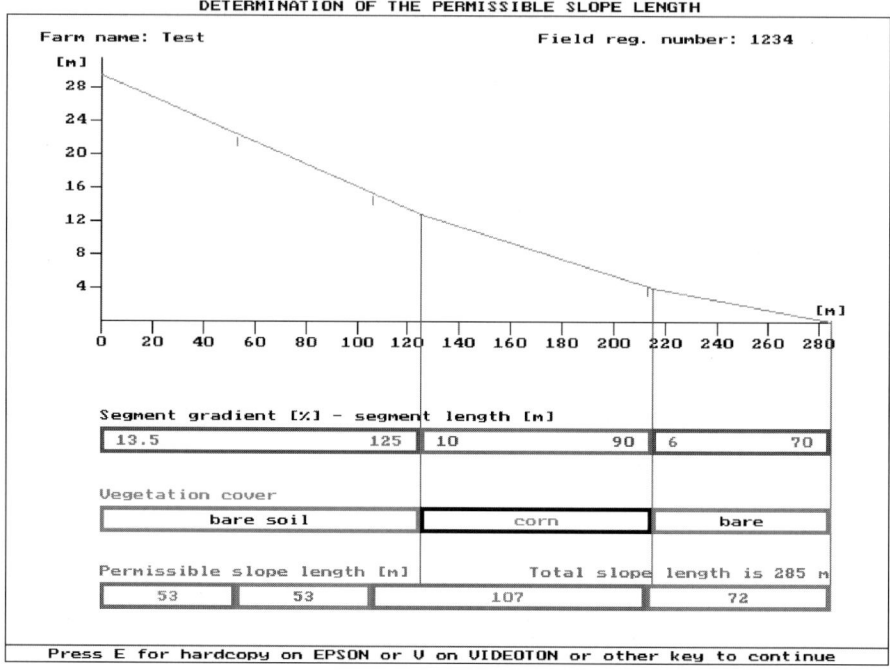

Fig. 8.4. SMODERP output-permissible slope length

8.4.2
Output Files

SMODERP allows specifying two types of output files. These provide information on the

- critical slope length
- overland flow and erosion data (volume of overland flow, peak flow, soil loss)

The data are given for the down slope end of each element and for each minute from the beginning of simulation.

Both output files are available in tabular and in graphic form. Examples of SMODERP output files are given in Fig. 8.4, 8.5, 8.6a and 8.6b.

8.5
Sensitivity Analysis

A sensitivity analysis was conducted for basic input parameters (Vrána 1991). The relative impacts of changes in the values of the different input parameters on the simulation results are shown in the Table 8.1. It also provides a guide to the accuracy required for input data determination.

Chapter 8 · SMODERP – A Simulation Model

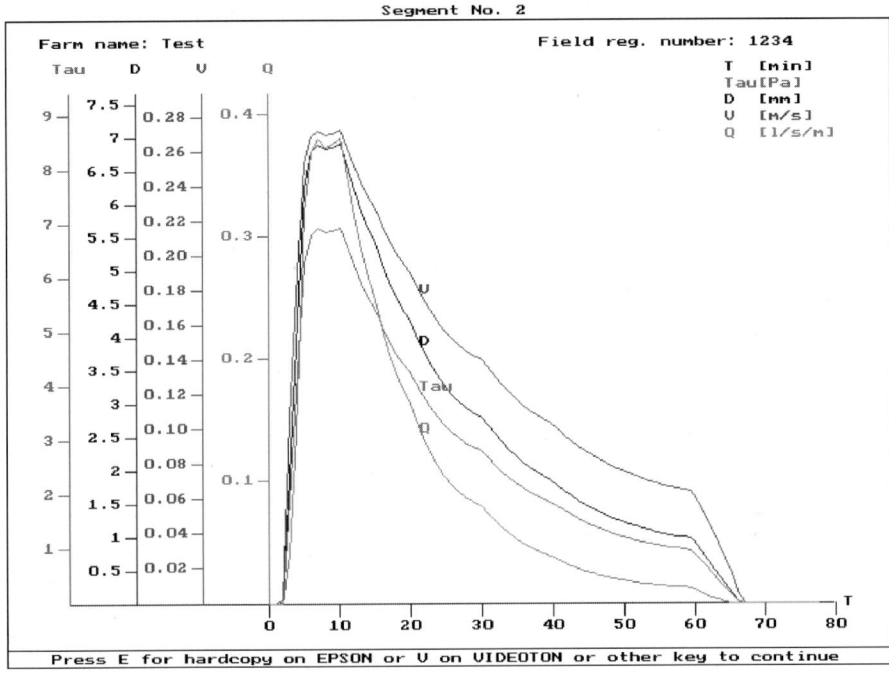

Fig. 8.5. SMODERP output-characteristics of overland flow (2nd segment)

Table 8.1. Results of sensitivity analysis

Input	Impact on outputs
Rainfall data (rainfall depth (mm) and duration (min))	★★★
Slope geometry – gradient (%)	★
Soil type	★★
Saturated hydraulic conductivity (cm min^{-1})	★★
Sorptivity (cm min$^{-0.5}$)	★
Manning's roughness coefficient	★★★
Depression storage (mm)	★★
Vegetation and crop type	★★
Potential interception (mm)	★
Relative leaf area	★
Vegetative and management factor (C-factor)	★

★ less important;
★★ important;
★★★ very important.

b

```
*****************************************************
* Total surface runoff volume:   24.02   (m3)       *
* Peak flow rate:                0.37    (l/s/m)    *
* Soil loss:                     0.4     (t/ha)     *
* Total soil loss:               40      (t)        *
*****************************************************
```

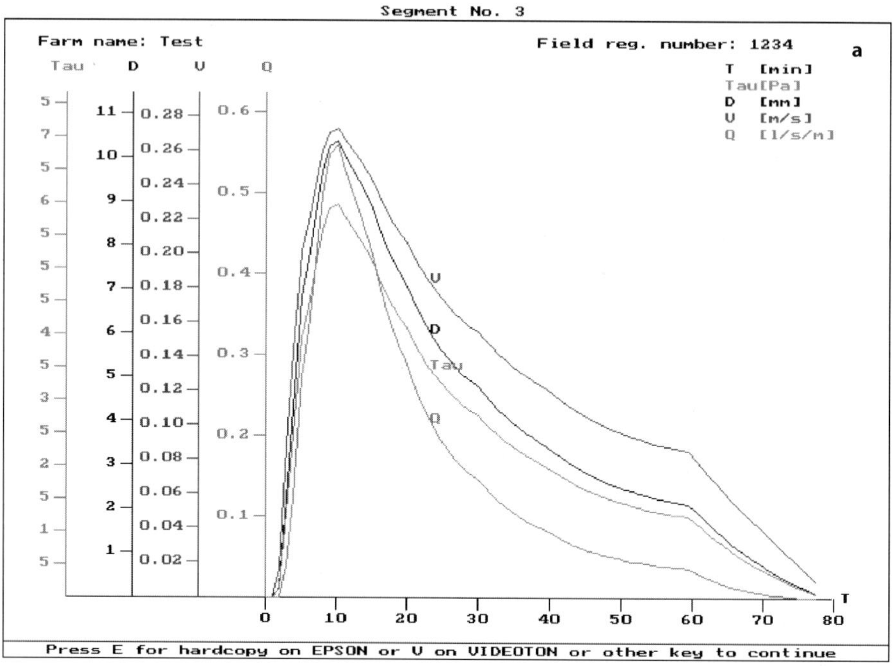

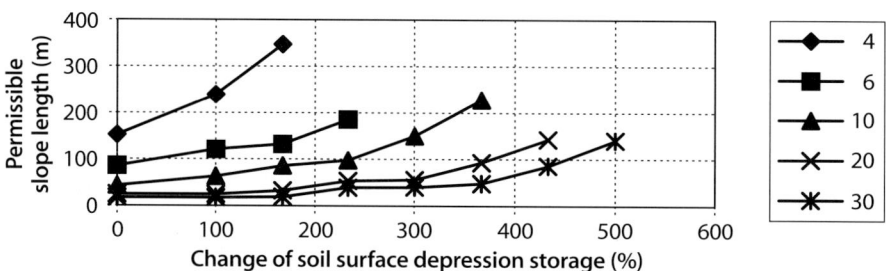

Fig. 8.6. a SMODERP output-characteristics of overland flow (end of the slope). **b** SMODERP output – overview of results of simulation

Fig. 8.7. Relation between critical slope length and relative change of soil surface depression storage (bare soil)

Sensitivity analysis studied the impact of change of selected input values on the resulting critical slope length. The tested parameters were

- soil surface depression storage
- relative leaf area and potential interception
- Manning's roughness coefficient
- hydraulic conductivity

Each input was analysed separately in such a way that its value was gradually increased and decreased from its average recommended value in a physically possible range. The permissible slope length was determined for each value of a tested parameter for a uniform slope with inclinations of 4, 6, 10, 20, and 30% by SMODERP simulations. Testing of all parameters was carried out for four basic soil types (sandy soils, sandy loam soils, loamy soils and clay loam soils) and for four types of vegetation (bare soils, row crops, small grain crops, and grass).

The results of sensitivity analysis are presented graphically as a relation between relative change of the tested parameter (the average recommended value was considered as 100%) and permissible slope length. The graphical relations for loamy soils, various parameters (soil surface depression storage, potential interception, Manning's roughness coefficient and hydraulic conductivity), and various types of surface cover are presented as examples: bare soil (Fig. 8.7, 8.10), row crops (Fig. 8.8), small grain crops (Fig. 8.9). The inclination of the lines in the graphs gives a measure of the influence of permissible slope length determination for incorrect choice of input parameter values.

It may be concluded that

- change of soil surface depression storage in a range between 0–5 mm has negligible influence, between 5 and 10 mm it is important, and for more then 10 mm it is very important.
- change of relative leaf area and potential interception has negligible influence, regardless of their values.
- change of Manning's roughness coefficient has great importance, regardless of its value.
- change of hydraulic conductivity and sorptivity values is very important for sandy loam soils and loam soils, important for clay loam soils, and less important for clay soils.

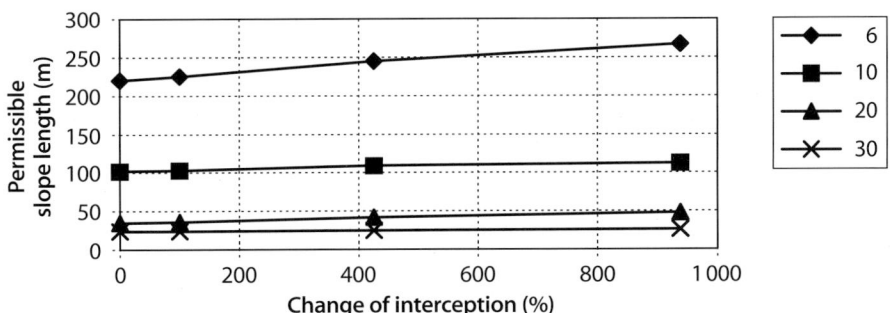

Fig. 8.8. Relation between permissible slope length and relative change of potential interception (row crops)

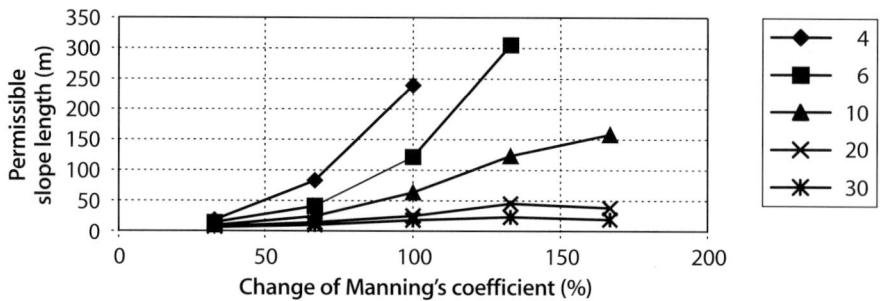

Fig. 8.9. Relation between critical slope length and relative change of Manning's coefficient (small grain crops)

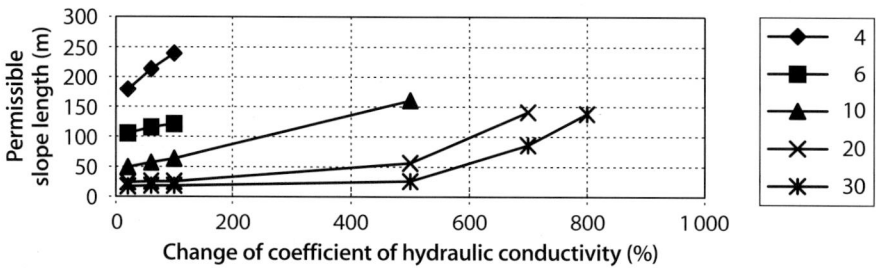

Fig. 8.10. Relation between critical slope length and relative change of coefficient of hydraulic conductivity (bare soil)

8.6
System Requirements

The code for the SMODERP model has been written in TURBO Pascal. SMODERP is applicable on IBM compatible computers with minimum RAM 640 kB. Two versions of SMODERP are available: an original freeware English version (1989) and an upgraded commercial version (1996, in Czech). A minimum hard disk capacity of 1.7 MB is required for installation and operation with files for the second version.

8.7
Use of SMODERP in Soil and Water Conservation

SMODERP outputs provide information both for decision making and for design in soil and water conservation. As the outputs of the model are both overland flow and erosion data, the model can be directly used for design of agronomic and mechanical conservation measures. The main applications of the model are for erosion hazard assessment and soil conservation planning for agricultural land; however it can also be used for landscape rehabilitation projects, land consolidation plans, restoration of waste dumps, protection of urban areas, etc. (Jansky et al.1992).

In the Czech Republic, the model has been used for more than 30 soil and water conservation projects on various scales and for various purposes. Several consulting firms have used the commercial version of the model.

Presently, a modified version for simulation of overland and concentrated flow is being tested for assessment of surface runoff and erosion in a small watershed.

8.8
Using the SMODERP Simulation Model for the Design of Conservation Measures as a Part of Landscape Revitalisation Studies

The changes in the political system in the Czech Republic after 1989 have led to significant changes in the national economy, in the system of land use, and to increasing environmental land use protection.

Efforts to correct the mistakes of the past period began soon after the revolution in parallel with the transition to the market economy and the restitution of fields to private owners.

To help in solving those problems a special fund was set up by two ministries the Ministry of the Environment and the Ministry of Agriculture (the Landscape Care Fund). Comprehensive studies of all problems in each region had to be made before measures could be undertaken. These studies included an assessment of the optimal use of fields, revitalisation of rivers and brooks, the design of new small reservoirs, field roads and the location of trees and bushes. Implementation of these measures will also influence water quality in a positive way. An investigation of all agricultural fields from the point of view of disposition to water erosion and design of the conservation measures forms a very significant part of these studies.

The SMODERP simulation model proved useful in many cases for studying erosion problems. The application of he SMODERP includes three steps:

- the determination of input data
- the simulation of erosion at each field
- the design of conservation measures

8.8.1
Input Data for the SMODERP Model

The input data to the SMODERP model are:

- design rainfall
- morphological characteristics
- soil characteristics

Tables of heavy rainfalls for 98 climatic stations have been made for all regions of the Czech Republic. Rainfalls of different probabilities are given in these tables. The probability of rainfall must be chosen according to the influence of surface runoff and transport of soil particles on water quality. For fields from which the surface water runs into the free landscape the one- or two-years probability is taken. For the fields with runoff into villages or cities it is appropriate to take rainfall with five- to ten-years probability. The length of the design rainfall is 2 hours.

Morphological characteristics (length and slope of the field under investigation) are taken from the map (scale 1 : 10 000) and modified according to the results of field investigation (e.g. the interruption of surface runoff by the ditches along field roads etc.). The number of characteristic profiles is given by the area, form and slope of the field blocks. In some field a single characteristic profile is enough, while other fields need two or more profiles.

Soil characteristics are given in the soil maps which have been made for agricultural land of the Czech Republic. As those maps were prepared about 15 years ago, it is appropriate to use the results from the field investigation to correct the data obtained from the map. The SMODERP model needs two soil characteristics – coefficient of hydraulic conductivity, and sorptivity. Both characteristics can be found find by laboratory analysis of soil samples.

8.8.2
Calculation of the Critical Slope Length

The critical slope length is calculated for each characteristic profile of the investigated area. The simulation is run for designed rainfall, morphological and soil characteristics and three possible types of plants (row plants, small grains and grass). The calculated critical slope length is compared with the actual slope length.

If the actual length of the investigated field for row plants is larger than the simulated critical slope length, a simulation for small grains or for grass must be repeated.

8.8.3
Design of Conservation Measures

The results of critical slope length calculation for all characteristic profiles form the basis for the design of conservation measures. Firstly, easier and less expensive measures (plant changing) should be preferred. If this type of conservation measure is not sufficient, the design of the technical measures (ditches, terraces) follows. In this case, it is suitable to use existing field road ditches to intercept surface runoff.

8.8.4
Conclusion

Case studies of landscape revitalisation provide a very sound basis for funding of revitalisation.

However, in the Czech Republic the design of erosion measures only leads to recommendations to the farmers about how to protect their fields from water erosion. The present laws do not oblige landowners to implement the design measures and state grant policies do not motivate farmers to prefer conservation farming technologies rather than market-based activities.

In spite of these problems, case studies of landscape revitalisation form a significant basis for decision-making.

8.9
Protection of the Urban Areas against Surface Runoff and Sediment from a Small Agricultural Watershed (Application of the SMODERP Model)

8.9.1
Description of Current Situation

The village Vinare is located 60 km east of Prague (in Kutná Hora district) (Vrána et Dostál 1995). A small watershed of about 130 ha with no permanent channel is adjacent to the village on its south side. The outflow point of the watershed is a very small road bridge, very near a housing area.

In the past, all the agricultural land in the watershed was divided into a few hundred small fields. Surface runoff was guide off through the road bridge to a non-lined channel and to the stream. However during the "collectivization" period, all the small fields were combined into large field. Small field trucks, trees and bushes were removed. The river bridge is in a very bad condition today due to lack of maintenance, clogging of the culvert with sediments and overloading of the bridge with heavy machinery.

Every year, high surface runoff volumes with severe soil erosion are observed in the watershed every year after heavy rainfall or snow melt. Since the road bridge is almost out its function, all the water from the watershed floods the adjacent fields and gardens, filling house basements and deposits the sediment in the adjacent areas.

The present problem is clearly caused by combination of unfavourable natural conditions (loess soils with low permeability, a slope up to 10%, the shape of the watershed) and unsuitable land use (removing all natural barriers, creating a single united field producing plants which provide low soil protection – e.g. maize).

8.9.2
Methodology

The study is guided by the following principles:

- It is necessary to prevent the concentration of surface runoff in order to achieve this objective. Crop management measures will be proposed in the first step (changes in crop rotation). In the next step, technical measures (diversion and drainage ditches) are suggested in such cases where crop management measures are not sufficient.
- The retention capacity of the watershed should be improved in order to slow down the runoff velocity and to reduce peak runoff rates.
- It is necessary to prevent soil erosion and the transport of the mobilised sediment.

8.9.3
The Solution

- The SMODERP model for used for various alternatives of land uses and spatial patterns of ditches to determine the critical slope lengths and characteristics of surface runoff. The whole watershed was divided into small planes, represented by characteristic profiles. The computation itself was provided for these profiles, and then the results used for the whole width of the planes.
- The information about critical slope lengths shows the need for soil erosion measures and the location any necessary ditches.
- The results of simulating surface runoff provide surface runoff hydrographs in outflow points of characteristic profiles. For the purposes of computation the ditches were divided into segments of length 100 m each, where the velocity, size and peak value of the flow wave were determined by the kinematic wave method.
- This is not a standard way of using the SMODERP model (and it has not yet been widely used by model users), but it is the only way to assess the influence of individual soil erosion measures designed on surface runoff volume and peak value.
- Basic mathematical simulations have been provided for 5 alternative space patterns of technical soil erosion measures (ditches), for various crop rotations and for design rainfall with periodicity 10 years. The results of total runoff, peak discharge and soil loss for individual characteristic profiles provides a good overview of the influence of various soil erosion measures on surface runoff and erosion risk. For all 5 alternatives, but only for a combination of the permanently grassed upper part of the watershed with all the rest of the land used for maize, a detailed computation was performed of the movement of flood waves, determining the total runoff and peak flow rate at the outflow point of the watershed (road bridge).

8.9.4
Input Data

Design Rainstorm

The design rainstorm from near raingauge station Sec with return period of 10 years was used for simulation (Table 8.2).

Morphological Data

Maps of the Czech Republic in the scale of 1 : 10 000 were used as a source of topography. An additional field survey was provided to obtain the most accurate information. The whole catchment was divided into 10–15 partial planes (depending on alternative), represented by characteristic profiles.

Soil Data

The basic source of data was 1 : 5 000 soil maps of the Czech Republic. Additional field survey and laboratory measurements were carried out to obtain more accurate data.

Table 8.2. Design rainfall

T (min)	H (mm)
5	14.4
10	20.9
15	24.2
20	26.2
30	28.8
40	30.7
60	33.4
90	36.1
120	38

Table 8.3. Guide values of vegetation cover and soil properties for the Vinare watershed

Soil type	Crop	K (cm min^{-1})	S (cm min$^{0.5}$)	LAI	Potential interception (mm)	Manning roughness
Loamy	Grass	0.016	0.130	1.0	0.40	0.100
	Small grain crops	0.015	0.125	0.30	0.20	0.040
	Row crops	0.014	0.115	0.16	0.13	0.035
Clay-loamy	Grass	0.0070	0.155	1.0	0.40	0.100
	Small grain crops	0.0060	0.145	0.30	0.20	0.040
	Row crops	0.0055	0.140	0.16	0.13	0.035

Land Use Data

The basic crop was a wide row crop (maize). The results of computation showed that they should be changed to small grain crops or even to permanent grass. The design values for vegetation cover and soil properties are displayed in Table 8.3.

8.9.5
Alternative Soil Protection Measures

Generally, agricultural management practices are considered first. Only if they have proved to be not sufficient, technical measures are evaluated in the next steps.

Here, the impact of crop rotation changes on soil erosion and surface runoff generation were first examined. Since even the change of the entire land use in the watershed to grassland did not reduce surface runoff volume, various patterns of ditches were examined in the following steps. For all simulations, a permanent grass strip with a width of 100–400 m was assumed in the upper part of the watershed.

The objective of this design option was lengthen the travel time of surface runoff in the ditches of the watershed. This will cause the move of peak flow rates from individual parts of the watershed what results in decrease of necessary capacity of the road bridge and the outflow channel.

8.9.6
Alternatives

Since it is not possible to reduce peak discharge rates and the amount of erosion by agricultural management practices to the required level, various patterns of ditches have been designed within the watershed.

Type A of diversion ditches has a depth of 0.5 m, a bottom width of 1.0 m, side slopes of 1 : 1, and a gradient of 1%. The ditches will be lined with a permanent grass cover. A path will be, built on the bank.

Type B of drainage ditches has a depth of 0.5 m, a bottom width of 3.0 m, and side slopes of 1 : 10. The slope will be equal to the slope of the terrain. The ditches will be lined with a permanent grass coverand will normally be cultivated.

Alternative 0

The so-called "zero variant" (Fig. 8.11) is an alternative without any technical measures. The peak flow rate for maize at the outflow point is 3 400 l s^{-1}.

Alternative 1

This alternative (Fig. 8.11) proposes diversion ditches at the level 250 m.a.s.l. The area above the ditches will be changed to permanent grassland.

This alternative provides good protection for the land in the upper part of the watershed. However, mainly in the lower part of the watershed, the situation remains unchanged.. Agricultural management practices results only in partial improvements due to the low permeability and high erodibility of the soil. With combined cultivation in the lower part of the watershed, the peak flow rate at the outflow point is 2 000 l s^{-1} for this alternative.

Alternative 2

This alternative (Fig. 8.11) is almost identical with alternative 1. The only difference is an additional ditch (P6) at the level about 240–246 m.a.s.l. in order provide better protection of this area. In this case, no erosion is observed in the lower part of the watershed even if it is entirely cover with corn.

The peak flow rate at the watershed outlet is 1 425 l s^{-1} for this alternative.

Alternative 3

Alternative 3 (Fig. 8.11) is a modification of alternative 2. This alternative considers the requirement to minimize the total length of all ditches. This variant provides

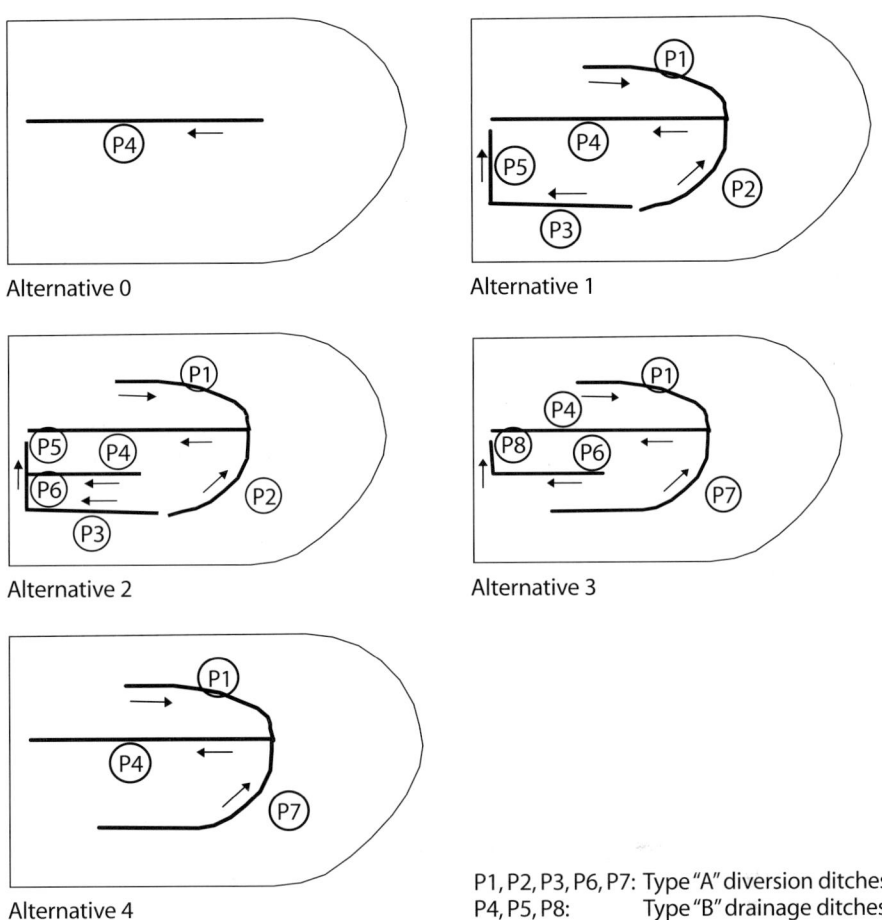

Fig. 8.11. Schemes of the spatial patterns of ditches for the individual alternatives

the best results in protection against soil erosion even for corn and decreases the peak rate at the watershed outlet. The peak flow rate at the outlet is $1350\ \mathrm{l\ s^{-1}}$ for this alternative.

Alternative 4

In this alternative (Fig. 8.11) ditches P6 and P8 are removed. This variant cannot be recommended, since even if the ditch along the road is removed, water will still flow here and will cause much greater damage. In addition, all the lower left part of the watershed is still affected by erosion.

With the land use being the same as in alternatives 2 and 3 the peak flow rate at the outlet is $2000\ \mathrm{l\ s^{-1}}$.

Road Bridge

The culvert of road bridge must be able to safely guide off the maximum peak flow rate that occur in the watershed. This will depend on the pattern of ditches that is chosen for implementation. The maximum possible diameter of tube is 700 mm, due to the terrain configuration. ($Q_c = 708$ l s^{-1}). It is possible to install more than one such tube, side-by-side, if necessary.

8.9.7 Conclusions

The current unsatisfactory situation in surface runoff and erosion events in Vinare, which leads to flooding of gardens, basements and road, several times a year, is caused mainly by a combination of unfavorable soil properties (low permeability, high erodibility) and unsuitable land use, together with a non-operating road bridge.

Corrective measures involve introducing changes in the agricultural land, which will prevent the development of erosion processes and will collect surface runoff out and safely remove it from the area.

The 5 alternative spatial patterns of ditches have been designed and evaluated in terms of exceeding critical slope lengths, total runoff volume, peak flow rate and total soil loss for a design rainfall with periodicity 10 years. The upper part of the watershed should be changed to permanent grass, and the lower part should be free for the production of both small grains and wide rows crops. For this combination, the advance of the flood wave in ditches was calculated with a spatial step of 50–100 m.

The drainage ditches (type B) are designed as a grass waterways and machines should be able to cross them at any point. The diversion ditches (type A) are designed as a narrow ditches.

The author considers the alternative 3 to be the optimum variant. It proposes to change the upper part of the watershed into permanent grass. On the lower border of the grass, ditches will be created, which will collect surface runoff from these areas and will take it to the valley bottom. The slope between the valley bottom and ditch P7 should be additionally divided approximately in the middle by a ditch that will target the road. In this way all the arable land can be cultivated and used even for

Table 8.4. Summary of peak discharges

Ditch No.	T (min)	Q (l s^{-1})
P1	23 – 25	153
P4	25	840
P6	22	562
P7	33	377
P8	29	604
Road bridge	25 – 30	1 350

Table 8.5. Results for individual profiles (alternative 3)

Profil No.	Crop	Slope length (m)	Slope width (m)	Runoff volume (m³/100 m)	Peak discharge (l s⁻¹/100 m)	Soil loss (t ha⁻¹)	Critical slope length (m)	Runoff volume (m³)	Soil loss (t)
1	Grass	185	180	269.0	30	0.3	×	484.2	1.0
2	Grass	220	400	299.5	40	0.4	×	1198.0	3.5
3	Grass	223	380	312.2	40	0.2	×	1186.4	1.7
4	Grass	395	350	953.5	90	0.4	×	3337.3	5.5
5	Grass	366	220	832.2	100	0.5	×	1830.8	4.0
6	Grass	111	210	176.2	30	0.2	×	370.0	0.5
7	Grass	102	160	156.7	30	0.1	×	250.7	0.2
12	Corn	115	140	211.0	50	0.5	×	295.4	0.8
13	Corn	320	320	506.6	70	0.4	278	1621.1	4.1
14	Corn	405	300	471.0	180	0.2	194	1413.0	2.4
21	Corn	170	320	172.1	50	0.6	74	550.7	3.3
22	Corn	220	480	626.4	160	1.6	85	3006.7	16.9
23	Corn	270	370	471.8	70	0.4	×	1745.7	4.0
24	Corn	190	230	426.9	70	0.4	×	981.9	1.7
25	Corn	240	320	291.6	70	0.7	128	933.1	5.4
								19205.0	55.0

corn with no limits. This variant is also documented in detail in tables (Table 8.4, 8.5, 8.6) and figures (Fig. 8.12).

The peak discharge at the watershed outlet is 1350 l s⁻¹. This discharge could be conduct off by two tubes of 700 mm diameter each ($Q_c = 1416$ l s⁻¹).

8.10
Numerical Simulation of Overland Flow on the Radovesice Waste Dump (North Bohemia)

8.10.1
Introduction

The reclamation of waste dumps is a serious problem associated with surface mining in the Northern Bohemian brown-coal district. The final period of reclamation is stipulated by the volume and layout of the dump body. The hydrological and physical parameters of the surface layer of the dump have also a significant influence on the success of the reclamation. If the dump surface is made up of unconsolidated, non-structural original material, intensive erosion processes will occur independently of the rainfall intensity. Increasing the infiltration capacity and improving the physical properties of the top layer is the only way to prevent the dump body from erosion before its final stabilisation and biological revitalisation.

Table 8.6. Summary for individual ditches for rainfall with periodicity 10 years and corn on the fields

Ditch No.	Pf No.	Q_{10} Corn V (m³)	G (t)
P1	5	1 830.8	4.0
	6	370.0	0.5
	7	250.7	0.2
		2 451.6	4.7
P4	12	295.4	0.8
	13	1 621.1	4.1
	14	1 413.0	2.4
	23	1 745.7	4.0
	24	981.9	1.7
	25	933.1	5.4
		6 990.2	18.5
P6	1	484.2	1.0
	21	550.7	3.3
	22	3 006.7	16.9
		4 041.6	21.2
P7	2	1 198.0	3.5
	3	1 186.4	1.7
	4	3 337.3	5.5
		5 721.6	10.7

Px Individual ditches (the important ones);
Pf Profile number;
V Total runoff volume;
G Total sediment transport.

8.10.2
Theoretical Considerations

The objective of decreasing erosion hazards is based on the concept of a "critical slope length"; if the actual slope length is greater than the critical slope length, the overland flow tends to concentrate into rills, and sheet erosion changes to more severe forms. Simulation methods that are based on the hydraulics of overland flow are used to determine the critical slope length.

The critical slope length was determined using the SMODERP simulation model for different reclamation technologies applied on a original waste dump surface and for various rainfall probabilities, with the aim to provide a recommendation for the selection of technology best suited to stabilise the dump surface.

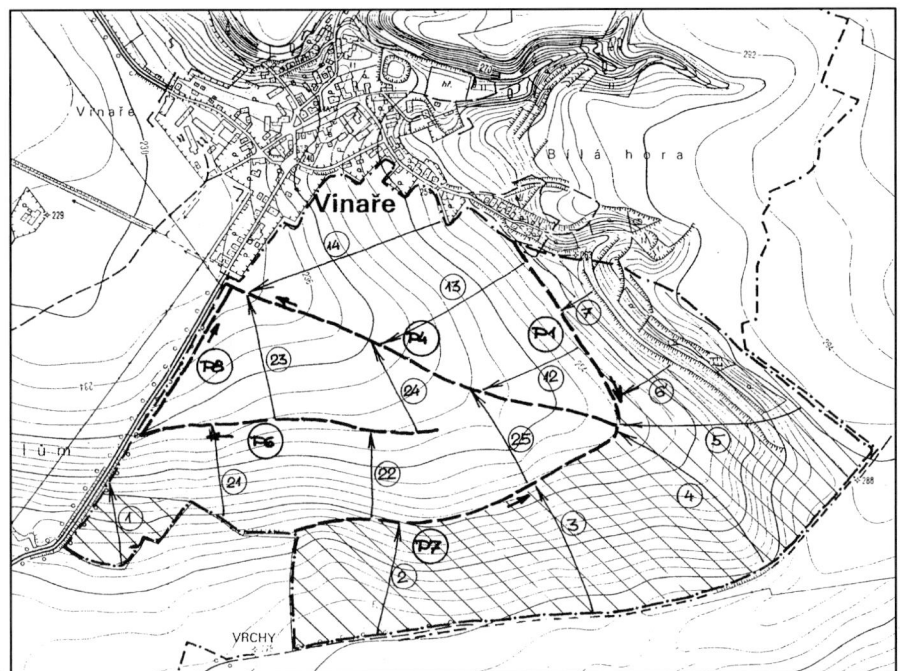

Fig. 8.12. Situation for alternative 3 (scale 1 : 10 000)

8.10.3
Methodology

The experimental plots are located on the Radovesice dump near the town of Bilina, in North Bohemia (Váška et al. 1993). The elevation is 232 m above sea level, and the annual precipitation is 488 mm. The following types of the dump surface reclamation technologies were evaluated by rainfall simulation:

I. Original material of the dump surface
II. Surface covered with a topsoil layer of marl, mixed with sandy loam
III. Surface covered with a layer of marl
IV. Surface covered with a layer of clay loam (loess)

The inputs were evaluated for running SMODERP for experimental plots that include e.g. design rainfall for its various probabilities ($p = 0.5, 0.2, 0.1$ and 0.02), depression storage, Manning's roughness coefficient for overland flow, etc. The most important inputs into the model are infiltration characteristics, because separation of surface and subsurface flow significantly influences the results of simulation. The Guelph permeameter method was used for field determination of hydraulic conductivity and sorptivity (Dolezal et al. 1990).

Table 8.7. Hydraulic conductivity (K_{fs}) and sorptivity (S) for experimental plots on the Radovesice dump

	Scenario			
	I	II	III	IV
K_{fs} (cm min^{-1})	0.0002	0.0176	0.0043	0.002
S (cm min$^{-0.5}$)	0.0249	0.2347	0.1154	0.0813

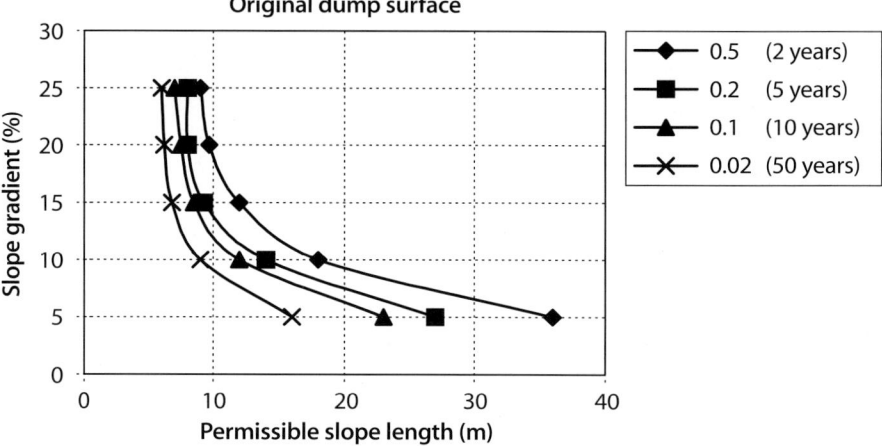

Fig. 8.13. Example of results of simulation of critical slope length

The results of the determination of the hydraulic conductivity and sorptivity values for the experimental plots are given in Table 8.7.

8.10.4
Results and Conclusions

From the conservation point of view, the most desirable reclamation measure for the initially sterile dump surface is to cover the surface with a layer of marl mixed with sandy loam of the thickness of 40–50 cm. By this measure optimal values of physical properties are achieved which positively influence the infiltration capacity of the surface layer and which subsequently create favourable conditions for the following biological rehabilitation (Fig. 8.13).

Covering the dump surface with marl only is suitable from the conservation point of view for higher rainfall probabilities (up to $p = 0.2$), while for lower probabilities ($p = 0.1$, resp. 0.02) the erodibility of the cover layer approaches the erodibility of the original dump material.

Overlaying the dump surface with clay loam (loess) is a less suitable measure because, its erodibility is nearly the same as the high erodibility of the original dump material.

The last two variants also have a low content of organic matter and undeveloped soil structure. This causes a significant deterioration of physical properties for high rainfall intensities when intensive wash-out of sealing colloidal particles occurs. These variants also create unfavourable conditions for biological rehabilitation.

The suitability of the reclamation measure for subsequent biological rehabilitation is – besides the conservation effect of the measure – an important condition for the final choice of a reclamation measure for waste dumps.

References

Dolezal F, Kuraz V, Zochova D (1990) Measurement of the field saturated hydraulic conductivity with the Guelph Permeameter and with other methods. Symposium IAHS, Vienna, Austria

Holy M, Mls J, Váška J (1982) Modelovani eroznich procesu. Studie CSAV c.6. pp. 81. Acad Praha.

Holy M, Váška J, Vrána K (1982a) Mathematisches Modell des oberirdischen Abflusses zur Bewertung von Erosionsprozessen. Z f Kulturtechnik und Flurbereinigung 23:269–279

Holy M, Váška J, Vrána K, Mls J (1982b) Analysis of surface runoff. CP-82-33. Int Inst for Applied Systems Analysis (IIASA), Laxenburg, Austria, pp 43

Holy M, Váška J, Vrána K (1986) Model of surface runoff and erosion processes. Proceedings of the conference "Hydrological Process in Catchments", 69–74, Krakow, Poland

Holy M, Váška J, Vrána K (1988) SMODERP: A Simulation Model for Determination of Surface Runoff and Prediction of Erosion Processes. Technical papers of the Faculty of Civil Engineering, CTU Prague, Series V8:5–42, Prague

Holy M, Váška J, Vrána K (1989) SMODERP: Simulation Model for Determination of Surface Runoff and Prediction of Erosion Processes. User's guide. Faculty of Civil Engineering CTU Prague, pp 24

Jansky L, Váška J, Furuya T, Okuyama T, Thoreson B (1992) Simulation Model of Surface Runoff and Erosion Process (SMODERP). Application and verification in Japan. Proceedings of the Int Symposium "Land Reclamation: Advances in Research and Technology". Younos T, Diplas P, Mostaghimi S (eds) ASAE, Nashville, USA, 24–29

Váška J, Vrána K (1993) Modeling of runoff and erosion from a small watershed. Proceedings "International Symposium on Runoff and Sediment Yield" (RSY-93), Warsaw, Poland, 101–107

Váška J, Kuraz V, Cermak P (1993) Numerical simulation of the surface runoff on the waste dump Radovesice (North Bohemia). Proceedings of the Symposium "Advances in Water Science", Vol II, Stara Lesna, UHH SAV Bratislava, Slovakia, 154–159

Chapter 9

Modeling Overland Flow and Soil Erosion for a Military Training Area in Southern Germany

R. Deinlein · A. Böhm

9.1 Objective

Soil erosion is a serious ecological and economical problem which will become more important in the future in European countries. Quantitative estimates of soil erosion can be essential to all issues of land management. Erosion models are particularly useful instruments for predicting soil erosion and for controlling the effects of land management techniques. This paper is divided into two sections. In the first section, several soil erosion models are evaluated in order to identify an erosion model for practical application. In the second section, the EROSION-3D model is chosen and applied to a military training area in southern Germany. Finally, the modeling results are presented and discussed for selected catchments.

Since soil loss predicted by the Universal Soil Loss Equation (USLE, Wischmeier and Smith 1978) provides only a rough annual estimate, it is not possible to simulate soil erosion for a single rainfall event. Further disadvantages of the USLE are the lacking representation of sediment transport and deposition processes, and its inherent deficit to address hydrological processes. For this reason, a more process-based erosion model was planned to be applied to a military training area in southern Germany. The chosen erosion model should have the following properties:

- single-event based
- interface to the Arc/Info GIS
- algorithms for modeling sediment transport, deposition and rainfall infiltration

9.2 Overview of Existing Erosion Models

The better a model simulates reality, the more parameters will be needed. Most of the following models include several submodules for climate, hydrology, transport of chemical substances, etc.

Generally, all existing erosion models can be related to one of the following three groups.

1. Erosion models which compute only the long-term soil loss, mostly using a modification of the USLE without any consideration of transport and deposition processes.
2. Erosion models which are based on empirical equations and which – in most cases – also use a modification of the USLE for calculating soil loss. Additionally, transport and deposition processes are simulated, based on topographic data inputs.

3. Erosion models which are process-oriented and which simulate the effects of raindrop splash, sheet runoff and further processes. Transport and deposition are calculated using topographic data inputs (e.g. EROSION-3D, WEPP).

9.2.1
RUSLE/MUSLE

The Revised Universal Soil Loss Equation (RUSLE, Renard et al. 1991) and the Modified Universal Soil Loss Equation (MUSLE, Williams and Berndt 1977) are modified versions of the USLE. They were developed to predict the long-term average annual soil loss carried off by runoff from field slopes with specific cropping and management systems. Both are empirical models which describe the relationship between the rate of erosion, slope, steepness, soil types, erodibility factors, rainfall, vegetative cover, and land use and management practices. RUSLE and MUSLE incorporate the extended knowledge about erosion dynamics by adding the following changes to the USLE (Renard et al. 1991):

- newly-developed rainfall-runoff erosivity term
- correction factors for high-climate erosivity
- seasonally variable soil erodibility term
- new method for calculating the cover and management term
- special terms for slope length and steepness
- newly-developed values for conservation practices

RUSLE and MUSLE are important erosion models because their algorithms are used in many erosion and hydrologic models which have been developed for small scale simulations (i.e. fields to small catchments) and for single rainfall events. The RUSLE and MUSLE estimates of rill and sheet erosion are given in tons per acre and year.

9.2.2
EPIC

The Erosion Productivity Impact Calculator (EPIC, Williams et al. 1983) simulates soil erosion and plant production. EPIC contains different submodels for hydrology, climate, erosion, nutrients, plant growth, soil temperature and economy. The submodel for hydrology is similar to that of CREAMS (Chemicals, Runoff and Erosion from Agricultural Management Systems, Knisel 1980). EPIC needs a great number of input data and computes horizontal and vertical flows of water, runoff, snowmelt and evapotranspiration. EPIC contains six alternative equations for simulating soil erosion:

- the USLE (Wischmeier and Smith 1978)
- the Onstad-Foster modification of the USLE (Onstad and Foster 1975)
- the MUSLE
- two further versions of the MUSLE
- a MUSLE version that uses input coefficients

The user specifies one of these equations which interacts with the other EPIC submodels. The six equations are identical except for their rainfall energy components:

- The USLE depends strictly upon rainfall intensity as an indicator of erosive energy and provides annual estimates only.
- The MUSLE and its modified versions use runoff variables only to simulate erosion and sediment yield. The runoff variables increase the prediction accuracy and eliminate the need for specifying the sediment delivery ratio (as used in the USLE to estimate sediment yield). By these variables, the equation is able to provide estimates of single-storm sediment yields.
- The Onstad-Foster equation contains a combination of the energy factors which are used in the USLE and MUSLE.

9.2.3
AGNPS

The AGricultural NonPoint Source (AGNPS, Young et al. 1987) model was developed to control water quality and how it is affected by soil erosion from agricultural and urban areas. AGNPS simulates runoff, sediment and nutrient transport from catchments ranging from a few to approximately 20 000 ha. However, the resolution is not very high in the latter case since the total number of grid cells is limited to 1 900 in AGNPS. The model contains modules for hydrology, erosion, sediment and chemical transport. They route sediments and contaminants through the catchment in a stepwise manner.

The hydrology module predicts the runoff volume and peak flow rate. The nutrient function estimates the nitrogen, phosphorus and chemical oxygen demand concentration in the runoff and sediment. The soil erosion function is based on the RUSLE model which considers soil erosion and deposition. Sediment is routed through the catchment from cell to cell using the relationship between sediment transport capacity and deposition based on a steady-state continuity equation developed by Foster (1976). A module that converts the output data into the Arc/Info GIS format is also available.

The AGNPS model requires a total of 22 input parameters for specifying the topography (digital elevation model for aspect and slope), soil characteristics and land use in the chosen catchment.

9.2.4
CREAMS

The Chemicals, Runoff and Erosion from Agricultural Management Systems (CREAMS, Knisel 1980) model was developed to analyse the water quality for field size areas. CREAMS also provides runoff and erosion estimates. Similar to AGNPS, CREAMS is a so-called field scale model for predicting the runoff, erosion and chemical transport from agricultural land. CREAMS contains physically-based equations and does not require calibration for each specific application. Estimates of percolation, runoff, erosion, dissolved and absorbed plant nutrients, and pesticides can be made for different management practices. CREAMS contains three submodels:

- The erosion submodel is based on elements of the MUSLE, including the modeling of the sediment transport capacity of overland flow.

- The hydrology submodel allows to specify two options. If only daily rainfall data are available, the SCS curve number model is used to estimate surface runoff. If hourly or breakpoint rainfall data are available, an infiltration-based model is used to simulate the generation of runoff. Water movement through the soil profile is modelled by using a simple capacity approach, with flow occurring if the water content of a soil horizon exceeds field capacity.
- The chemical submodel of CREAMS has a nitrogen component that considers mineralisation, nitrification and denitrification processes. Nitrate losses from the root zone due to leaching and plant uptake are calculated by special algorithms. Both the nitrogen and phosphorus parts of the nutrient component use enrichment ratios to estimate that portion of the two nutrients that is transported with the eroded sediment. The pesticide component considers foliar interception, degradation, and wash-off, as well as adsorption, desorption, and degradation in the soil.

9.2.5
GLEAMS

The Groundwater Loading Effects of Agricultural Management Systems modeling system (GLEAMS, Knisel et al. 1983) was developed for single fields. The main area of application is to assess the effects of different agricultural management systems on the movement of agricultural chemicals within and through the plant root zone. GLEAMS simulates the daily water balance, the sediment yield caused by erosion, the pesticide losses in water and sediment, and the loss of plant nutrients to water and sediment. Basically, the model is an updated and slightly modified version of the CREAMS model which contains an additional component for simulating the vertical flux of pesticides. Input requirements are daily rainfall, temperature, solar radiation, wind velocity, soil characteristics, pesticide information, and fertiliser and tillage data. The output can be presented as daily, monthly or annual values, with a maximum length of the climatic record of up to 50 years. The output includes estimates for runoff, sediment, pesticide mass and concentration, percolation volume, and plant nutrient mass and concentrations. The model is written in FORTRAN and runs on a PC.

9.2.6
ANSWERS

The Areal Nonpoint Source Watershed Environmental Response Simulation model (ANSWERS, Beasley and Huggins 1982) was developed to simulate the surface runoff and erosion on agricultural catchments. ANSWERS uses a grid-cell representation of the catchment and simulates the effects of a single storm event. Three basic erosion processes are considered:

- detachment of soil particles by raindrop impact (splash)
- detachment of soil particles by overland flow
- transport of soil particles by overland flow

The maximum number of grid cells for which the soil loss is simulated by ANSWERS is limited to 1 700. The resolution for large catchments is therefore very low. The following parameters are needed for calculating the erosion rate and deposition in each cell:

- precipitation data
- topography data (elevation, slope, aspect)
- soil data (porosity, moisture content, field capacity, infiltration capacity, USLE K-factor)
- land cover data (percent cover, interception, USLE C- and P-factor, surface roughness and retention)

9.2.7
EROSION-2D and EROSION-3D

EROSION-2D (Schmidt 1996) is a physically-based computer model for simulating the sediment transport on slopes. Three processes can be simulated separately:

- detachment of soil particles
- transport of the particles
- surface runoff

Soil erodibility is described by a parameter which is determined from the minimum discharge depth that initiates erosion. For quantitative estimates the model was calibrated with experimental data. The input parameters of the EROSION-2D model are the elevation co-ordinates of the initial slope profile, the surface and soil properties, and the vegetation cover. The parameter values may vary in space (e.g. changes along the slope profile) and time (e.g. seasonal variations). Sediment transport is always calculated for single precipitation or erosion events. These are characterised by their time of occurrence (day and month), their duration, and the temporal variation of the precipitation intensity. At the end of each simulation, the initial slope profile and the rates of erosion and deposition are presented in on-screen graphics and tables.

EROSION-3D is the three-dimensional version of EROSION-2D. The model simulates the effects of single rainfall events. EROSION-3D uses a temporal resolution of 1–15 min and can be applied to catchments with a maximum size of 30 000 ha. The topographical parameters are generated automatically from a digital elevation model. The model has a user interfaces for MS-Windows and geographical information systems (i.e. Arc/Info Grid and Grass) applications. Since it has been particularly developed for applications of limited data availability, the model requires the following input data only:

- digital terrain model
- soil moisture
- content of organic content
- hydraulic roughness (Manning's *n*)
- vegetation cover
- correction factor for hydraulic conductivity
- soil texture
- bulk density
- rainfall intensity
- rainfall duration

9.2.8
KINEROS

The KINematic Runoff and EROSion model (KINEROS, Woolhiser et al. 1990) is a process-oriented simulation model which predicts infiltration, surface runoff and erosion from small agricultural and urban catchments. KINEROS is a deterministic-conceptual model that simulates overland flow and surface erosion. The event-based model can be applied to catchments of approximately 10–20 km^2 in size. The terrain is represented by a cascade of planes and channels describing topography, catchment morphometry, soil properties, and land cover. The model requires the catchment topography, channel geometry, vegetation cover, soil texture and hydraulic characteristics as input data. The output includes summary information on sedimentology and hydrography. The model considers the spatial variability of rainfall, of saturated hydraulic conductivity, of runoff, and of the erosion parameters.

KINEROS has four main components:

1. a saturated hydraulic conductivity algorithm that allows the event-based calculation of excess rainfall
2. a kinematic routing algorithm that transforms rainfall events generated on the overland flow planes into a lateral inflow hydrography in the stream channel
3. a kinematic routing algorithm for routing the streamflow hydrography through the channel system and
4. an erosion algorithm that routes the sediment through the system and which is based upon detachment and transport rates for each soil type

9.2.9
OPUS

OPUS (no acronym, Smith 1992) is a model to simulate the transport of nitrogen, phosphorus and carbon in the soil and the surface water on and from a small catchment. It is mainly intended as a simulation tool for studying the potential pollution from different agricultural management practices. Opus includes models for simulating

- plant growth
- soil cover changes
- human water use
- uptake of nutrients
- cycling of soil nitrogen, phosphorus and carbon
- transport of adsorbed pesticides and nutrients
- interaction of surface water and soil water
- runoff and erosion

The main time interval for all simulations is one day, but an annual summary of the simulation results is also provided if specified by the user. OPUS allows the user to choose between two types of simulation: The simulation may be based on time-intensity data of rainfall or, for more general questions, it may be based either on recorded daily rainfall, or on stochastically generated rainfall data. OPUS is a potential model for global change research since it may help to understand how the effects of changes in precipitation and temperatures may affect non-point source pollutants.

9.2.10
SPUR

SPUR (Simulating Production and Utilization of Range Land, Carlson and Thurow 1991) is a simulation and process model to assess the effects of different management and climate change scenarios on rangeland sustainability. The soil erosion is computed with the MUSLE without calculating transportation and deposition separately. Two versions of the model (SPUR I and SPUR II) are currently existing.

SPUR I contains four components for hydrology, plant growth, animal physiology and harvesting (for both domestic and wildlife), and economics. The driving variables of this model are daily precipitation, maximum and minimum temperature, solar radiation and wind velocity.

SPUR II is a grassland ecosystem simulation model which has been developed to determine beef cattle performance and production by simultaneously simulating the production of up to 15 plant species on 36 heterogeneous grassland sites. SPUR II simulates grassland hydrology, nitrogen cycling, and soil organic matter on grazed ecosystems. This model also simulates the rangeland production under different climatic regimes, environmental conditions and management alternatives. The model requires fairly detailed parameterisation when new plant species are to be included in the simulation. The erosion and runoff routines used in SPUR II may not be valid under all rangeland conditions.

9.2.11
WEPP

The Water Erosion Prediction Project (WEPP, Lane and Nearing 1989) is a physically-based model which considers the spatial and temporal variability of the main erosion processes. It was developed to replace the USLE and RUSLE as a soil erosion prediction system. The model has different submodels for climate, evaporation, plant growth, sediment transport and accumulation. Modeling inputs include generated or observed climate data, topography and plant and tillage characteristics. The model output is provided as the daily sum of runoff, biomass, soil detachment, sediment and soil water, etc. The user may specify other levels of temporal aggregation for the model output, depending on the particular problem to be investigated. The erosion submodel calculates both interrill (caused by splash) and rill erosion.

The current version of the WEPP model runs on a personal computer. The model can be used either for modeling hillslope erosion processes, or for simulating hydrologic and erosion processes on small catchments. The main areas of application of the WEPP model are the investigation of land-use changes, and the conservation of agricultural resources. The model needs a large number of input parameters which are, in most cases, not available for European conditions. Long-term scenarios of climate change may be used as input if the user wishes to assess its effects on future soil erosion.

9.2.12
EUROSEM

The European Soil Erosion Model (EUROSEM, Morgan et al. 1993) has been developed as a counterpart to the WEPP model. Fifteen European research groups co-operated to validate EUROSEM for European conditions. EUROSEM provides a procedure for assessing the risk to soil erosion and designing soil protection measures. The single-event and process-based model predicts soil erosion by water for fields and small catchments. Unlike similar models, EUROSEM simulates the deposition of detached material even if the maximum transport capacity of overland flow has not been reached. The model also explicitly computes both rill and interrill erosion, and the transport of sediment from interrill areas to rills. These calculations are based on the dynamic mass balance equation of erosion (Bennett 1974; Kirkby 1980; Woolhiser et al. 1990). EUROSEM can be used applied to the following three situations:

1. The erosion is predicted for a single plane or element which represents a small field with reasonably uniform slope, soil and land cover conditions.
2. The erosion is calculated for a consecutive series of multiple planes or cascading elements which represent a heterogeneous slope, with each plane having uniform slope, soil and land cover characteristics.
3. The erosion is predicted for a small catchment which is conceptualized as a cascading series of multiple planes and branching channels, with the plane elements contributing their runoff to the channels.

EUROSEM requires two input files, one containing the rainfall parameters of the given storm event, and the other containing the catchment characteristics such as topography, soil and cover.

9.3
Assessment of Model Applicability

Most of the described models have not been developed to exclusively predict erosion since, in fact, they simulate very complex processes of nutrient cycling or water pollution (e.g. AGNPS, EPIC, GLEAMS, OPUS). The following models are simulating soil erosion only:

- EROSION-3D
- KINEROS
- WEPP
- EUROSEM
- USLE, RUSLE and MUSLE

For the practical application to be conducted in this study, three main aspects were identified as the most important ones when selecting the erosion model:

1. Most of the models have been developed for agricultural catchments.
2. Most of the models have been developed and validated for North American conditions.
3. In many cases the general availability of soil data is low.

Table 9.1. Comparison of the main model characteristics

Erosion model	Event based	Empirical equation for simulating erosion	Interface to Arc/Info	Simulation of transport, deposition, etc.	Required input data	Maximum number of grid cells	User friendliness
USLE	No	Yes	No	No	+	n.i.	Yes
RUSLE/MUSLE	Yes	Yes	No	No	+	n.i.	Yes
EPIC	No	Yes	No	No	+++	n.i.	No
AGNPS	Yes	Yes	Yes	Yes	++	1 900	Yes
CREAMS	No	Yes	No	Yes	++	n.i.	No
GLEAMS	No	Yes	No	Yes	++	n.i.	No
ANSWERS	Yes	No	No	Yes	++	1 700	Yes
EROSION-2D/3D	Yes	No	Yes	Yes	++	50 000	Yes
KINEROS	Yes	N.I.	No	Yes	+++	n.i.	No
OPUS	Yes	No	No	Yes	+++	n.i.	No
SPUR I/II	Yes	Yes	No	No	+++	n.i.	No
WEPP	Yes	No	Yes	Yes	+++	n.i.	No
EUROSEM	Yes	No	Yes	Yes	+++	n.i.	No

n.i.   No information;
+   Few;
++   Moderate;
+++   Many.

A further aspect to be considered during the selection procedure is the fact that, despite of the more accurate representation of the relevant processes of erosion, the more research-oriented models such a EUROSEM or WEPP usually require very cost-demanding methods for the determination of input parameters. Table 9.1 shows a comparison of all erosion models and requirements.

With respect to the scope of the to-be-tackled erosion problem, the available input data and the model characteristics, the following erosion models were identified as most appropriate:

1. *AGNPS* is the only model that meets all requirements. On the other hand, one has to consider that the catchment representation of AGNPS is limited to 1 900 gridcells. The low number will result in a low resolution for larger catchments in particular. The model also needs the relatively high number of 22 input parameters.
2. *EROSION-3D* has been specially validated for German conditions and requires the low number of 7 soil input parameters.
3. *EUROSEM* has the advantage of having been developed especially for European conditions, but it needs more than 60 input parameters for execution.

The chosen erosion models are compared in Table 9.2 by listing the required input parameters.

Table 9.2. Summary of input parameters required for AGNPS, Erosion-3D and EUROSEM

Parameter	AGNPS	Erosion 3D	EUROSEM
Digital terrain model	Yes	Yes	No
Slope	Yes	No	No
Aspect	Yes	No	No
Rill slope	No	No	Yes
Interrill slope	No	No	Yes
Slope shape	Yes	No	No
Slope length	Yes	No	Yes
Channel slope	Yes	No	Yes
Channel sideslope	Yes	No	No
Left channel sideslope	No	No	Yes
Right channel sideslope	No	No	Yes
Side slope of concentrated flow paths	No	No	Yes
Width of channel bottom	No	No	Yes
Gully source level	Yes	No	No
Channel indicator	Yes	No	No
Manning roughness	Yes	Yes	Yes
Across-slope roughness	No	No	Yes
Erodibility factor	Yes	Yes	No
Cohesion of soil-root matrix	No	No	Yes
Detachability of soil particles	No	No	Yes
Gravity of sediment particles	No	No	Yes
Soil texture	Yes	Yes	No
Median particle diameter	No	No	Yes
Standard deviation of sediment diameter	No	No	Yes
Rock content	No	No	Yes
Soil porosity	No	No	Yes
Infiltration recession factor	No	No	Yes
Bulk density	No	Yes	Yes
Initial soil moisture	No	Yes	Yes
Maximum soil moisture	No	No	Yes
Landcover value	Yes	Yes	No
Percentage canopy cover	No	No	Yes
Percentage basal area of vegetation	No	No	Yes
Average acute angle of plant stems	No	No	Yes
Effective canopy height	No	No	Yes
Plant leaf shape factor	No	No	Yes
Interception storage	No	No	Yes
Landmanagement value	Yes	No	No
Support practice factor	Yes	No	No
Surface condition constant	Yes	No	No
Fertilisation level	Yes	No	No
Fertiliser availability factor	Yes	No	No
Oxygen demand factor	Yes	No	No
Impoundment factor	Yes	No	No
SCS curve number	Yes	No	No
Intensity of precipitation	Yes	Yes	Yes
Duration of precipitation	Yes	Yes	Yes
Air temperature	No	No	Yes

Table 9.2 shows that AGNPS model requires many parameters especially for describing the transport of chemically active substances and the relief (e.g. slope geometry). Most of these parameters are often not available and have to be determined from field investigations. Since it is the most complex model included in this comparison, EUROSEM needs a extremely large number of parameters for execution. Most of them are usually not available from standard databases such as maps or guidevalue lists. EROSION-3D requires only 10 parameters through which the main processes of soil erosion are described. Similar to AGNPS, EROSION-3D is initialised by a digital terrain model from which the main relief parameters such as slope and aspect are derived.

9.4
Applying EROSION-3D to a Subcatchment

In spring 1998 it was decided to apply the EROSION-3D model to a military training area in southern Germany. The required input parameters (see Sect. 9.2.7) were determined from existing databases such as landcover and soil maps. Rainfall and runoff data had been monitored since 1990 for 6 catchments within the military training area. We selected one of these areas with a size of 275.6 ha for the simulation, so that the existing data could be used for verifying the modeling results. A precipitation event on 29–30 July 1996 was chosen for the simulation since its runoff and precipitation data had been monitored and analysed. The precipitation data with a resolution of 10 minutes are shown in Table 9.3.

The EROSION-3D pre-processor was used for computing a "topology" grid from a digital terrain model with a resolution of 10 m. This grid file contains topology data such as aspect and slope, but also hydrological data such as flow directions, the stream network and the catchment outlets. The EROSION-3D pre-processor displays the complete topology dataset for each gridcell simply by clicking on it (Fig. 9.1).

For each simulation run, a grid file has to be created first for each input parameter. The results of the simulation for the 275.6 ha catchment are shown in Fig. 9.2. Areas

Table 9.3. Erosion-3D input data for the rainfall event on 29–30 July 1996

Time from start of rainfall (min)	Time interval (min)	Rainfall depth (mm)
10	0 – 10	0.93
20	10 – 20	0.28
30	20 – 30	0.01
40	30 – 40	0.36
50	40 – 50	0.07
60	50 – 60	0.06
70	60 – 70	0.04
80	70 – 80	0.04
90	80 – 90	0.03
100	90 – 100	0.01
	Total depth	1.83

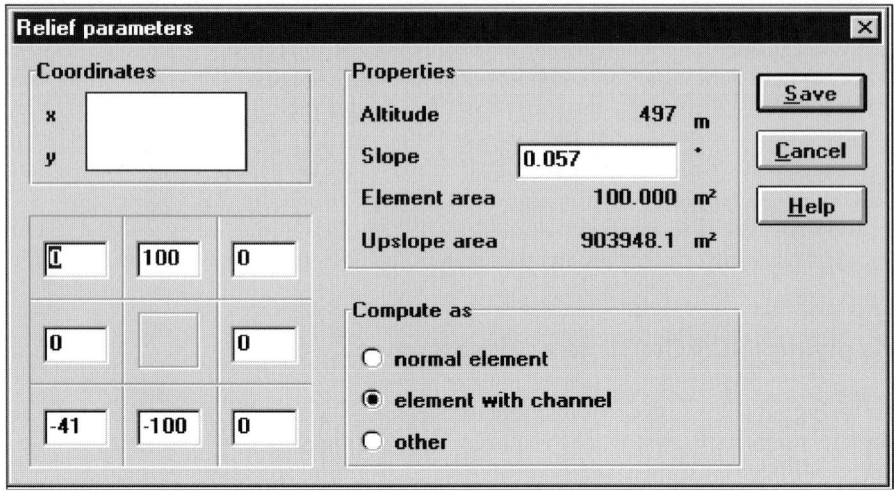

Fig. 9.1. Example data sheet of a selected grid cell

of deposition are plotted in green and blue, and areas of erosion are indicated by the yellow to red colours. All values are expressed in kg per square meter. The stream network is indicated by the black lines.

The EROSION-3D main program displays the simulation data for every gridcell by clicking on it. The detailed dataset for a gridcell located in the stream channel close to runoff monitoring station is shown in Fig. 9.3.

Since the runoff data for the event on 29–30 July 1996 had been recorded by the monitoring station, the simulation result could be easily compared to the measured values. The measured net erosion from the entire catchment (including forested areas) is 69.9 kg ha^{-1}. This value amounts to 149.6 kg ha^{-1} if related to the nonforested (i.e. agricultural) areas only. EROSION-3D predicts a mean net erosion of 55 kg ha^{-1} for the entire catchment (see Fig. 9.3), which equals an amount of 117.7 kg ha^{-1} for the nonforested areas.

9.5
Simulation of Different Soil Moisture Scenarios

The following two scenarios were developed for a further subcatchment within the military training area. Both simulations are based on the same precipitation event on 29–30 July 1996. The only parameter which has been changed was the initial soil moisture. The two scenarios were developed in order to demonstrate the great influence of the soil moisture content on erosion. Table 9.4 shows the initial soil moisture values for both scenarios, depending on soil texture and landcover. The values of the "worst-case scenario" are printed in bold. The lower values are moderate soil moisture contents. The high values of the worst-cases scenario represent the soil moisture contents of saturated soils. The high soil moisture values are usually observed after rain periods.

Table 9.4. Low and high soil moisture values (%) for different types of soils and land use. The higher values printed in bold refer to the "worst-case" scenario

Soil texture	Type of land use													
	Forest		Grassland 0% cover		Grassland 21–40% cover		Grassland 41–60% cover		Grassland 61–80% cover		Grassland 81–100% cover		Secondary roads	
Clayey loam ($Lt_{2/3}$)	29	48	27	47.5	27	47.5	28	47.5	28	47.5	28	47.5	22	39
Sandy loam ($Ls_{2/3}$)	33	41	31	40	31	40	32	40	32	40	32	40	25	33.5
Silty loam (Lu)	32	43.5	30	43	30	43	31	43	31	43	31	43	24	36.5
Loamy sand (Sl_4)	26	34	24	33	24	33	25	33	25	33	25	33	19	28.5

After having created the grid file for each input parameter, the soil erosion was simulated first for the subcatchment and normal soil moisture conditions. In the second step, the worst-case scenario was modelled using the high soil moisture values. Figure 9.4 (see at page 178) shows the areas of deposition (green to blue) and erosion (yellow to red) for the normal (left) and worst-case conditions (right).

The two scenarios demonstrate that the extent of soil erosion greatly depends on the initial soil moisture content. For normal conditions, soil erosion occurs only in a rather small area. On the other hand, the area affected by erosion is much larger if worst-case conditions are assumed. The right map shows that soil erosion occurs in areas which are otherwise not affected by erosion. The total amount of sediment delivery from the entire catchment is therefore also much higher.

The EROSION-3D model can be easily used for simulating a great variety of scenarios not only on military training lands. These scenarios may be developed for assessing the effects of different types of landcover and/or land use. Areas which are affected by soil erosion can easily be identified, so that conservation measures can be initiated and implemented.

9.6
Conclusions

The practical application of the EROSION-3D model shows that it is an easy-to-use tool for erosion prediction and scenario simulation. The modeling results agree reasonable well with measured data of sediment yield.

In future, EROSION-3D will be used in particular for the following purposes:

- simulating erosion and sediment deposition before and after military training periods, so that the impacts of training activities on sediment production can be demonstrated.
- simulating "worst case scenarios", i.e the model results provide an answer to more specific "what happens if …" questions.
- identifying erosio-susceptible areas and assisting in the planning of appropriate counter measures such as the declaration of "off-limits-areas", areas scheduled for re-seeding areas and other conservation measures.

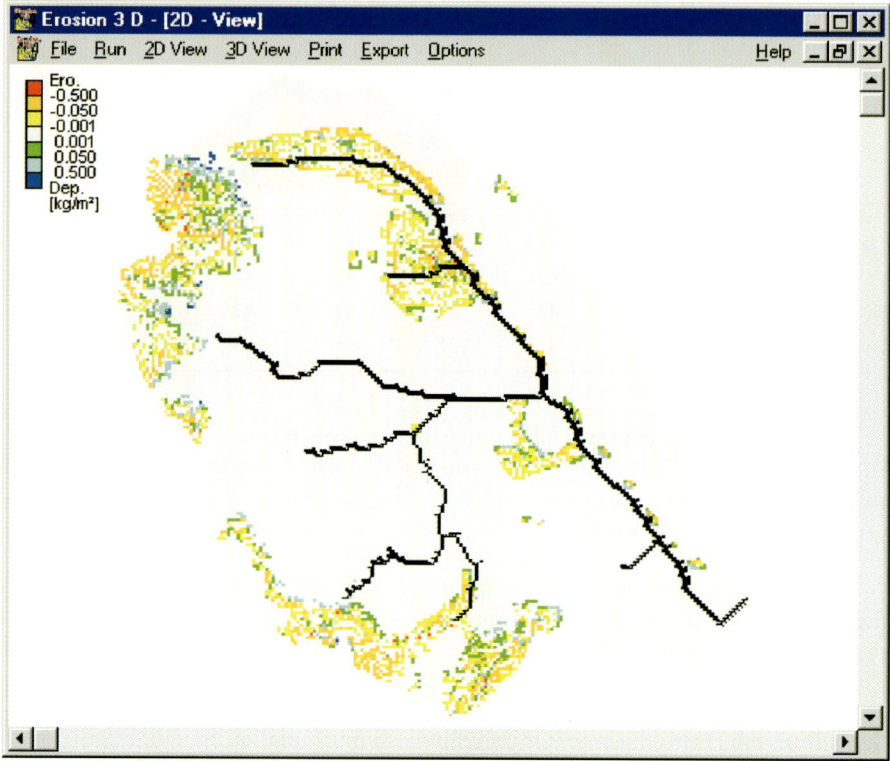

Fig. 9.2. Spatial distribution of deposition (*blue*) and erosion (*red*) in the modelled subcatchment

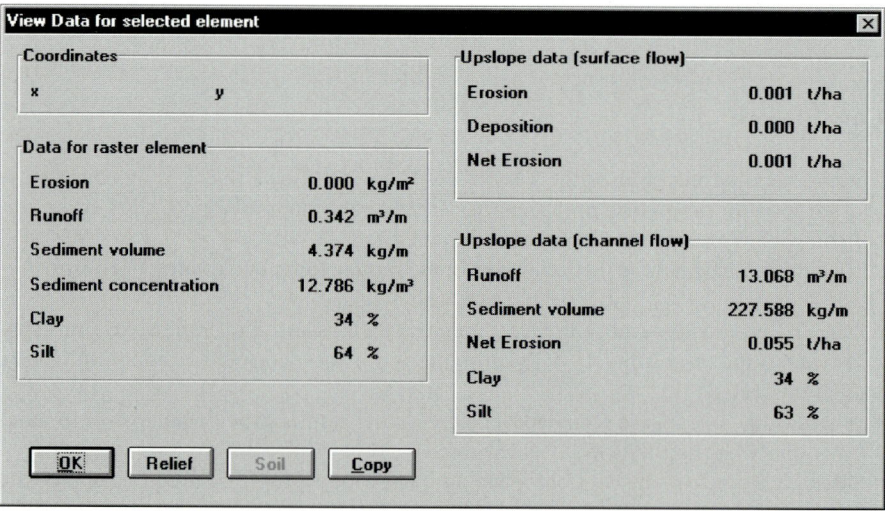

Fig. 9.3. Detailed dataset for a grid cell located in the stream channel close to the runoff monitoring station

CHAPTER 9 · Modeling Overland Flow and Soil Erosion for a Military Training Area 177

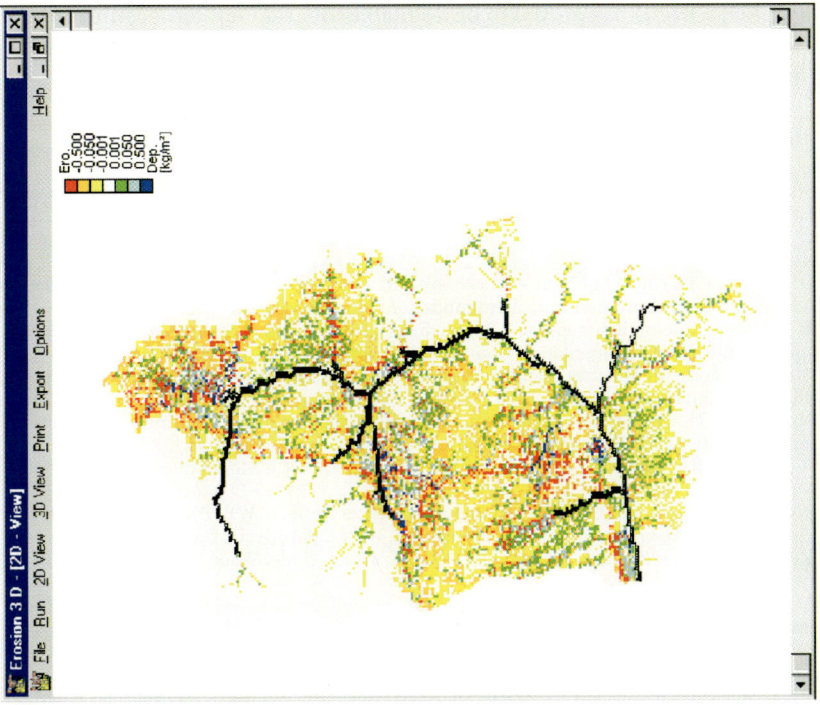

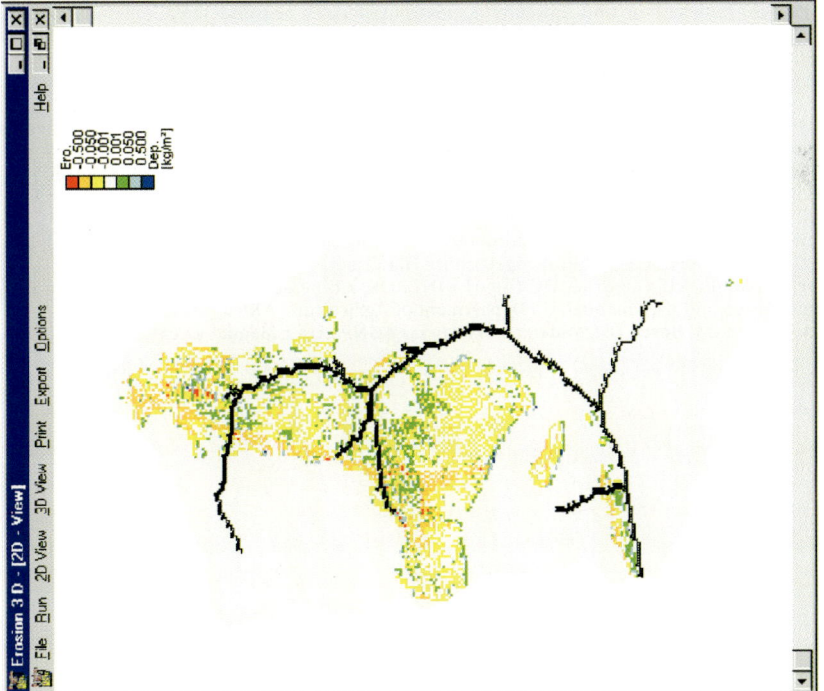

Fig. 9.4. Spatial distribution of deposition (*blue*) and erosion (*red*) in the modelled catchment. The maps show the simulation results for normal (*left*) and "worst-case" soil moisture conditions (*right*)

Finally, EROSION-3D has proved to be a suitable tool for long-term estimations of soil erosion. The primary purpose of such assessments may be to evaluate the effects of land management techniques.

References

Beasley DB, Huggins LF (1982) ANSWERS (Areal Nonpoint Source Watershed Environmental Response Simulation). User's manual. Environmental Protection Agency Report No. 905/9-82-001, Chigaco, Ill
Bennett JP (1974) Concepts of mathematical modeling of sediment yield. Water Resources Research 10:485-492
Carlson DH, Thurow TL (1991) SPUR-91 - Workbook and user guide. 1991 upgrade of the simulation of production and utilization of rangeland models. Auston, Texas A&M University
Foster GR (1976) Sedimentation, general - Proceedings of the National Symposium on Urban Hydrology, Hydaulics and Sedimentation Control, University of Kentucky, Lexington
Kirkby MJ (1980) Modeling water erosion processes. In: Kirkby MJ, Morgan RPC (eds) Soil erosion, 183-216, Wiley, Chichester
Knisel WG (1980) CREAMS: A field-scale model for chemicals, runoff and erosion for agricultural management systems. US Department of Agriculture, Conservation Research Report 26
Knisel WG, Leonard RA, Davis FM (1983) GLEAMS Version 2.1, Part I: Model documentation. UGA-CPES-BAED, Pub 5, Nov 1993
Lane LJ, Nearing MA (1989) USDA-Water Erosion Prediction Project (WEPP): hillslope profile version. Profile model documentation. NSERL Report 2, West Lafayette, USDA-ARS, 269 p
Morgan RPC, Quinton JN, Rickson RJ (1993) EUROSEM: a user guide. Silsoe College, Cranfield University, UK
Nearing MA, Lane LJ, Lopes VL (1994) Chapter 6: Modeling Soil Erosion. Soil Erosion Research Methods, 2nd edn., 127-156
Onstad CA, Foster GR (1975) Erosion modeling on a watershed. Transactions of the American Society of Agricultural Engineers 18(2):288-292
Renard KG, Foster GR, Weesies GA, Porter JP (1991) RUSLE - Revised universal soil loss equation. Journal of Soil and Water Conservation 46:30-33
Schmidt J (1996) Entwicklung und Anwendung eines physikalisch begründeten Simulationsmodells für die Erosion geneigter, landwirtschaftlicher Nutzflächen. Berliner Geographische Abhandlungen 61, Berlin, Institut für Geographische Wissenschaften, 148 p
Smith RE (1992) OPUS: an integrated simulation model for transport of nonpoint-source pollutants at the field scale. Vol I, documentation (USDA-ARS 98). Washington, DC, USDA-ARS, 120 p
Williams JR, Berndt HD (1977) Sediment yield prediction and utilization of rangelands. Documentation and user guide. US Department of Agriculture, ARS 63, Washington, DC
Williams JR, Dyke PT, Jones CA (1983) EPIC: a model for assessing the effects of erosion on soil productivity. In: Laurenroth WK et al. (eds) Analysis of ecological systems. State-of-the-Art in Ecological Modeling 553-572, Amsterdam
Wischmeier WH, Smith DD (1978) Predicting rainfall erosion losses: A guide to conservation planning. US Department of Agriculture, Agriculture Handbook 537, Washington, DC
Woolhiser DA, Smith RE, Goodrich DC (1990) KINEROS, a kinematic runoff and erosion model: documentation and user manual. US Department of Agriculture, ARS 77, Washington, DC
Young RA, Onstad CA, Bosch DD, Anderson WP (1994) AGNPS User Manual v4.03

Part II
Model Validation

Chapter 10

A Process-Based Evaluation of a Process-Based Soil-Erosion Model

A. J. Parsons · J. Wainwright

10.1
Introduction

In recent years, efforts to achieve better prediction of soil erosion by overland flow have emphasized the development of process-based models. It is claimed that process-based erosion models provide several advantages over empirically based erosion-prediction technology, amongst which is the capability to estimate spatial and temporal distributions of soil loss (Nearing et al. 1989). Evaluation of such process-based models has, however, been generally made in terms of their outputs rather than in terms of their correct representation of the erosion processes (see, for example, the evaluations reported in Boardman and Favis-Mortlock 1998). Such evaluation is limited for two reasons. First, a match between the output of the model and observed values for erosion does not demonstrate that the model is successfully representing the operation of processes. As Grayson and Moore (1992) have convincingly demonstrated in the context of overland flow hydrographs, the same output can be generated from a variety of sources and types of overland flow. These authors showed that a match of the runoff hydrograph is insufficient to demonstrate that the model is adequately representing the hillslope processes that generate the runoff. Secondly, where model output and observed runoff and erosion rates do not agree, an evaluation only in terms of model output provides no basis for identifying an explanation for the disparity that would lead to improvements to the model.

One likely reason why evaluations of process-based models have not been made in terms of their representation of the erosion processes is the difficulty of obtaining suitable data for such evaluations. Typically, data from runoff plots or larger areas comprise no more than runoff and sediment yield, often with poor temporal resolution. In this paper we present an evaluation of a process-based model of soil erosion for a site for which information on runoff hydraulics and sediment detachment rates are available, in addition to data on runoff and sediment yield with a high temporal resolution. Thus we are able to undertake a more process-based evaluation of the model. The model evaluated is the European Soil Erosion Model, EUROSEM (Morgan et al. 1993) and the test site is a large runoff plot located on a semi-arid grassland hillslope in southern Arizona.

10.2
EUROSEM

EUROSEM is a single-event, process-based model for predicting soil erosion by water (Morgan et al. 1993). In common with other process-based models for soil erosion, EUROSEM consists of coupled runoff and erosion models. A summary of the model, taken from Morgan et al. is given in Fig. 10.1. EUROSEM can be used to predict erosion from a single plane within which the soil, slope and land cover conditions are reasonably uniform, cascading planes, or multiple planes and channels. In the present study it is used to predict erosion over three cascading planes.

The model requires two input files, one describing the rainfall of the storm event, and the other the characteristics of the area for which the predictions are to be made. The latter is extensive, even for simple interrill surfaces such as those being examined in this study. A summary of these input parameters is given in Table 10.1. For many situations parameter values are likely to be unknown, and Morgan et al. (1993) provide information on which

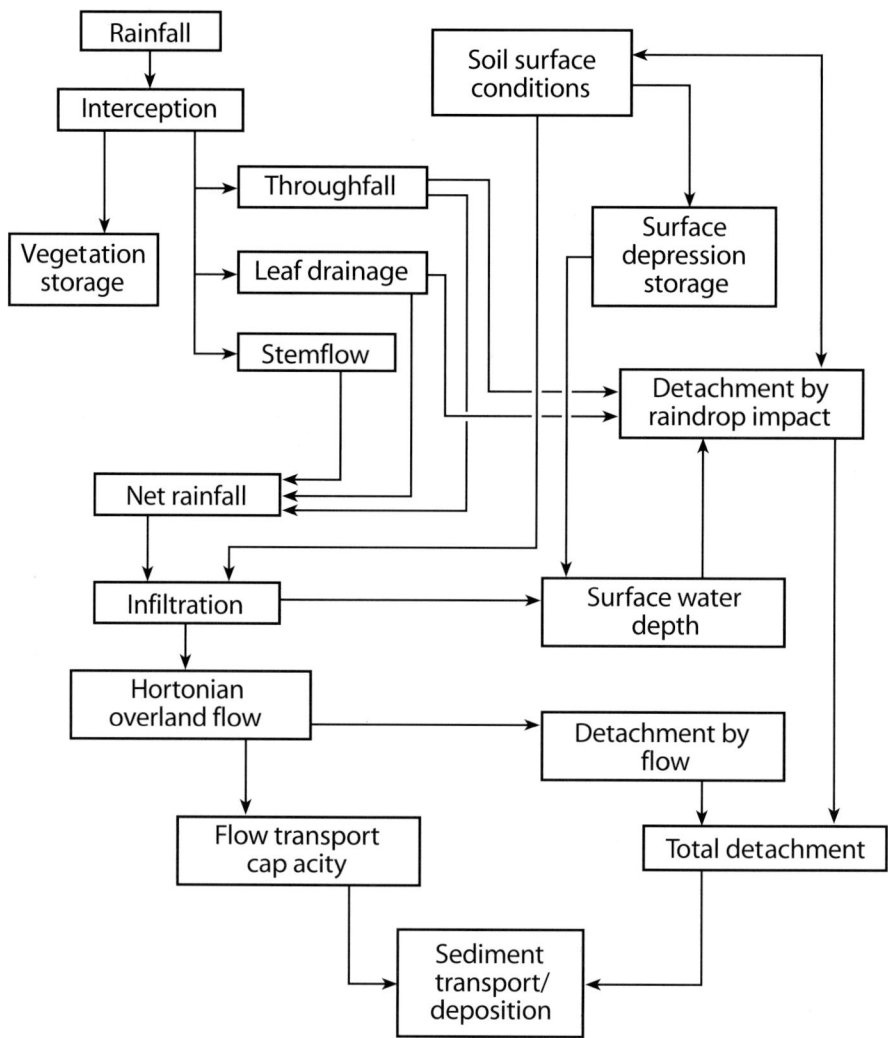

Fig. 10.1. Flow chart of EUROSEM (after Morgan et al. 1993)

CHAPTER 10 · A Process-Based Evaluation of a Process-Based Soil-Erosion Model

Table 10.1. Input parameters for EUROSEM

Parameter	Measured (m) or estimated (e)
Maximum length of cascading planes	m
Cohesion of the soil-root matrix	e
Percentage canopy cover	m
Median particle diameter of the soil	m
Maximum depth to which erosion can occur	m
Maximum interception storage	e
Detachability of soil particles by raindrop impact	m
Saturated hydraulic conductivity	m
Effective net capillary drive	e
Value of Manning's n	m
Proportion of surface covered by impermeable materials	m
Percentage basal area of vegetation	m
Average acute angle of the plant stems to the soil surface	e
Effective canopy height	m
Soil porosity	e
Across-slope roughness	m
Infiltration recession factor	m
Downslope roughness	m
Specific gravity of sediment particles	m
Proportion of rock in the soil by volume	m
Slope	m
Plant leaf shape factor	m
Standard deviation of sediment diameter	m
Air temperature at time of rainfall	m
Initial volumetric moisture content of the soil	m
Maximum volumetric moisture content of the soil	e
Width of plane	m
Length of plane	m

estimates can be based. Table 10.1 distinguishes between those parameters for which field measurements were obtained in the present study and those which were estimated, based on a combination of measured field data and guidance from Morgan et al.

10.3
The Test Site

The test site consists of a runoff plot 18 m wide by 29 m long located on a grassland hillslope within the Walnut Gulch Experimental Watershed, southern Arizona. The slope of the plot, and much of the hillslope on which it was located, was about 7.5°. It was underlain by deep, calcareous, gravelly soil of the Hathaway-Bernardino association (Gelderman 1970) developed on well cemented, coarse, Quaternary alluvium. The vegetation of the plot was dominated by grass species (*Bouteloua* spp., *Andropagon bardinodis* and *Hilaria belangeri*). Scattered among the grasses were a number of other species, notably *Calliandra eriophylla*, *Gutierrezia lucida* with occasional, small *Prosopis juliflora* and *Opuntia* spp. In total, the vegetation covered 33% of the ground. Between the plants and groups of plants the soil surface was exposed. This surface was characterized by scattered stones (>2 mm) covering 32% of the surface and set in a matrix of 70% sand, 13% silt and 17% clay. The ground surface had a microtopography characterized by small treads and risers which resulted from trapping of finer sediment behind clumps of grass (Parsons et al. 1996). This microtopography had an amplitude 2 to 5 cm and a wavelength 20 to 50 cm.

The plot was equipped with a sprinkler system to provide artificial rainfall at a constant intensity of 80 mm h^{-1}. The rainfall has a median drop size of 1.70 mm and provides 0.438 J m^{-2} s^{-1} of kinetic energy. Further details of the design of this system is described by Parsons et al. (1990). Three rainfall simulation experiments (lasting

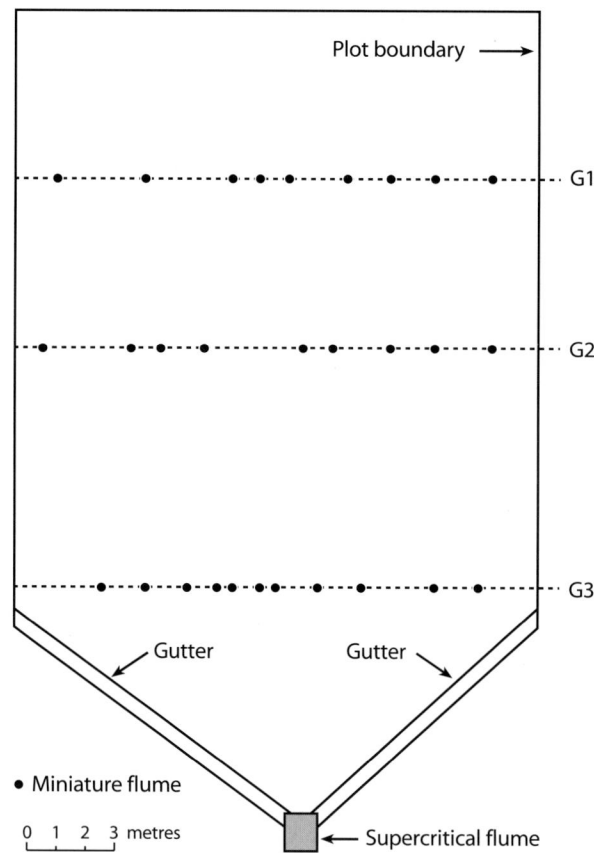

Fig. 10.2. The runoff plot showing cross sections G1, G2 and G3 that provide the lower boundaries of the three cascading planes of the EUROSEM model

56, 50 and 77 min) were performed on the plot, of which only the first is discussed here. Runoff from the plot outlet was recorded using a supercritical flume. Discharge through the flume was sampled at frequent intervals to obtain data on sediment yield from the plot. Within the plot, 3 cross-sections G1, G2 and G3, located 6 m, 12 m and 20.5 m from the upslope boundary, respectively, were established for the determination of within-plot runoff, flow hydraulics and sediment yield (Fig. 10.2). At each of these cross-sections several miniature flumes (Parsons and Abrahams 1989) were installed at which discharge and sediment load were sampled every few minutes during the experiments. At five-minute intervals during the experiments, the depth of overland flow was sampled at 50-cm spacings across each of the cross-sections. For experiment 1, the first of these measurements was made at minute 4 and the last was made at minute 39. The discharge measurements made at the plot outlet, which continued throughout the experiment, indicate that equilibrium runoff was attained at about this time. Using these depth data, together with depth-discharge rating curves for the miniature flumes, discharge estimates were obtained for the cross-sections at five-minute intervals throughout the experiments. Details of the computational method used to obtain these discharge estimates are given in Parsons et al. (1990). Using these discharge estimates together with the sediment load and discharge data obtained from the miniature flumes, total sediment load passing through the cross-sections was estimated for each of the discharge estimates. Details of the methods used to obtain these sediment-load estimates are given in Abrahams et al. (1991). The cross-sections G1, G2 and G3 define the lower boundaries of the three cascading planes in the EUROSEM model representation of the upper 20.5 m of the plot.

In addition, data were collected on rates of splash transport which, it is generally agreed (e.g. Ellison 1947; Schultz et al. 1985), are closely linked to total raindrop detachment. Indeed, the EUROSEM User's Guide (Morgan et al. 1993) uses measurement of splash rate to determine the rate of soil detachment by rainfall. These data were obtained using four splash kites (Parsons et al. 1994) deployed at five locations on the plot for five-minute periods in running sequences. Further details of the experimental method and the data obtained from these splash kites are given in Parsons et al. (1994).

10.4
Results

10.4.1
Runoff

Input parameters that affect runoff, but which were not measured, are maximum interception storage, maximum volumetric moisture content of the soil, average acute angle of the plant stems to the soil surface, effective net capillary drive and soil porosity (Table 10.1). For the first two of these parameters a single guide value is given in Morgan et al. (1993); for the others a range of values is given. For these latter parameters the mean values appropriate to the vegetation or soil type were used initially. The result of running the model with these parameter values is that no runoff is produced until 52 minutes after the commencement of rainfall, compared to the first recording of observed runoff 4 minutes after the commencement of rainfall. Reducing the values of the parameters to the minimum values given in Morgan et al. results in the runoff

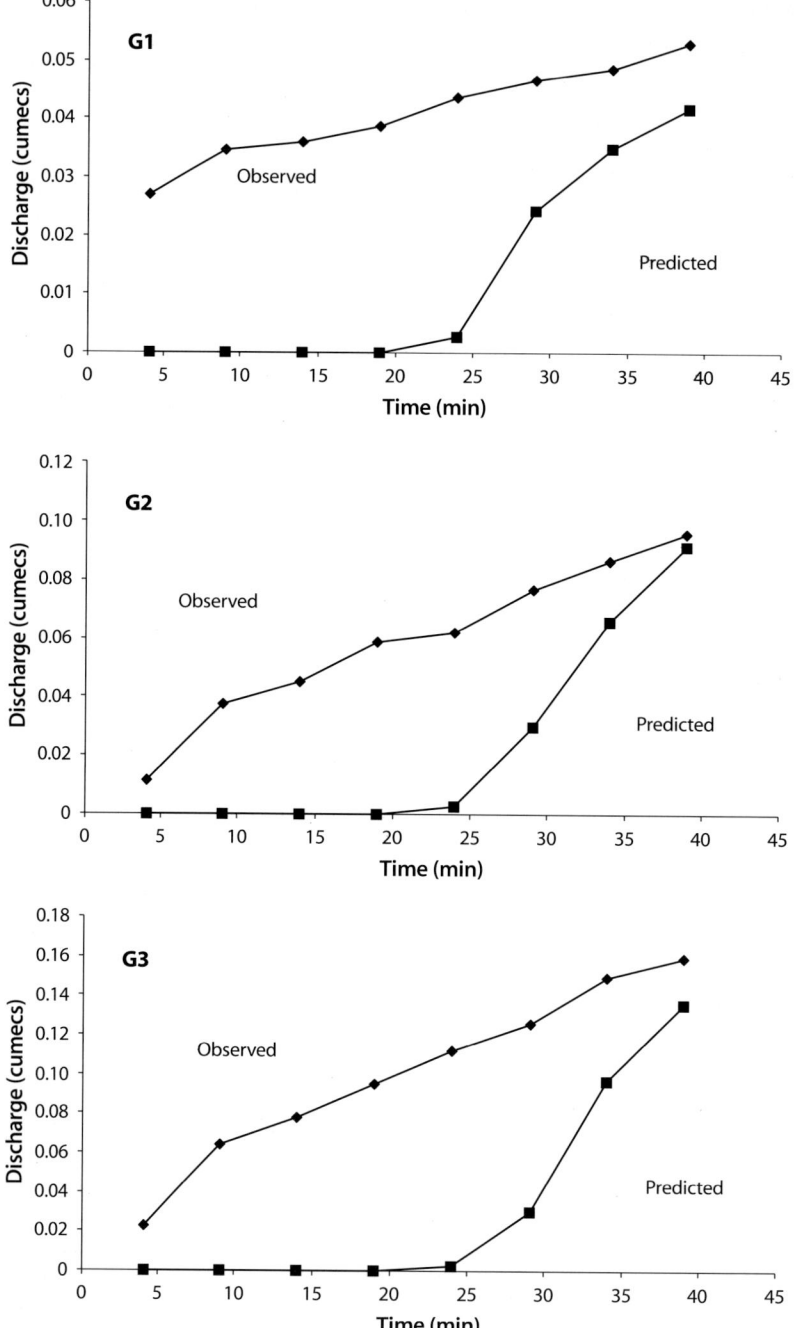

Fig. 10.3. Observed and predicted hydrographs for G1, G2 and G3, using mimimum recommended values for unmeasured parameters

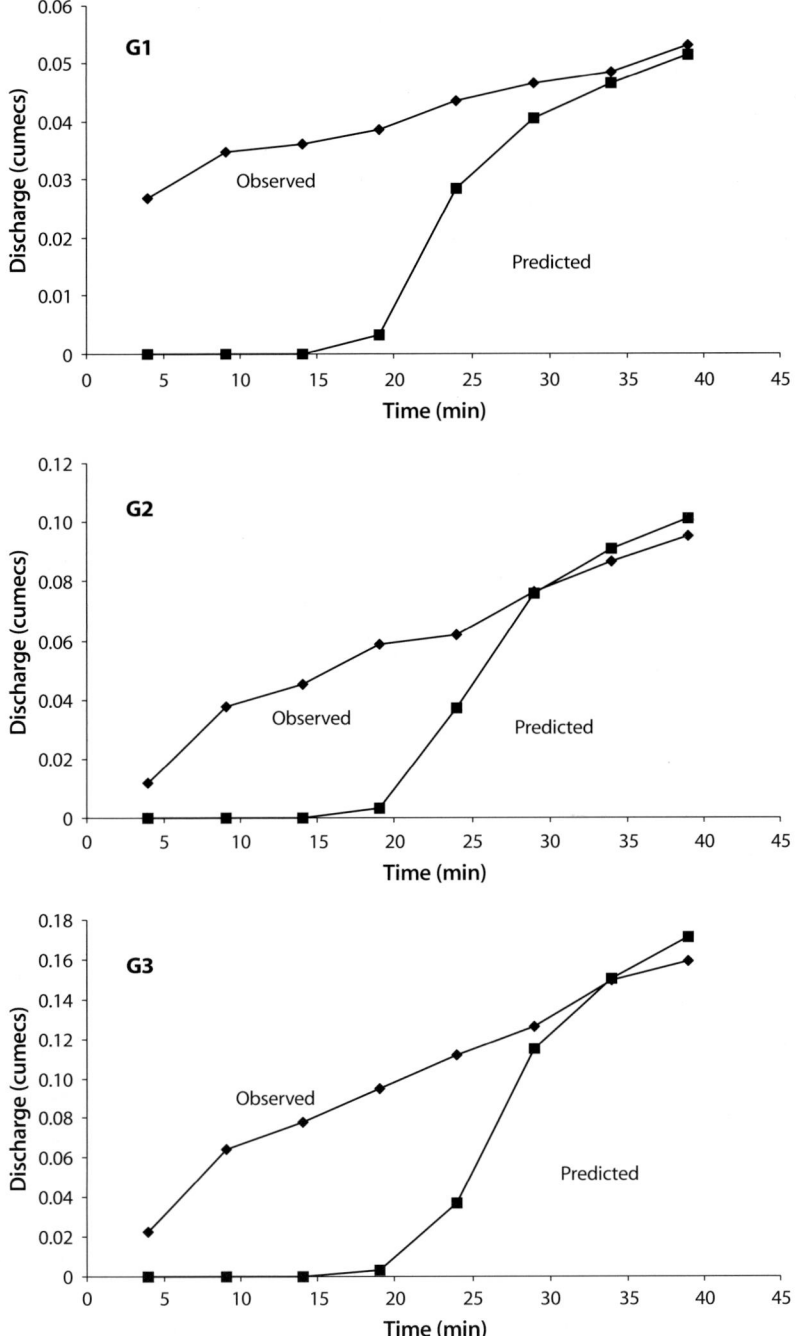

Fig. 10.4. Observed and predicted hydrographs for G1, G2 and G3, using optimum values for unmeasured parameters (Table 10.2)

Table 10.2. Optimum values of unmeasured parameters for best-fit hydrographs

Maximum interception storage (DINTR)	Effective net capillary drive (G)	Soil porosity (POR)	Maximum volumetric soil moisture content (THMAX)	Average acute angle of plant stems to the soil surface (PLANGLE)
1.5 mm	105 mm	30%	30%	60°

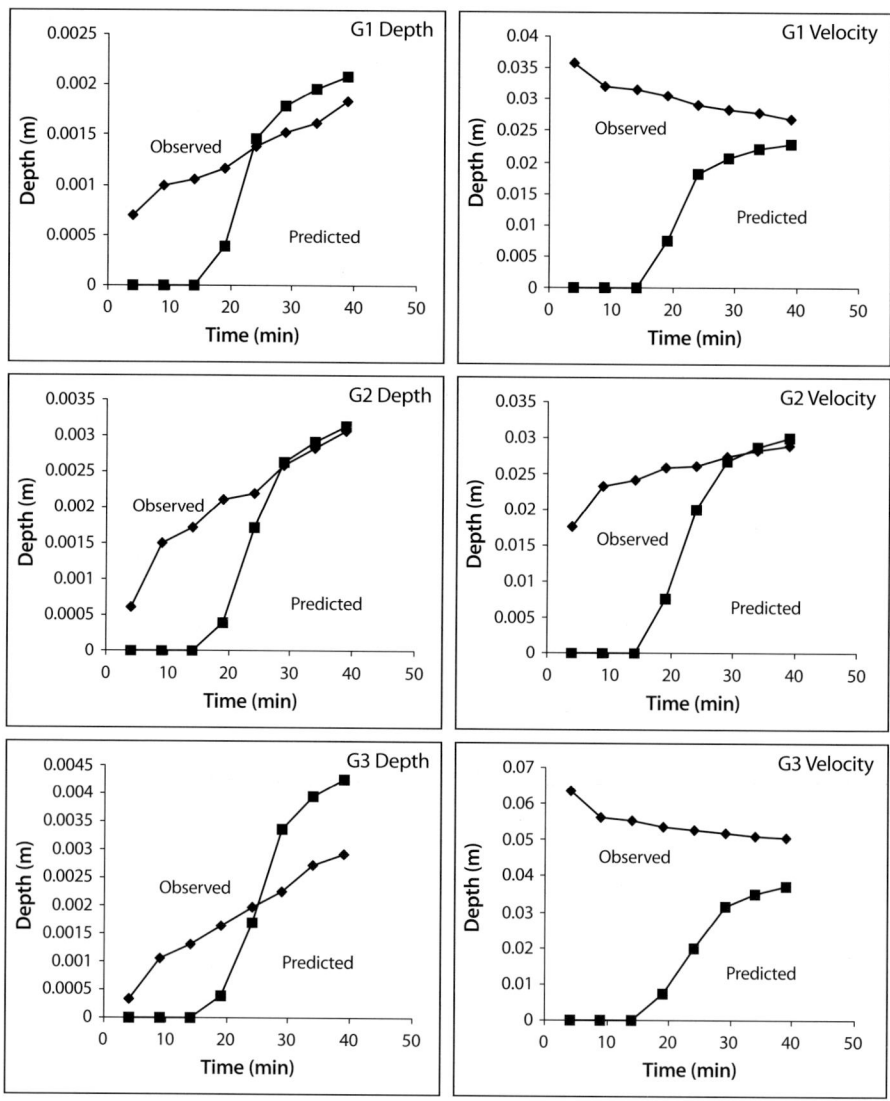

Fig. 10.5. Runoff hydraulics at G1, G2 and G3 for the hydrographs shown in Fig. 10.3

hydrographs shown in Fig. 10.3. Runoff at minute 39 of the experiment is still too low, and the commencement of runoff is still much delayed compared to that observed.

Without a reasonable prediction of runoff, it is impossible for a process-based model to produce predictions of erosion that are accurate in terms of both the amounts of erosion and the processes causing that erosion. To obtain suitable predictions of runoff, the model was run repeatedly, using a range of values for the unmeasured parameters (not necessarily within the ranges given in Morgan et al.) in order to obtain the best possible fits to the observed hydrographs. Optimum values for these parameters are given in Table 10.2 and the resultant hydrographs for G1, G2 and G3 are presented in Fig. 10.4. Even with these optimised values, the predicted start of runoff is much later than that observed and the hydrographs are much steeper, but a very close match to the observed runoff for the last 10 minutes of the measurement period at G1 to G3 is achieved at all three cross-sections.

A comparison of predicted runoff hydraulics for the three cross-sections with those observed for the period when observed and predicted discharge are in close agreement is shown in Fig. 10.5. In general, the comparison is quite favourable, although there is a tendency for the model to overestimate depth (and, correspondingly, to underestimate velocity) in all cases. The difference is greatest at G3 where the predicted depth is almost 50% higher than that observed. However, one additional consideration that needs to be mentioned is that these comparisons are with mean depth of flow across the entire 18 m of the cross-sections including, therefore, parts of the cross-sections where there was no flow. In fact, on average, flow was present over about 15 m of the three cross-sections so that the actual mean depths of flow over the inundated parts of the cross-sections were somewhat higher than shown in Fig. 10.5 (see Parsons et al. 1997).

10.4.2
Erosion

For erosion, the only additional input parameter for which direct measurement was not available is soil cohesion. Because this parameter affects only the equation used by the model to calculate detachment by flow, and because flow detachment should be relatively unimportant in interrill flow, its value should be equally unimportant. A value of 4 kPa was selected, based upon table 6 in Morgan et al. (1993).

Comparison of observed and predicted erosion rates for the three cross-sections can be usefully made only for the last 10 minutes of the measurement period, when observed and predicted discharge were in close agreement. These comparisons are displayed in Fig. 10.6. As this figure shows, the predicted erosion rates are typically an order of magnitude too high.

10.5
Discussion

10.5.1
Runoff

The processes generating runoff are rainfall input, infiltration and flow routing. Of these, the one that is given least explicit treatment in the model is flow routing.

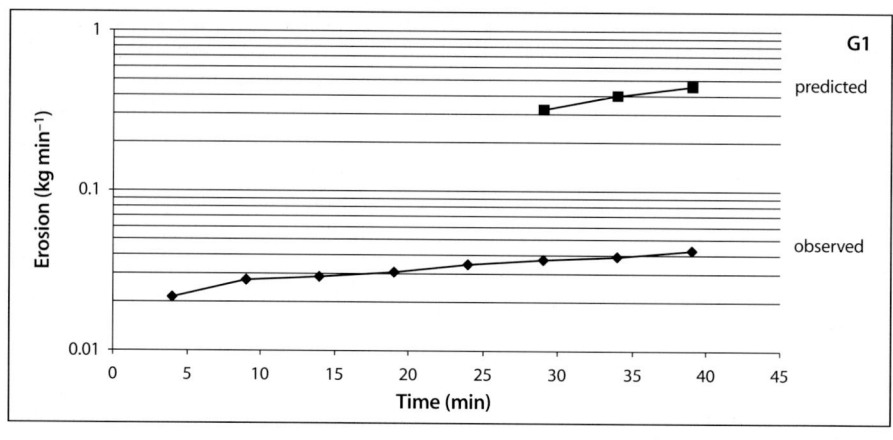
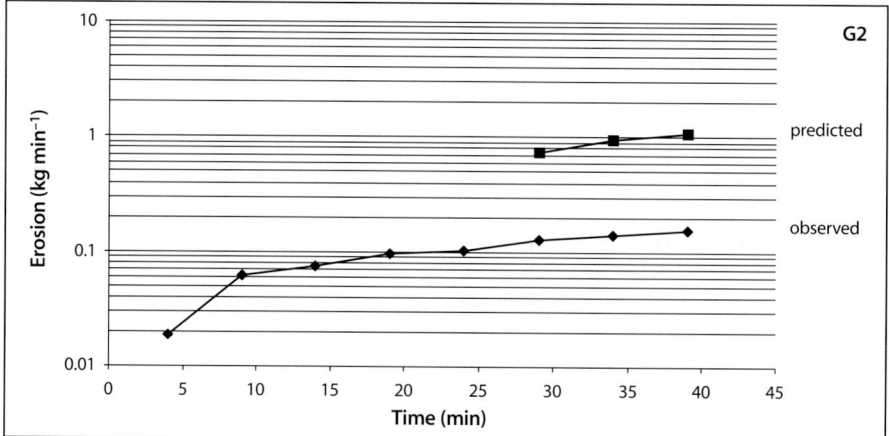
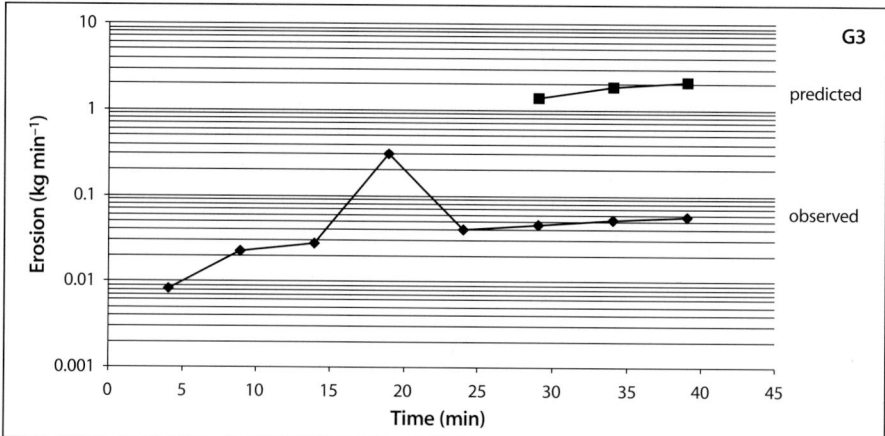

Fig. 10.6. Observed and predicted erosion rates for minutes 29 to 39 obtained from model hydrographs and hydraulics of Fig. 10.3 and 10.4, respectively

This routing depends upon both spatial variations in infiltration and surface microtopography.

On stony soils, such as those present on the plot used in this study, runoff is generated from the stone-covered portion of the ground surface almost as soon as rainfall begins. Initially, this runoff will be absorbed by the adjacent soil, but at the same time it will hasten the time to ponding of that soil. To the extent that this runoff follows selective paths over the ground surface and thereby promotes heterogeneity in the infiltration characteristics of the soil surface, it will advance the appearance of runoff at any selected measuring point. Because EUROSEM does not consider these small-scale heterogeneities in the infiltration characteristics of soil, it fails to capture this effect of stones, or other local runoff-producing areas of the soil surface. Such local runoff-producing areas are almost certainly responsible for the rapid appearance of runoff on the plot and, consequently, a hydrograph less steep than that predicted. Without representation of spatial heterogeneity in runoff production within the planes used by EUROSEM it would seem inevitable that predicted hydrographs would show the commencement of runoff later than observed and a hydrograph that is too steep.

The treads and risers that characterise the surface microtopography of the plot also have a significant impact on runoff routing, and hence the runoff hydrographs. Parsons et al. (1997) showed that water ponding on the treads typically exited laterally and downslope in the ratio of 2:1. Because EUROSEM, in common with other soil-erosion models, treats the ground surface as a plane on which flow depth is uniform, such flow divergence by the microtopography cannot be modelled. Although the effect of microtopography on hydraulic roughness is recognised in EUROSEM, its effect in controlling variations in flow depth is not.

10.5.2
Erosion

In EUROSEM, prediction of interrill erosion depends upon four processes: soil detachment by raindrop impact, soil detachment by flow, transport by flow and deposition by flow (Morgan et al. 1992). Both flow transport and deposition potentially reduce erosion below the sum of detachment by raindrops and flow. Therefore, inasmuch as the predicted erosion is too high, it is overprediction of the processes of detachment that would seem to be likely sources of the error.

Detachment by raindrop impact DET (g m^{-2} s^{-1}) is determined from the equation

$$DET = kKE^{1.0}e^{-bh} \tag{10.1}$$

where KE is the kinetic energy of the rainfall (J m^{-2}), k is an index of the detachability of the soil that depends on soil texture, b is an exponent, taken to have a value of 2.0 (Morgan et al. 1992), and h is water depth (m). As is evident from this equation, the tendency of the model to overpredict flow depth can be expected to lead to significant underprediction, rather than overprediction, of the detachment rate. The value for k was obtained from the measurements of splash rates according to the method of calculation given in Poesen (1985) and Poesen and Torri (1988). The value so calculated is 0.14 g J^{-1}, which is an order of magnitude smaller than the value recommended in Morgan et al. (1993) for a soil of the type on the plot in a dry initial condition. However, it should be noted that our value was obtained from the undisturbed surface of the plot, which includes its vegetation cover, whereas the values given in Morgan *et al.* and based on Poesen (1985) are for bare soil surfaces. Unfortunately, there are no data which would enable us to judge whether or not the difference between the recommended detachability of the bare soil and that observed for the undisturbed (vegetated) surface is reasonable, nor to adjust our value to an equivalent for a bare soil. Although bare soils are common on agricultural fields and hence of major importance in modeling soil erosion, quite evidently, however, it would be better if the model could make use of measurements of soil detachability on a range of undisturbed surfaces, since no other measurement has any real meaning in terms of the erosion processes.

The modelled detachment rate is, additionally, modified by the presence of non-erodible materials at the soil surface so that

$$DET' = DET(1 - PAVE) \tag{10.2}$$

where $PAVE$ is the proportion of the surface covered by non-erodible materials.

Using the calculated value of k and taking into account the effect of plant canopy cover on kinetic energy of the rainfall, the calculated detachment rate is 2.168 g m^{-2} min^{-1} at zero depth of runoff. This value compares to the observed rate of splash transport (corrected for area of the collecting surface according to Poesen and Torri 1988), of 1.668 g m^{-2} min^{-1}. Given the findings of Schultz et al. (1985) that the total weight of soil loosened by raindrop impact is 14 to 20 times greater than that of the soil carried in splash droplets, the calculated detachment rate would seem to be too low. Consequently, it does not appear likely that this is the explanation for the overprediction of the erosion rate. Using the recommended value for k of 1.7 g J^{-1} given in Morgan et al. (1993) yields a predicted detachment rate of 26.332 g m^{-2} min^{-1}, which is much more in keeping with the ratio for total detachment to splash transport claimed by Schultz et al. (1985). However, quite surprisingly, this much higher value for k does not lead to an even greater overprediction of the erosion rate. This finding further suggests that it is not an error in the estimation the rate of soil detachment by raindrops that is responsible for the high predicted erosion rate.

Although it would be expected that soil detachment by runoff would be negligible in interrill areas, overprediction of such detachment is one possible explanation for the overprediction of erosion rate. Soil detachment by runoff DF is given by the equation

$$DF = ywv_s(TC - C) \tag{10.3}$$

where y is a flow-detachment efficiency coefficient, w is the width of the flow, v_s is the settling velocity of the particles, TC is the transport capacity of the flow, and C is the sediment concentration of the flow. The efficiency coefficient y is given by

$$y = \frac{u_{gmin}}{u_{gcrit}} \tag{10.4}$$

where u_{gmin} has a recommended value of 1 cm s^{-1} and u_{gcrit} is given by the equation

$$u_{gcrit} = 0.89 + 0.56 COH \tag{10.5}$$

where COH is soil cohesion (kPa), estimated from the soil texture as 4 kPa.

Transport capacity for interrill flow TC is determined from equations developed by Everaert (1991) such that

$$TC = \frac{b}{<r>_s q}\left((<O> - <O>_c)^{\frac{0.7}{n}} - 1\right)^k \tag{10.6}$$

where $k = 5$, $<r>_s$ = the sediment density, n is Manning's n and b is a function of particle size defined by:

$$b = \frac{19 - \frac{d_{50}}{30}}{10^4} \tag{10.7}$$

$<O>$ is the modified stream power ($g^{1.5}$ cm$^{-2/3}$ s$^{-4.5}$) defined by Bagnold as:

$$<O> = \frac{(U^*u)^{\frac{3}{2}}}{h^{\frac{2}{3}}} \tag{10.8}$$

in which U^* is shear velocity (m s^{-1}). $<O>_c$ is the critical value of $<O>$ found by using the Shields critical shear velocity (White 1970):

$$U_c^* = \sqrt{y_c(<r>_s - 1)gd_{50}} \tag{10.9}$$

where y_c is the modified Shields critical shear velocity based on particle Reynolds number (m s^{-1}) and g is the acceleration due to gravity (m s^{-2}).

Because, as stated above, it is generally agreed that detachment by flow is negligible in interrill area, it might be expected that EUROSEM would calculate zero detachment by flow for the three planes in the present application of the model. Such an assumption, however, is at odds with results from running the model. If detachment by flow is zero, then varying soil cohesion should have no effect on the predicted erosion rate. As Table 10.3 shows, this is not the case. Selecting values of COH from the estimated value of 4 kPa up to 40 kPa shows that as COH increases so the erosion rate decreases, levelling off when COH reaches a little over 20 kPa. Evidently, EUROSEM is attributing a very large proportion of the observed erosion to detachment by overland flow.

If the model is run with a value of COH of 30 kPa to prevent flow detachment and the recommended value of k (the index of soil erodibility) of 1.7 g J^{-1} is used, the model predicts erosion rates that are quite close to those observed (Fig. 10.7). However, it should be noted that in this case the model predicts declining erosion rates as discharge increases. This prediction disagrees with the observed increasing erosion rates for the same part of the experiment. In all probability, the predicted decline in ero-

Table 10.3. Predicted erosion rates (kg min^{-1}) at G1, G2 and G3 for a range of values of soil cohesion

Plane	Time (min)	Cohesion (kPa)	4.0	8.0	12.0	16.0	20.0	30.0	40.0
1	29		0.3231	0.2655	0.0448	0.0053	0.0037	0.0037	0.0037
	34		0.3963	0.3208	0.0496	0.0041	0.0024	0.0023	0.0023
	39		0.4489	0.3584	0.0521	0.0037	0.0189	0.0018	0.0018
2	29		0.7488	0.6202	0.1029	0.0117	0.0082	0.0082	0.0082
	34		0.9356	0.7626	0.1187	0.0083	0.0040	0.0039	0.0039
	39		1.064	0.8548	0.1248	0.0071	0.0026	0.0024	0.0024
3	29		1.359	1.102	0.1746	0.0218	0.0160	0.0158	0.0158
	34		1.857	1.507	0.2303	0.0171	0.0090	0.0088	0.0088
	39		2.135	1.708	0.2472	0.0133	0.0044	0.0041	0.0041

sion rate is associated with the effect of increasing flow depths on raindrop detachment. In reality, the spatial variability of depths of overland flow on the three runoff planes (Parsons et al. 1996) means that simple relationships among mean overland flow depths, detachment rates and erosion rates do not exist, as has been demonstrated by Abrahams et al. (1991) and Parsons et al. (1994).

10.6
Conclusions

In this study we have compared the predictions of runoff and erosion made by EUROSEM with observed data for a large runoff plot in southern Arizona, paying particular attention to the processes responsible for the predicted values. With only a few exceptions, the values of parameters needed for running the model have been obtained from field measurements. Neither runoff nor erosion is well predicted by the model. In the case of runoff, for which several of the parameter values had to be estimated based upon guidance given by Morgan et al. (1993), EUROSEM predicts much later initiation times than observed. Using parameter values outside the ranges given it Morgan et al., it is possible to achieve a close match between observed and predicted runoff for the last 10 minutes of the measurement period. At this time, runoff hydraulics also show reasonably close agreement with those observed, indicating that hydraulic roughness has been reasonably well dealt with in the model. However, such calibration is wholly at odds with the notion of process-based modeling. Furthermore, as Parsons et al. (1997) have demonstrated, it is quite possible to obtain reasonable matches between observed and predicted hydrographs provided sufficient understanding of the processes controlling the hydrographs is included in the modeling.

Correct runoff hydraulics are a pre-requisite for a correct representation of the processes of sediment detachment and transport. Even then, small errors in modeling hydraulics may be magnified in calculations of erosion rates (see Wainwright and Parsons 1998). Comparisons between observed and predicted erosion for the 10-minute

CHAPTER 10 · A Process-Based Evaluation of a Process-Based Soil-Erosion Model

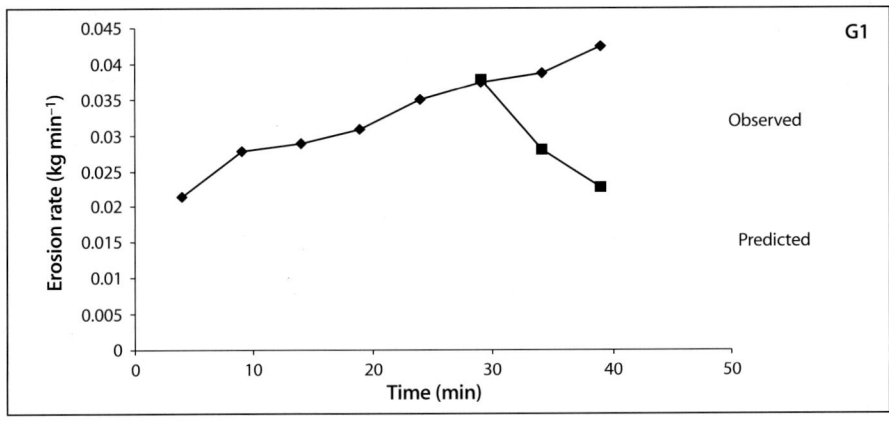

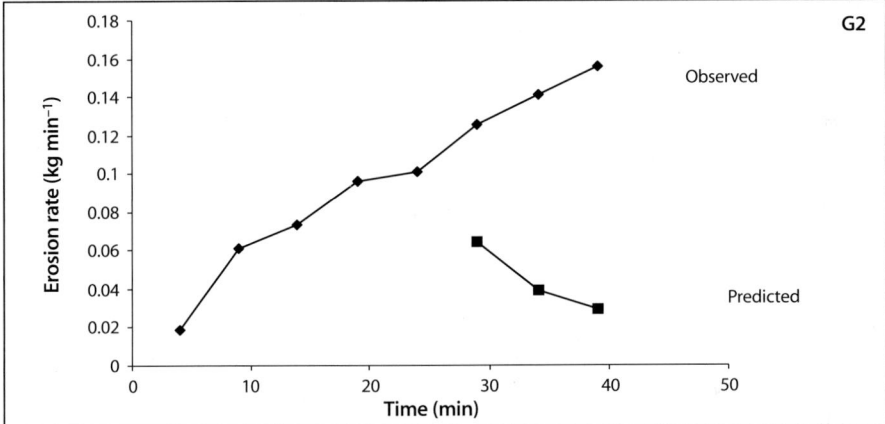

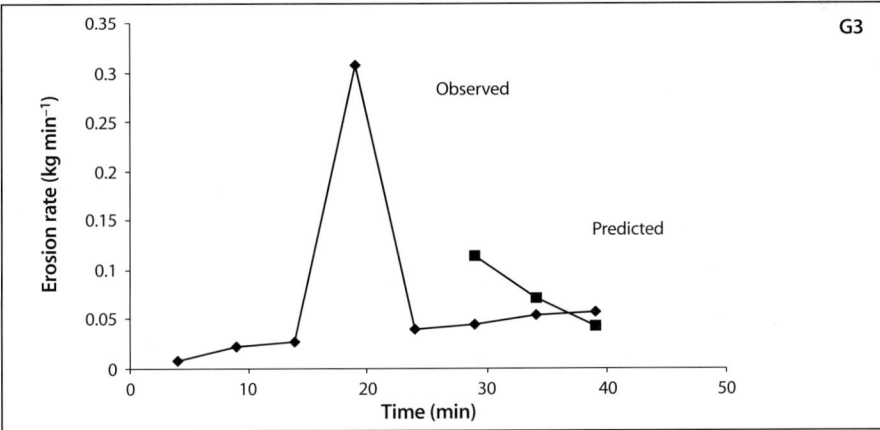

Fig. 10.7. Observed and predicted erosion rates for minutes 29 to 39 obtained from model hydrographs and hydraulics of Fig. 10.3 and 10.4, respectively, and using values of 30 kPa for soil cohesion (COH) and 1.7 g J^{-1} for soil detachability (k)

period for which predicted runoff hydraulics were reasonable shows that the model calculates a substantial amount of flow detachment even though, according to the description given of the model (Morgan et al. 1992), it should not. Forcibly removing this component to erosion, and using a recommended value for soil detachability that is an order of magnitude greater than that measured yields reasonably accurate erosion values.

The study demonstrates that detailed information on the processes responsible for runoff and erosion permits a more rigorous evaluation of the results of an erosion model. In the present case, the evaluation shows that much of the predicted erosion is being attributed to flow detachment which contrary both to the accepted view of erosion processes in interrill areas, and to the supposed equations used in the model. Furthermore, the study demonstrates the significance of EUROSEM's failure to take account of spatial variations in runoff production within the planes for which runoff and erosion are calculated. These small-scale variations have important effects on the processes of infiltration and sediment detachment, affecting both the timing and amount of runoff and erosion. The study raises the question of the extent to which erosion models can claim to be truly process-based without including explicit representation of such spatial variability.

Acknowledgements

We thank John Quinton for his assistance in answering queries in our use of the EUROSEM program. The field data were obtained with the support of the Natural Environment Research Council Grant GR3/7999.

References

Abrahams AD, Parsons AJ, Luk SH (1991) The effect of spatial variability in overland flow on the downslope pattern of soil loss on a semi-arid hillslope, southern Arizona. Catena 18:255–270
Boardman J, Favis-Mortlock D (eds) (1998) Modeling soil erosion by water. Springer-Verlag, ASI-NATO Series I-55, Berlin
Ellison WD (1947) Soil erosion studies – Part II. Agricultural Engineering 28:197–201
Evereart W (1991) Empirical relations for the sediment transport capacity of interrill flows. Earth Surface Processes and Landforms 16:513–522
Gelderman FW (1970) Soil survey of Walnut Gulch Experimental Watershed: a special report. US Depart Agricult, Soil Conserv Service and Agricultural Res Service, Portland, Ore, 54 p
Grayson RB, Moore ID (1992) Effect of land-surface configuration on catchment hydrology. In: Parsons AJ, Abrahams AD (eds) Overland flow hydraulics and erosion mechanics. UCL Press, London, 147–175
Nearing MA, Foster GR, Lane LJ, Finkner SC (1989) A process-based soil erosion model for USDA-water erosion prediction technology. Transactions of the American Society of Agricultural Engineers 32:1587–1593
Morgan RPC, Quinton JN, Rickson RJ (1992) Eurosem documentation manual. Silsoe College, 34 p
Morgan RPC, Quinton JN, Rickson RJ (1993) Eurosem: A user guide. Silsoe College, 84 p
Parsons AJ, Abrahams AD (1989) A miniature flume for sampling interrill overland flow. Phys Geogr 10:96–105
Parsons AJ, Abrahams AD, Luk SH (1990) Hydraulics of interrill overland flow on a semi-arid hillslope, southern Arizona. Journal of Hydrology 117:255–273
Parsons AJ, Abrahams AD, Wainwright J (1994) Rainsplash and erosion rates in an interrill area on semi-arid grassland, southern Arizona. Catena 22:215–226
Parsons AJ, Wainwright J, Abrahams AD (1996) Runoff and erosion on semi-arid hillslopes. In: Anderson MG, Brooks SM (eds) Advances in hillslope processes. Wiley, Chichester, 1061–1078
Parsons AJ, Wainwright J, Abrahams AD (1997) Distributed dynamic modeling of interrill overland flow. Hydrological Processes 11:1833–1859

Poesen J (1985) An improved splash transport model. Z Geomorph 29:193–211
Poesen J, Torri D (1988) The effect of cup size on splash detachment and transport measurements. Part I: Field measurements. Catena Suppl 12:113–126
Schultz JP, Jarrett AR, Hoover JR (1985) Detachment and splash of a cohesive soil by rainfall. Transactions of the American Society of Agricultural Engineers 28:1878–1884
Wainwright J, Parsons AJ (1998) Sensitivity of sediment-transport equations to errors in hydraulic models of overland flow. In: Boardman J, Favis-Mortlock D (eds) Modeling soil erosion by water. Springer-Verlag, ASI-NATO Series I-55, Berlin
White SJ (1970) Plane bed thresholds of fine grained sediments. Nature 228:152–153

Chapter 11

WEPP, EUROSEM, E-2D:
Results of Applications at the Plot Scale

A. Schröder

11.1
Introduction

Agriculture has a major influence on the state of the environment. Simply due to their share of the land surface, their effects on the regional budgets of energy and matter, and the limited availability of soil fertility as the main resource, agricultural land use systems should be the subject of continuous evaluation and improvement. In the context of the growing importance of soil conservation as a means of sustainable development, fundamental soil erosion research which guides the development and application of predictive simulation systems is a *desirable* issue because it helps to optimise agricultural production techniques with respect to energy input, soil and nutrient losses, and food security.

Since the mid-1970s several soil erosion modeling systems have been developed mainly to predict the effects of different growing techniques and crop rotations on runoff and soil loss from agricultural land. CREAMS (Knisel 1980), ANSWERS (Beasley et al. 1980), AGNPS (Young et al. 1987), EPIC (Sharpley and Williams 1990a, b) and OPUS (Ferreira and Smith 1992; Smith 1992) are probably the most important models which should be mentioned here. The main erosion components of these models are based on a concept originally formulated in the Universal Soil Loss Equation (Wischmeier and Smith 1965, 1978) which relates the total amount of soil loss to six factors (rainfall erosivity, soil texture, slope length and grade, soil cover and conservation practice) which were – based on empirical analyses – identified as the most important ones. The main improvements incorporated in the models mentioned above are mathematical formulations for describing the *processes* of soil water movement, infiltration, runoff generation, retention and concentration, particle detachment and deposition, transport capacity, and nutrient and pesticide leaching. The spatial and temporal variability of the processes of soil erosion is usually accounted for by square grid or slope segment representations of the field surface, and explicit routing algorithms for the prediction of the peak rates and duration of the runoff and sediment hydrographs. More comprehensive models such as EPIC and OPUS use special equations for the simulation of daily weather changes, plant growth, and the impacts of tillage on soil surface and structure. In general, modern erosion models have become complex computer systems which should be applicable to all questions of agricultural soil conservation.

11.2
Objectives and Methodology of the Study

In this study the hillslope version of the Water Erosion Prediction Project (WEPP) (Flanagan and Nearing 1995; Flanagan and Livingston 1995), the European Soil Erosion Model (EUROSEM) (Morgan et al. 1992, 1993), and the EROSION-2D (E-2D) model (Schmidt et al. 1996) are investigated. These computer models were developed during the last ten years and represent the "state of the art" of physically-based soil erosion prediction technology.

The study was conducted mainly for the following reasons. First, a computer model – which is a complex system by itself – is used to simulate another complex system, i.e. a compartment of the natural soil-water-plant system within which soil erosion occurs. Due to the causal, spatial and temporal complexity of soil erosion processes any computer model is a simplification of the natural system. Thus, the first objective of the investigation is to assess the *structure*, *quality* and *relevance* of the soil erosion processes which are simulated in the model. This part of the study consists mainly of a description and comparison of the model structure and components. The second objective is to make statements on the models' *behaviour*, i.e. to identify those input parameters which have a strong impact on the model predictions. In this study it was decided to conduct a one-dimensional sensitivity analysis which means that the dependence of changes in the predicted amount of soil loss on changes in one input parameter are investigated. The last objective of this study is to assess the predictive capability or *performance* of each model by comparing measured to predicted amounts of runoff and soil loss. For this analysis eight rainfall simulator experiments were carried out on several plots on loessy and sandy soils near Leipzig and Berlin.

The basic motivation for conducting this study was to obtain "first and fundamental" information on currently used soil erosion models. The findings and conclusions of this study are largely influenced by the limited time and the specific conditions under which the models were applied. Due to the large number of unvalidated components which are used in all models any fundamental statements on the general applicability of each model and the transferability of the findings of this study should be made with caution.

11.3
Model Comparison

The Water Erosion Prediction Project (WEPP) (Flanagan and Nearing 1995; Flanagan and Livingston 1995) soil erosion model is a joint development of several agencies of the U.S. Department of Agriculture (Agricultural Research Service, Natural Resource Conservation Service – former Soil Conservation Service –, Forest Service), and of the Bureau of Land Management of the U.S. Department of the Interior. The main objectives of the research and development project which was allocated at the National Soil Erosion Research Laboratory (NSERL) at Purdue were the replacement of the Universal Soil Loss Equation (USLE, Wischmeier and Smith 1978) and its revised form (RUSLE, Renard et al. 1997) by a physically-based simulation model which could account for various types of agricultural land use, crop production and conservation techniques by representing the fundamental processes of soil erosion. The model can be applied both to single hillslopes of a maximum length of about 100 m and to small watersheds up to about 2.5 km^2 (Foster and Lane 1987). In the latter case the watershed is divided into hillslope elements and channels ('open-book method'). WEPP is the only model included in this comparison which can be run in a continuous simulation mode in

CHAPTER 11 · WEPP, EUROSEM, E-2D: Results of Applications at the Plot Scale

order to simulate the effects of complex crop rotations. In this case daily weather data are produced by a stand-alone stochastic climate generator which is included with the model. The findings of this study are based on the hillslope version 94.7 from August 1994 which was run in the single-event mode only. The simulation model and further information on the WEPP programme can be obtained from the USDA-NSERL homepage (http://soils.ecn.purdue.edu/wepp).

The concept of the European Soil Erosion Model (EUROSEM) (Morgan et al. 1992, 1993) was developed after a series of workshops organised by the E.U. in the 1980's (Prendergast 1983; Morgan and Rickson 1988). The main research was conducted at the Silsoe College (G.B.), the Leuven University (Belgium) and the Soil Conservation Research Institute at Firenze (Italy). The model is applicable to the same problems as described for the WEPP model although it was decided to simulate single rainstorms only. Compared to the WEPP model EUROSEM puts more emphasis on the detailed simulation of the temporal changes in the runoff generation and sediment discharge (sedigraphs). The representation of hillslopes and catchments – which is similar to that used in WEPP – and several other model components were adapted from the KINEROS model (Woolhiser et al. 1990). In this study, the EUROSEM version 3.1 from June 1993 was used. Information on the latest EUROSEM version can be obtained from the Silsoe College homepage (http://www.silsoe.cranfield.ac.uk/eurosem).

The EROSION-2D model (E-2D) (Schmidt 1996) was developed by J. Schmidt during the mid-1980 at the Department of Geography at the Free University Berlin. The main objective of the research programme which was funded by the German Ministry for Education and Research was the implementation of a single-event hillslope erosion model which should be primarily used for the assessment of agricultural soil protection measures such as strip cropping, check dams, ditches, roads or tree lines. The model should require a minimum number of input parameters for easy and wide use in agricultural planning offices and authorities. E-2D was extensively validated between 1993 and 1996 in a joint erosion research programme of the Saxonian State Office for Environment and Geology and the Saxonian Research Institute for Agriculture. The results of this study a based on the model version 2.06 which was issued in June 1994. A grid-based watershed version ('EROSION-3D') has been developped by von Werner (1995). E-2D is available from the two institutions mentioned above.

11.3.1
Input and Output Parameters

The number of input parameters required for the hillslope profile versions of WEPP, EUROSEM and E-2D are listed in Table 11.1. The large number of input parameter number needed for the WEPP model is due to the boundary and starting conditions which have to be specified in the management file for the minimum simulation period of one year. With its complicated file building screen and structure this file is the weakest point in the input parameter estimation procedure. A special problem with the WEPP model is the insufficient declaration of those parameters which are *actually used* in the event simulation.

About 60% of the parameters in the climate and management files are monthly weather data, tillage, crop rotation and plant parameters which are used to drive the

Table 11.1. Minimum number of input parameters for the single-event hillslope versions of WEPP, EUROSEM and E-2D. The number increases with increasing within-segment and soil profile variability

WEPP		EUROSEM		E-2D	
Climate file[a] (*.CLI)	60	Precipitation file (*.PCP)	2	Rainfall file (*.REG)	2
Slope file (*.SLP)	5	Hillslope characteristics file (*.PAR)	35	Relief file (*.REL)	2
Soil file (*.SOL)	20			Soil file (*.BOD)	15
Management file[a] (*.MAN)	80				
Total	165		37		19

[a] File includes parameters which are needed for the continuous simulation only but which have to be specified by the user in order to run the single-event model without problems. The total minimum number of parameters actually needed for a single event simulation with WEPP is approximately 100.

long-term subroutines (e.g. climate generator, plant growth, soil disturbance, compaction) for the continuous simulation. An advantage of WEPP are the soil type, plant parameter and tillage implement databases which assist the user in the building of the soil and management files.

The high number of input parameters in the EUROSEM hillslope file is needed for the specification of the interception characteristics of the plant canopy, the hydraulic properties of the soil surface and the soil itself. All models use ASCII input files which can be accessed, built or stored via file building screens from the model interface only in WEPP and E-2D.

The spatial and temporal resolutions of the main model inputs, internal calculations and outputs are given in Tables 11.2 and 11.3. An important aspect of model applicability is that in some cases fundamental information such as the maximum numbers of homogenous elements, the internal resolution of the model calculations or the type of interpolation algorithms is not or only insufficiently given in the model documents or the on-screen help features. A common aspect of all models is the spatial discretization scheme of the hillslope profile as a series of homogeneous sections. The soil profile can be divided into several layers in all but the EUROSEM model. On the other hand this model is the only one which allows the user to define an above-ground canopy height for the calculation of interception storage and throughfall. The question arising from the different vertical discretization schemes is to what extent the infiltration and runoff characteristics are influenced by the soil properties in a depth of more than 30 cm during a rainfall event.

The horizontal distribution of soil erosion which helps to quickly identify the 'critical' points on a hillslope is plotted graphically in WEPP and E-2D only. This option should become a standard feature in soil erosion models because it helps the user in the identification of 'critical' slope sections and in the assessment of the efficiency of the suggested conservation measures.

Table 11.2. Spatial resolution of main model inputs and calculations

	WEPP	EUROSEM	E-2D
Rainfall	Homogenous over profile	Homogenous over profile[a]	Homogenous over profile
Slope profile description	Distance-slope pairs, min. number: 2, max. number: 30, automatic interpolation, algorithm: ?	Mean slope, straight	Distance-elevation pairs, min. number: 2, min. spacing: 1 m, automatic spline interpolation with 2nd order polynomial
Max. slope length	? Max. model input: 999 999 m	?	1 000 m
Max. slope width	? Max. model input: 999 999 m	?	Default: 1 m
Horizontal soil and surface properties	Homogenous OFEs, max. number: 20	Homogenous single plane, max. number: ?	Homogenous profile sections, max. number: 1 000
Soil profile description	Homogenous layers, max. number: 8, min. thickness: 0.01 m, max. thickness: 5.00 m	Homogenous layer, max./min. number: 1, "infinite" thickness or depth to non-erodible layer	Homogenous layer, max./min. number: 1, "infinite" thickness
Distribution of runoff along hillslope profile	Internal: at points with variable default spacing: Slope length (m) / 100, output: OFE total	Internal: ? output: plane total	Internal and output: fixed intervals with default spacing of 1 m, expressed per unit width, plotting: min. spacing: 1 m, max. spacing: 999 m[b]
Distribution of erosion along hillslope profile	Internal and output: at points with Variable default spacing: Slope length (m) / 100, expressed per unit area and OFE total[b]	Internal: ? output: rill and interrill totals, plane totals	Internal and output: fixed intervals with default spacing of 1 m, expressed per unit area[b], plotting: min. spacing: 1 m, max. spacing: 999 m

[a] Profile can be split into more than one plane with individual rain gauge weighting factors;
[b] Distribution along hillslope profile is plotted graphically;
? Unknown or not explicitly stated in the user documents or the file building screens.

A clear lack of all models is the insufficient representation of spatial rainfall variability. Only EUROSEM allows the user to specify individual raingauge weighting factors which can be assigned to different hillslope sections. Topographic and high-intensity raincell effects which may influence the sediment delivery processes and erosion budgets especially of larger catchments cannot be assessed by these models.

The temporal resolution of the main model components (Table 11.3) depends on the available resolution of the rainfall data and the maximum length of the time steps which are used for the numerical solution of the kinematic wave equations to produce

Table 11.3. Temporal resolution of main model inputs and calculations

	WEPP	EUROSEM	E-2D
Rainfall	Breakpoint data, min. number: 2 datapairs/day, max. number: 50 datapairs/day	Internal: breakpoint data, min. number: 2, max. number ?	Input: 10-min default average, internal: 1 min, output: 10-min default average
Infiltration	Internal: 1 min, output: 1 min and event total	Internal: user-specified[a], output: event total	Internal: 1 min, output: 10-min default average
Runoff	Internal: 1-min default, output: runoff hydrograph and event total[b]	Internal: user-specified[a], output: runoff hydrograph and event total[b]	Internal: 10-min default average, output: event total
Erosion	Internal and output: event total	Internal: user-specified[a], calculated together with runoff[a] output: sedigraph and event total[b]	Internal: 10-min default average, output: event total

[a] Minimum length of time increment of kinematic wave calculations is total length of simulation (in min) divided by 500;
[b] Hydrograph and/or sedigraph is plotted graphically;
? Unknown or not explicitly stated in the user documents or the file building screens.

the runoff hydrograph. When breakpoint data are available a resolution of one minute is used in WEPP and E-2D for the infiltration excess calculations. These values are aggregated in E-2D to 10-min averages from which the erosion totals of the event are calculated. In WEPP the erosion calculations for the rainfall events are based on the peak discharge rate, the runoff duration and a modified 'effective' runoff volume. These values are derived from a semi-analytical solution or an approximation of the kinematic wave model. The first option is used for the single event simulation with breakpoint rainfall data. The second scheme is applied to the 'artificial' rainstorms produced by the weather generator for the continuous simulation in order to save computational time. The erosion calculations are thus based on a simplified, steady-state runoff regime. The 'most dynamic' model included in this comparison is EUROSEM because it uses an explicit numerical solution scheme both for the kinematic wave equations and the sediment concentration calculations. This is the only model which produces a sedigraph (sediment discharge hydrograph) with a time step of usually less than one minute. The spatial distribution of soil erosion is unfortunately not calculated.

11.3.2
Model Structure

A list and description of the main model components which are used in the simulation of single rainstorms are given in Table 11.4. Only selected components are briefly described and assessed in the following sections. It can be inferred from Table 11.4 that all models are based on the same working scheme which is illustrated in Fig. 11.1: First, rainfall is assigned a certain erosivity value which characterises the ability of the rain-

Table 11.4. Main model components of the event simulation versions

Component	WEPP	EUROSEM	E-2D
Rainfall erosivity	Rainfall intensity	Rainfall kinetic energy (Brandt 1989), leaf drainage kinetic energy (Brandt 1990)	Rainfall momentum flux (Schmidt 1996)
Interception	Cover factor (no storage)	Storage model (Merriam 1973)	Cover factor (no storage)
Infiltration	Matrix infiltration (Green and Ampt 1911; Mein and Larson 1973; Chu 1978), pedotransfer functions for parameter estimation (WEPP project)	Matrix infiltration (Smith and Parlange 1978)	Matrix infiltration (Green and Ampt 1911), pedo-transfer functions for parameter estimation (van Genuchten 1980; Campbell 1985; Vereecken et al. 1989)
Surface runoff generation	Infiltration excess (Horton overland flow), surface retention	Infiltration excess (Horton overland flow), surface retention	Infiltration excess (Horton overland flow)
Surface runoff routing	Kinematic wave, semi-analytical solution (Stone et al. 1992) or approximate solution	Kinematic wave, numerical solution (Woolhiser et al. 1990)	Steady state (10-min average)
Particle detachment by rainfall	Proportionality constant: 'interrill detachability' (Foster and Meyer 1975), pedotransfer functions for parameter estimation (WEPP project)	Proportionality constant: 'soil detachability index' (Poesen 1985; Govers 1991; Everaert 1992)	Proportionality constant: 'soil resistance' (Schmidt 1996)
Particle detachment by runoff	Proportionality constant: 'rill detachability' (Foster and Meyer 1975), pedotransfer functions for parameter estimation (WEPP project)	Proportionality constant: 'flow detachment efficiency coefficient' (Rose et al. 1983; Styczen and Nielsen 1989)	Not distinguished from rainfall particle detachment
Sediment routing	Steady-state sediment continuity equation (Foster and Meyer 1975)	Dynamic mass balance equation (Bennett 1974; Woolhiser et al. 1990)	Steady-state mass balance equation (Schmidt 1996)
Rills	Characteristics to be specified by user	Characteristics to be specified by user, simulation of rill incision	Not considered
Transport capacity of interrill flow	Not considered	Modified stream power equation (Bagnold 1966; Everaert 1991)	Momentum flux equation (Schmidt 1996)
Transport capacity of rill flow	Modified Yalin bedload equation (Yalin 1977; Finkner et al. 1989)	Modified unit stream power equation (Yang 1979; Govers 1990)	
Sediment composition	Sand, silt, clay, organic matter (Foster et al. 1985)	Median soil particle diameter (Campbell 1985)	Sand, silt, clay (Schmidt 1996)
Sediment enrichment	Yes	Not considered	Yes

drops to remove the soil particles from the bare surface. The rainfall impact is reduced by a plant cover or canopy coefficient. Second, soil erodibility which describes the ability of the soil surface to withstand the erosive impact of raindrops is usually derived from the surface and soil characteristics. Third, the infiltration approaches used

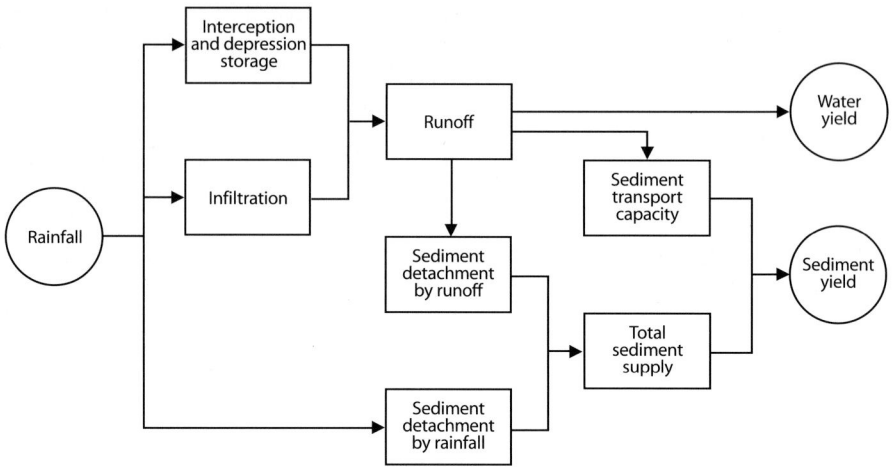

Fig. 11.1. Simplified representation of the main factors and interactions causing soil erosion by water (Hartley 1987)

in all models are based on the 'Horton overland flow' concept (Horton 1933). This means that the total amount of water which is available at the soil surface during each time step of the simulation is the positive difference between the actual rainfall intensity and infiltration rates. Fourth, the overland flow which is the input to the runoff routing routines is the thus generated 'rainfall excess' reduced by surface storage. Fifth, the ability of the overland flow to detach soil particles from the soil surface is calculated from the overland flow volume and velocity, and a further soil erodibility factor. Sixth, the total amount of sediment which can be moved by the overland flow along a hillslope section per unit time is limited by its available transport capacity. Seventh, the sediment amount which is deposited on a certain hillslope section per unit time is usually calculated from the positive difference between the sum of the sediment input from the upslope segment, the sediment eroded by raindrop and flow impact, and the available transport capacity of the overland flow in the actual hillslope segment. The main differences among the approaches in the model components are the spatial and temporal resolutions of the calculations, the number of sub-processes, the degree of simplification, and the accuracy with which the boundary conditions are represented.

11.3.2.1
Rainfall

Accumulated time-rainfall volume pairs (breakpoint data) are used in WEPP and EUROSEM for the description of the rainfall intensity distribution during rainstorms. These data enter the infiltration and rainfall erosion calculations directly in WEPP. In EUROSEM, the rainfall kinetic energy value is calculated for the throughfall as (Brandt 1989):

$$E_{kin(DT)} = 8.95 + 8.44 \log I \tag{11.1}$$

where
- $E_{kin(DT)}$ = rainfall energy of direct throughfall (J m^{-2} mm^{-1})
- I = rainfall intensity (mm h^{-1})

This equation is based on the raindrop size investigations of Marshall and Palmer (1948). The kinetic energy of the leaf drainage is calculated from the plant canopy height by the following equation (Brandt 1990):

$$E_{kin(LD)} = 15.8\sqrt{PH} - 5.87 \tag{11.2}$$

where
- $E_{kin(LD)}$ = energy of leaf drainage (J m^{-2} mm^{-1})
- PH = effective height of plant canopy (m)

The rainfall kinetic energy produced by Eq. 11.1 is approximately 20% lower than the rainfall erosivity estimates of the R-factor in the Universal Soil Loss Equation (Wischmeier and Smith 1978). The maximum rainfall energy which can be reached with Eq. 11.1 is approximately 25 J m^{-2} mm^{-1}. This value corresponds to a canopy height PH of about 4 m.

In E-2D the rainfall erosivity is expressed in terms of its momentum flux (momentum divided by unit time) (Schmidt 1996):

$$\varphi_{r,\alpha} = r_\alpha \rho_r v_r \sin\alpha \left(1 - \frac{A_{leaf}}{A}\right) \tag{11.3}$$

where
- $\varphi_{r,\alpha}$ = momentum flux exerted by falling droplets (kg m^{-1} s^{-2})
- r_α = effective rainfall intensity related to slope inclination (m s^{-1})
- ρ_r = fluid density of rainfall (kg m^{-3})
- v_r = mean fall velocity of droplets (m s^{-1})
- A_{leaf} = portion of ground covered by plants or crop residue (m^2)
- A = total area (m^2)
- α = slope (°)

The crucial factors on which both the kinetic energy and the momentum flux equations depend are the size and velocity distributions of the raindrops. The E-2D and the USLE rainfall erosivity estimates are based on the fall velocity and intensity-size distribution investigations of Laws (1941), Laws and Parsons (1943) and Gunn and Kinzer (1949) which were carried out under laboratory conditions. Due to the differences in the raindrop size and fall velocity distributions which arise from the various natural precipitation generation mechanisms (i.e. convective, advective), and the exclusion of the impact of squalls which may accelerate the fall velocity it is likely that the rainfall erosivity indices used in the models seriously misrepresent the erosive potential of natural rainfall.

11.3.2.2
Infiltration

The Green-Ampt approach (Green and Ampt 1911) is used in the WEPP and the E-2D model in order to simulate the temporal changes in the infiltration rate during the rainstorm. The differential form of this equation for a soil matrix of infinite depth is:

$$i = K_e \left(1 + \frac{(\phi - \theta_o)\psi_c}{I}\right) \tag{11.4}$$

where
- i = actual infiltration rate (m s^{-1})
- K_e = effective hydraulic conductivity of the wetted zone (m s^{-1})
- ϕ = effective porosity (m^3 m^{-3})
- θ_o = initial saturation (m^3 m^{-3})
- ψ_c = effective capillary tension or wetting front suction potential (m)
- I = cumulative infiltration depth (m)

Equation 11.4 describes the exponential approach of the actual infiltration rate i to the hydraulic conductivity K_e when I approximates infinity. The most important assumptions of this approach are the piston-like entry of the water into the soil and a sharply defined wetting front which separates the fully saturated (above) and unsaturated (below) zones. The driving parameters of the Green-Ampt model are the wetting front suction potential ψ_c, the soil moisture deficit $\phi - \theta_o$, and the effective saturated conductivity K_e.

The wetting front suction term is calculated in WEPP from the soil type, the soil water content and the soil bulk density. The empirical relationship is a modified version of an equation originally developped by Rawls and Brakensiek (1983) and which has been changed several times since its first use in the early WEPP versions (see Rawls et al. 1989). Unfortunately the currently used function is not specified in the latest WEPP model documentation (Flanagan and Nearing 1995). The soil moisture deficit is calculated in a similar manner from empirical functions which were developed during the extensive WEPP rainfall simulation studies. The effective saturated conductivity is calculated from the sand and clay contents, and the cation exchange capacity of the topsoil. The thus estimated 'baseline' conductivity is adjusted in order to account for the influences of plant residue, rock and canopy cover, the soil crust resulting from rainfall impact since the last tillage, and macropores such as wormholes on the actual value.

A similar approach to the calculation of the Green-Ampt parameters from pedotransfer functions is used in the E-2D model. The relationship between the actual soil water content and the wetting front suction potential (soil water retention curve) is estimated from the soil texture, the bulk density and the carbon content on the basis of empirical functions developped by Van Genuchten (1980) and Vereecken et al. (1989). The hydraulic conductivity is calculated from an approach developped by Campbell (1985). It requires the soil texture, the bulk density, and the contents of clay and silt as input. The effects of rocks, wormholes or buried residue on the infiltration parameters have to assessed by the user.

The EUROSEM infiltration model is the Smith and Parlange equation (Smith and Parlange 1978) which is also used in the KINEROS model (Woolhiser et al. 1990):

$$i_{max} = K_s \frac{e^{(I/B)}}{e^{(I/B)} - 1} \tag{11.5}$$

with

$$B = G(\theta_o - \theta_s) \tag{11.6}$$

where
- i_{max} = maximum infiltration rate or infiltration capacity (cm min^{-1})
- K_s = saturated hydraulic conductivity (cm min^{-1})
- I = cumulative rain depth already absorbed by the soil (cm)
- B = an integral capillary and water deficit parameter of the soil (cm)
- G = effective net capillary drive (cm)
- θ_o = initial soil water content (m^3 m^{-3})
- θ_s = maximum soil water content (m^3 m^{-3})

Similar to the Green-Ampt infiltration model, Eq. 11.5 describes the exponential decay of the initial maximum infiltration rate until it approximates the saturated hydraulic conductivity or final infiltration rate K_s. The net effective net capillary drive G conceptually equals the suction front potential term in the Green-Ampt approach; it is actually this 'scaling parameter' which decides how fast the difference between the initial and the maximum soil water content is reduced. In EUROSEM the effective net capillary drive, the soil moisture deficit parameter and the final infiltration rate are not derived from soil or surface properties by pedotransfer functions; they have to be selected by the user with the help of guide value tables in the model documentations.

Because of the unknown pedotransfer functions of the WEPP model it was not possible to compare the estimates resulting from the different approaches described in this study. A general shortcoming of all approaches is the consideration of matrix flow and saturation only. Unsaturated hydraulic conductivities are usually more than one order of magnitude higher than saturated ones. This effect is mainly due to preferential flows in biopores and cracks produced by tillage, and the inhomogeneities in the soil structure of the unsaturated zone which produce a rather irregular vertical infiltration pattern and wetting front ('fingering') during a rainfall event.

In WEPP and EUROSEM the infiltration excess is reduced by the depression storage before becoming the input to the routing routine. In EUROSEM the storage term is derived from the ratio of the straight line distance to the actual length of a chain which is laid on the soil surface between two points. In WEPP the storage term is calculated from the random roughness of the soil surface and the slope based on an empirical equation originally developped by Onstad (1984). The roughness measures are treated as constant during the rainfall event which ignores the influence of relief lowering due to raindrop impact and soil subsidence.

The portion of the rainfall which enters the soil is considered lost in all models. This concept ignores the effects of return or saturation excess flow (Dunne 1978). In all models the infiltration excess is calculated first for each time step and hillslope segment first and then routed downslope. By this approach the infiltration excess is not allowed to re-infiltrate on a downslope segment if its potential infiltration rate exceeds the net rainfall rate.

11.3.2.3
Surface Runoff Routing

If the potential surface storage depression storage is completely satisfied, the positive difference between the net rainfall intensity at the ground surface and the infiltration rate becomes the input to the overland flow calculations. The basic equations which describe the movement of a water wave through arbitrarily shaped cross-sections along an inclined surface are simplified forms of the Saint-Vernant equations which are based on the laws of mass and momentum conservation (Fread 1993):

$$\frac{\partial A}{\partial t} + \frac{\partial Q}{\partial x} = r(t) - i(t) = q(x,t) \tag{11.7}$$

and

$$Q = \alpha P R^{m-1} \tag{11.8}$$

with

$$\alpha = C\sqrt{s} \quad \text{(WEPP)} \tag{11.9}$$

or

$$\alpha = \frac{\sqrt{s}}{n} \quad \text{(EUROSEM, E-2D)} \tag{11.10}$$

where
- A = cross-sectional area (m²)
- t = time (s)
- Q = discharge (m³ s⁻¹)
- x = downslope distance (m)
- r = rainfall intensity (m s⁻¹)
- i = local infiltration rate (m s⁻¹)
- q = lateral inflow rate (m s⁻¹)
- R = hydraulic radius (m)
- P = wetted perimeter (m)
- m = depth-discharge exponent Chezy: $m = 3/2$, Manning: $m = 5/3$ (–)
- α = depth-discharge coefficient (WEPP: m$^{1/2}$ s⁻¹, EUROSEM: m$^{1/3}$ s⁻¹) (–)
- C = Chezy coefficient (m$^{1/2}$ s⁻¹)
- s = average slope (m m⁻¹)
- n = Manning roughness coefficient (s m$^{-1/3}$)

The so-called rating equation (Eq. 11.7) describes the proportionality between the storage on a plane and the one-dimensional flux of overland flow. Each of the three models uses different methods for calculating the actual flow rate at various points along the hillslope.

In WEPP the overland flow is conceptualized as plane runoff which means that A is substituted by the average flow depth h (expressed in m). Equations 11.7 and 11.8 are solved analytically by the methods of characteristics (Eagleson 1970) which requires the rewriting of these equations as differential equations on characteristic curves on the x-t plane:

$$\frac{dh}{dt} = v(t) \tag{11.11}$$

and

$$\frac{dx}{dt} = \alpha m h(t)^{n-1} \tag{11.12}$$

where
- h = flow depth (m)
- v = runoff or rainfall excess rate (m s^{-1})

These equations are solved together with the infiltration calculations by using a Runge-Kutta iteration scheme with as spatial resolution of one hundredth of the total hillslope length and a time step of one minute. The recession limb of the hydrograph is calculated until the routed runoff volume equals 95% of the total infiltration excess volume, or the discharge rate equals 10% of the peak discharge rate. The approximate method which is used for the calculation of the runoff volume, the peak runoff rate and the runoff duration of the artificial rainstorms produced by the weather generator is based on empirical relationships among these parameters which were developped from kinematic wave simulations. The Chezy friction coefficient C is an area-weighted average of the rill and interrill area factors. They are derived from geometric surface roughness parameters, the plant and rock cover, and the plant type.

The kinematic wave equations are solved numerically in EUROSEM by the Newton-Raphson iteration technique (finite-difference grid, four-point implicit method, see Woolhiser and Liggett 1967; Woolhiser et al. 1990) which calculates explicit runoff depth values for each x-t pair along the hillslope. The time steps and spacing between the hillslope points is kept sufficiently short in order to ensure the computational stability of the routing algorithms. The decision as to whether flow is routed as rill or interrill runoff depends on the existence and the spacing of rills. Hillslopes with no rills are modelled as shallow overland flow areas like in WEPP and E-2D. For surfaces with rills of a spacing of more than one meter the flow length and direction is calculated from the vector sum of the hillslope inclination and the interrill slope towards the rills. The interrill flow routing is abandoned if the rill spacing is less then one meter. The Manning rill and interrill roughness coefficients have to be estimated from guide values in the EUROSEM user documents.

The kinematic wave equations are not solved in E-2D. The model calculates the infiltration excess for every minute of the simulation and every meter of the hillslope first. In a second step the momentum flux of the thus generated overland flow volume is determined by the following equations:

$$\varphi_q = \frac{w_q v_q}{\Delta x} \tag{11.13}$$

with

$$w_q = q\rho_q = [(r_\alpha - i)\Delta x + q_{in}]\rho_q \tag{11.14}$$

and

$$v_q = \frac{s^{1/2} h^{2/3}}{n} \tag{11.15}$$

with

$$h = \left(\frac{qn}{\sqrt{s}}\right)^{3/5} \tag{11.16}$$

where
- φ_q = momentum flux of flow (kg m^{-1} s^{-2})
- w_q = mass rate of flow (kg m^{-1} s)
- v_q = mean flow velocity (m s^{-1})
- Δx = length of slope segment (m)
- ρ_q = fluid density (kg m^{-3})
- r_α = effective rainfall intensity on slope segment (m s^{-1})
- i = infiltration rate (m s^{-1})
- q = discharge of actual segment (m^3 m^{-1} s^{-1})
- q_{in} = inflow from upslope segment (m^3 m^{-1} s^{-1})
- h = mean flow depth (m)
- n = Manning hydraulic roughness (s m$^{-1/3}$)
- s = slope (m m^{-1})

The thus generated overland flow depth and velocity enters the erosion calculations which are made for every minute. This approach was chosen because of the rather small contributions of the rising and the falling limbs to the total amount of erosion, compared to the amount that occurs during the steady state of the runoff hydrograph. The overland flow calculations are made for non-rilled surfaces only in order to avoid the difficulties associated with the specification of rill characteristics. The increasing overland runoff depth and velocity along the hillslope is seen as the main flow variable which controls erosion (Schmidt 1996). The hydraulic roughness n has to be specified by the user with the help of guide values.

11.3.2.4
Erosion and Deposition

The movement of the sediment along the hillslope is described in WEPP on the basis of the steady-state sediment continuity equation (Foster and Meyer 1972) which is applied rill flow conditions:

$$\frac{dG}{dx} = D_r + D_i \qquad (11.17)$$

where
- G = sediment load (kg s^{-1} m^{-1})
- x = distance downslope (m)
- D_r = rill erosion rate (kg s^{-1} m^{-2})
- D_i = interrill erosion rate (kg s^{-1} m^{-2})

Equation 11.17 uses the peak runoff rate as input. The rill and interrill erosion rates are determined from the following relationships:

$$D_i = K_i I_e \sigma_{ir} SDR_{RR} F_{nozzle} \left(\frac{R_s}{w}\right) \qquad \text{(Interrill areas)} \qquad (11.18)$$

and

$$D_c = K_r (\tau_f - \tau_c) \qquad \text{(Rill areas)} \qquad (11.19)$$

where
- D_i = interrill sediment delivery rate to rill (kg s^{-1} m^{-2})
- K_i = interrill soil erodibility parameter (kg s^{-1} m^{-4})
- I_e = effective rainfall intensity (m s^{-1})
- σ_{ir} = interrill runoff rate (m s^{-1})
- SDR_{RR} = sediment delivery ratio (–)
- F_{nozzle} = adjustment factor to account for sprinkler irrigation nozzle energy variation (–)
- R_s = spacing of rills (m)
- w = rill width (m)
- D_c = detachment capacity in a rill (kg s^{-1} m^{-2})
- K_r = rill soil erodibility parameter (s m^{-1})
- τ_f = flow shear stress (Pa)
- τ_c = critical shear stress to initiate particle detachment (Pa)

The independent variables in Eq. 11.18 were changed several times in the different versions of WEPP due to unclear impacts of the rainfall intensity and the overland flow depth on the particle detachment (see Kinnell 1991 and Ghidey and Alberts 1994 for discussion). The critical shear stress, and the rill and interrill soil erodibility parameters are predicted from soil texture and organic matter content which were developed from the comprehensive rainfall simulation experiment conducted during the WEPP research programme (Elliot et al. 1989). The sediment delivery ratio SDR_{RR} describes the ratio between the total amount of detached interrill sediment and that which actually reaches the rill; it is thus an indirect measure of the transport capacity of the shallow interrill flow. The delivery ratio is estimated from the random surface roughness, the tillage row side-slope and the soil texture.

The interrill erosion rate is always greater than zero and added to the rill erosion rate. A rill spacing of 1 m is assumed if no rills are specified by the user. Whether erosion or deposition occurs in a rill segment is decided on the basis of the sign of the rill erosion rate D_f in Eq. 11.17:

$$D_f = D_c\left(1 - \frac{G}{T_c}\right) \quad \text{if } t > t_c \text{ and } G < T_c \quad \text{(Erosion)} \quad (11.20)$$

and

$$D_f = \frac{\beta v_f}{q}(T_c - G) \quad \text{if } G \geq T_c \quad \text{(Deposition)} \quad (11.21)$$

where
- D_f = net detachment or deposition (kg s⁻¹ m²)
- τ = flow shear stress (Pa)
- τ_c = critical shear stress to initiate particle detachment (Pa)
- D_c = detachment capacity of rill flow (kg s⁻¹ m²)
- G = sediment load (kg s⁻¹ m²)
- T_c = transport capacity (kg s⁻¹ m⁻¹) (= G_{max})
- β = 0.5 = raindrop-induced turbulence coefficient (–)
- v_f = effective fall velocity for sediment particles (m s⁻¹), calculated by Stokes' law
- q = discharge per unit width (m² s⁻¹)

A fundamental problem of Eq. 11.17 is that all excess sediment calculated by Eq. 11.21 is deposited within the limits of the rill section dx (Δx). This may cause large jumps in the amount of deposited sediment at the end of each segment (Schramm 1994). This problem is avoided in WEPP by calculating the amount of erosion and deposition for many points on the slope. For this, normalised versions of Eq. 11.20 and 11.21 are used which are inserted into Eq. 11.17. The resulting equations are solved numerically in case of detachment (Runge-Kutta method), and by a closed-form solution in case of deposition. The spatial distribution of the total storm soil loss which is calculated for every slope point is calculated from the accumulated sediment load G over the effective runoff duration.

In EUROSEM the erosion calculations are based on the dynamic mass balance equation (Bennett 1974; Woolhiser et al. 1990) in which the changes in the surface runoff volume and the sediment concentration per time step and slope segment are considered simultaneously:

$$\frac{\partial(AC_s)}{\partial t} + \frac{\partial(QC_s)}{\partial x} - e_f(x,t) = q_s(x,t) \tag{11.22}$$

where
- A = cross-sectional area of flow (m²)
- C_s = sediment concentration (m³ m⁻³)
- t = time (s)
- Q = discharge (m³ s⁻¹)
- x = distance downslope (m)
- e_f = net detachment rate or rate of erosion of the bed per unit length of flow (m³ cm⁻¹ s⁻¹)
- q_s = external input or extraction of sediment per unit length of flow (m³ cm⁻¹ s⁻¹)

Equation 11.22 is applied to rill and interrill areas, thus accounting simultaneously for both the raindrop and runoff particle detachment. In WEPP the different soil particle detachment regimes are confined to the rill (flow erosion) and interrill areas (raindrop erosion). The net detachment rate e_f is calculated from

$$e_f = DET + DF \tag{11.23}$$

with

$$DET = k(KE)e^{-bh} \tag{11.24}$$

and

$$DF = \eta w_q v_s (TC - C) \tag{11.25}$$

where
- e_f = net detachment rate or rate of erosion of the bed per unit length of flow (m³ cm⁻¹ s⁻¹)
- DET = soil detachment by raindrop impact (g cm⁻²)
- DF = balance between rate of soil particle detachment by flow and particle deposition rate (m³ cm⁻¹ s⁻¹)
- k = soil detachability index by raindrop impact (g J⁻¹)
- KE = total kinetic energy of the rain (J m⁻²)
- b = 2 = empirical exponent (–)
- h = depth of surface water layer (mm)
- η = flow detachment efficiency coefficient (–)
- w_q = width of flow (m)
- v_s = settling velocity of particles (m s⁻¹)
- TC = volumetric transport capacity (m³ m⁻³)
- C = volumetric sediment concentration of flow (m³ m⁻³)

Similar to the D_f term in WEPP (Eq. 11.17) the sign of the soil particle detachment rate DF – which depends on the difference between the transport capacity TC and the sediment concentration C – decides upon whether flow erosion or deposition occurs. The efficiency coefficient η accounts for the reduction of the pick-up rate of the overland flow through soil cohesion whenever C is less than TC. η is calculated from the soil cohesion which has to be measured with a torvane or to be estimated from guide values. The soil detachability index k describes the soil erodibility due to raindrop impact; it is the unit mass of soil mass which is removed per unit kinetic rainfall energy from the soil surface. The coefficient has to be estimated from the soil texture and the surface condition by the user. The surface flow depth h accounts for the damping effect of the runoff layer on the raindrop impact.

For small values of the raindrop detachment DET or decreasing slope gradients which cause a drop in the transport capacity TC, the net detachment rate e_f in Eq. 11.22 may become negative if the sediment concentration C exceeds the transport capacity TC. In this case the excess sediment is routed downslope over short distances which results in a smoothing out of the deposition process.

In E-2D particle detachment occurs if the sum of the momentum fluxes exerted by rainfall (Eq. 11.3) and overland flow (Eq. 11.13) exceeds a critical momentum flux or soil resistance φ_{crit}:

$$E = \frac{\varphi_q + \varphi_r}{\varphi_{crit}} \tag{11.26}$$

with

$$E = 0 \quad \text{if } \varphi_q = 0 \tag{11.27}$$

where
- E = erosion ($E > 1$) or deposition ($E < 1$) coefficient (–)
- φ_q = momentum flux exerted by flow (kg m^{-1} s^{-2})
- φ_r = momentum flux exerted by droplets (kg m^{-1} s^{-2})
- φ_{crit} = critical momentum flux to initiate soil erosion (kg m^{-1} s^{-2})

Rill and interrill areas are not distinguished in Eq. 11.26. The critical momentum flux is defined by the following equation:

$$\varphi_{crit} = q_{crit} \Delta y \rho_q v_q \tag{11.28}$$

where
- φ_{crit} = critical momentum flux to initiate particle detachment (kg m^{-1} s^{-2})
- q_{crit} = critical volume rate of flow to initiate particle detachment (m^3 m^{-1} s^{-1})
- Δy = width of slope segment (m)
- ρ_q = fluid density (kg m^{-3})
- v_q = mean flow velocity (m s^{-1})

Equation 11.28 relates φ_{crit} solely to a threshold flow intensity only. Unlike the erodibility concepts of WEPP and EUROSEM, the rainfall intensity has no effect on the initiation of the motion of soil particles. Values for φ_{crit} have to be estimated by the user with the help of guide values which are given for various soil textures and conditions in the E-2D handbook. The dimensionless erosion index E is correlated to the sediment discharge by the following relationship:

$$q_s = 0.0001(1.75E - 1.75) \tag{11.29}$$

where
- q_s = potential sediment discharge (kg m^{-1} s^{-1})

The actual amount of sediment that is eroded or deposited on a slope segment is determined from a mass balance check in which the maximum possible sediment discharge (transport capacity) of the calculated runoff, and the difference between the upslope sediment input and the downslope output are considered. The resulting amounts of soil erosion or deposition are accumulated over all 10-min intervals of the simulation to yield the respective values for the every meter of the hillslope for the entire storm.

The accuracy and appropriateness of the erosion and deposition concepts which are used in WEPP, EUROSEM and E-2D can be assessed in rather general terms only. From a 'physical' point of view the distinction between rills and interrill areas is a reasonable one because it allows the separate description of the relevant processes, i.e. the detachment dominated by raindrop impact in the interrill areas, and the detachment and transport by flow in rills (Meyer and Wischmeier 1969; Foster and Meyer 1972). In WEPP, the interrill areas are treated as the main delivery regions for the sediment, and the transport capacity and deposition calculations are made for rill flow only. The results are converted to per unit area values which describe the spatial distribution of erosion along the hillslope profile. Like the EUROSEM calculations this approach requires the specification of the rill geometry. If this is not done by the user, WEPP estimates the rill spacing and characteristics from empirical functions. The problem associated with this concept can be seen when comparing the counteracting impacts of runoff on the interrill sediment delivery: Whereas in Eq. 11.18 the interrill detachment increases with an increasing flow rate due to the increasing transport capacity, the interrill splash detachment in Eq. 11.24 decreases due to the shielding effect of the surface water depth. It is thus useful to distinguish rill and interrill erosion, transport capacity and deposition regimes – as it is done in EUROSEM – in order to account for the different boundary conditions, but this does not overcome the problem of a comprehensive and unequivocal description of the relevant processes. The approach in E-2D to calculate erosion and deposition on a discharge- or sheet flow-depth basis only can be seen as the most realistic one because it requires no specification of the rill characteristics. However, the high erosion and sediment dislocation potential of concentrated rill flow – especially on long hillslopes – might be underestimated by this model.

A particular problem is the way in which fast changes in the erosion and deposition regimes – which can arise mainly from changes in the transport capacity along the hillslope – are dealt with in each model. The deposition equations in WEPP (Eq. 11.21) and EUROSEM (Eq. 11.25) contain special particle sink or pick-up parameters (efficiency coefficients) which – together with the estimates made at points and not for discrete segments – avoid abrupt temporal or spatial changes in the erosion and deposition patterns. Rather steep gradients at the hillslope segment boundaries can be expected for the spatial erosion-deposition pattern calculated with the E-2D model because of the (event-accumulated) mass balance which is determined for each segment width for each 10-min time step.

11.3.2.5
Sediment Transport Capacity of Surface Runoff

The transport capacity of overland flow is the maximum capability of a defined runoff volume to carry a certain amount of suspended sediment per unit time beyond which no further increase in the sediment concentration is possible due to deposition.

WEPP uses a simplified form of the Yalin (1977) transport capacity equation which was developped by Finkner et al. (1989). Based on the research of Julien and Simmons (1985) who identified slope gradient, unit discharge, rainfall intensity and shear stress as the variables with the highest influence on the transport capacity of overland flow, the modified Yalin transport capacity is calculated from the following expression:

$$T_c = k_t \tau_f^{3/2} \tag{11.30}$$

where
- T_c = transport capacity (kg m^{-1} s^{-1})
- k_t = transport coefficient (m$^{1/2}$ s^2 kg$^{-1/2}$)
- τ_f = hydraulic shear acting on soil (N m^{-2})

In the original form of the Yalin equation the movement of the channel bedload particles is 'driven' by the positive difference (excess shear) between the shear stress exerted on the ground by flow, and a threshold bed shear stress to initiate the motion of the particles. The flow shear stress τ_f is calculated from the following equation:

$$\tau_f = \rho g R s \tag{11.31}$$

where
- τ_f = flow shear stress (N m^{-2})
- ρ = fluid density (kg m^{-3})
- g = 9.8 = gravity constant (m s^{-2})
- R = hydraulic radius (m)
- s = slope gradient (m m^{-1})

In order to avoid a numerical solution scheme for the estimation of the flow shear stress and transport capacity at every point of the hillslope and for every timestep, a simplified approach was developed which can be described only briefly here. The hydraulic shear in Eq. 11.31 is calculated as a function of the discharge and the slope gradient at the end of complex slopes first. The hydraulic shear value then becomes the input to the transport capacity calculation at the end of the slope using the original Yalin equation. With this method the transport coefficient k_t in Eq. 11.30 is obtained from the expression

$$k_t = \frac{T_{co}}{\tau_{so}^{3/2}} \tag{11.32}$$

where
- T_{co} = transport capacity computed using τ_{so}
- τ_{so} = representative shear stress for entire slope

k_t is used for the calculation of the effective transport capacity at each point on the hillslope. Thus, the transport coefficient has to account for the changes in T_c according to the hillslope geometry. The k_t-values calculated from the actual slope at the end of the profile and the uniform reference slope were found to yield the best estimates of k_t for the entire hillslope. All flow is treated as rill flow in the transport capacity calculations.

The transport capacity calculations are made for each particle class (fine and coarse sand, silt, clay, organic matter) of the eroded soil. Excess load in one class is distributed among the others. A mass balance check at the end of each deposition region ensures that the total mass of the particle fraction leaving the section does not exceed that which enters it plus the interrill contribution in the region. The new surface soil composition in the deposition region becomes the input to the erosion calculations of the following rainstorm.

The transport capacity calculations in EUROSEM are based on the investigations of Govers (1990) and Everaert (1991, 1992). The rill and interrill transport capacities are estimated from different formulas which are based on stream power theory (Bagnold 1966; Yang 1973). The expression for the rill flow transport capacity is:

$$TC_{RF} = a(\omega - \omega_{cr})^b \qquad (11.33)$$

with

$$a = \left(\frac{d_{50} + 5}{0.32}\right)^{-0.6} \qquad (11.34)$$

$$b = \left(\frac{d_{50} + 5}{300}\right)^{0.25} \qquad (11.35)$$

$$\omega = vs \qquad (11.36)$$

$$\omega_{crit} = 0.4 \text{ cm s}^{-1} \qquad (11.37)$$

where
- TC_{RF} = transport capacity of rill flow (m³ m⁻³)
- ω = stream power (cm s⁻¹)
- v = mean velocity (cm s⁻¹)
- s = slope gradient (m m⁻¹)
- ω_{crit} = critical stream power to initiate particle detachment (cm s⁻¹)
- a, b = empirical constants (–)
- d_{50} = median particle diameter (μm)

The transport capacity of the sheetlike interrill flow is calculated from the following equations:

$$TC_{SF} = \frac{m}{\rho_s q}\left(\left(\Omega - \Omega_{crit}\right)^{\frac{0.7}{n}} - 1\right)^n \tag{11.38}$$

with

$$n = 5 \tag{11.39}$$

$$m = 0.0001 \frac{19 - d_{50}}{30} \tag{11.40}$$

$$\Omega = \frac{\omega^{3/2}}{h^{2/3}} \tag{11.41}$$

$$\Omega_{crit} = \frac{\left(0.5 u_c^{*2} v\right)^{3/2}}{h^{2/3}} \tag{11.42}$$

with

$$u_c^* = \sqrt{y_c(\rho_s - 1)g d_{50}} \tag{11.43}$$

where
- TC_{SF} = transport capacity of interrill sheet flow (m³ m⁻³)
- ρ_s = specific gravity of sediment particles (g cm⁻³)
- q = specific discharge (m³ cm⁻¹ s⁻¹)
- Ω = effective stream power (cm s⁻¹)
- ω = stream power as defined in Eq. 11.36 (cm s⁻¹)
- h = flow depth (cm)
- Ω_{crit} = critical Bagnold stream power to initiate particle detachment (cm s⁻¹)
- d_{50} = median particle diameter (µm)
- v = mean flow velocity (cm s⁻¹)
- u_c^* = critical shear velocity (cm s⁻¹)
- y_c = modified Shields critical shear velocity (cm s⁻¹), based on particle Reynolds number
- g = 9.81 = Gravity constant (m s⁻²)
- m, n = empirical constants (–)

The reason for using the stream power concept is given with its better representation of the rather laminar runoff conditions which usually occur on agricultural land. These are seen to be better accounted for by the product of excess shear stress and the mean flow velocity (Govers 1990). The Yalin equation considers only the amount of excess shear which is usually the main factor controlling particle detachment in fully turbulent channel flow only. A fundamental problem of the stream power and transport capacity calculations for sheetlike overland flow is the estimation of the effective flow depth. Due to the roughness of the soil surface even the runoff on the interrill areas becomes concentrated after short flow lengths in more or less clearly defined rills or slightly incised flow paths.

The transport capacity calculations in E-2D are based on the idea that the maximum sediment concentration c_{max} of the overland flow is determined by the equilibrium between the downward-directed momentum flux of the settling sediment particles, and a counteracting upward momentum flux component exerted by turbulence:

$$\varphi_{p,crit} = \varphi_{q,vert} \tag{11.44}$$

with

$$\varphi_{p,crit} = c_{max} \rho_p A v_p^2 \tag{11.45}$$

$$\varphi_{q,vert} = \frac{\varphi_q \varphi_r}{\kappa} \tag{11.46}$$

where
- $\varphi_{p,crit}$ = momentum flux below which falling sediment particles cannot be held in suspension (kg m^{-1} s^{-2})
- $\varphi_{q,vert}$ = vertical momentum flux component exerted by flow (kg m^{-1} s^{-2})
- φ_r = momentum flux of rainfall (kg m^{-1} s^{-2})
- φ_q = momentum flux of flow (kg m^{-1} s^{-2})
- κ = 0.001 = coefficient defining the effective upward component of total momentum flux (–)
- c_{max} = maximum sediment concentration in flow (m^3 m^{-3})
- ρ_p = particle density (kg m^{-2})
- A = area of slope segment (m^2)
- v_p = settling velocity of soil particles in stagnant fluid (m s^{-1}), calculated from Stokes' law

Rearranging Eq. 11.44 to 11.46 and solving it for the maximum sediment concentration yields the transport capacity $q_{s,max}$ as:

$$q_{s,max} = c_{max} \rho_p q \tag{11.47}$$

where
- $q_{s,max}$ = transport capacity (kg m^{-1} s^{-1})
- q = volume rate of flow (m^3 m^{-1} s^{-1})

The value of the deposition factor κ was derived from flume experiments conducted by Schmidt (1988). The factor is assumed to be independent of the flow conditions and the discharge depth. The transport capacity (Eq. 11.47) is calculated for every main particle class of the parent soil. The particle size distribution of the eroded soil is then determined from a comparison between the potential sediment discharge and the individual transport capacity of that class. Similar to the WEPP model the initial texture of the eroded soil is assumed to be the same as the parent soil. The relative increase in the share of the finer particles along the transport distance is thus caused by the excess transport capacity in the coarser particle classes which is 'used up' in the finer classes.

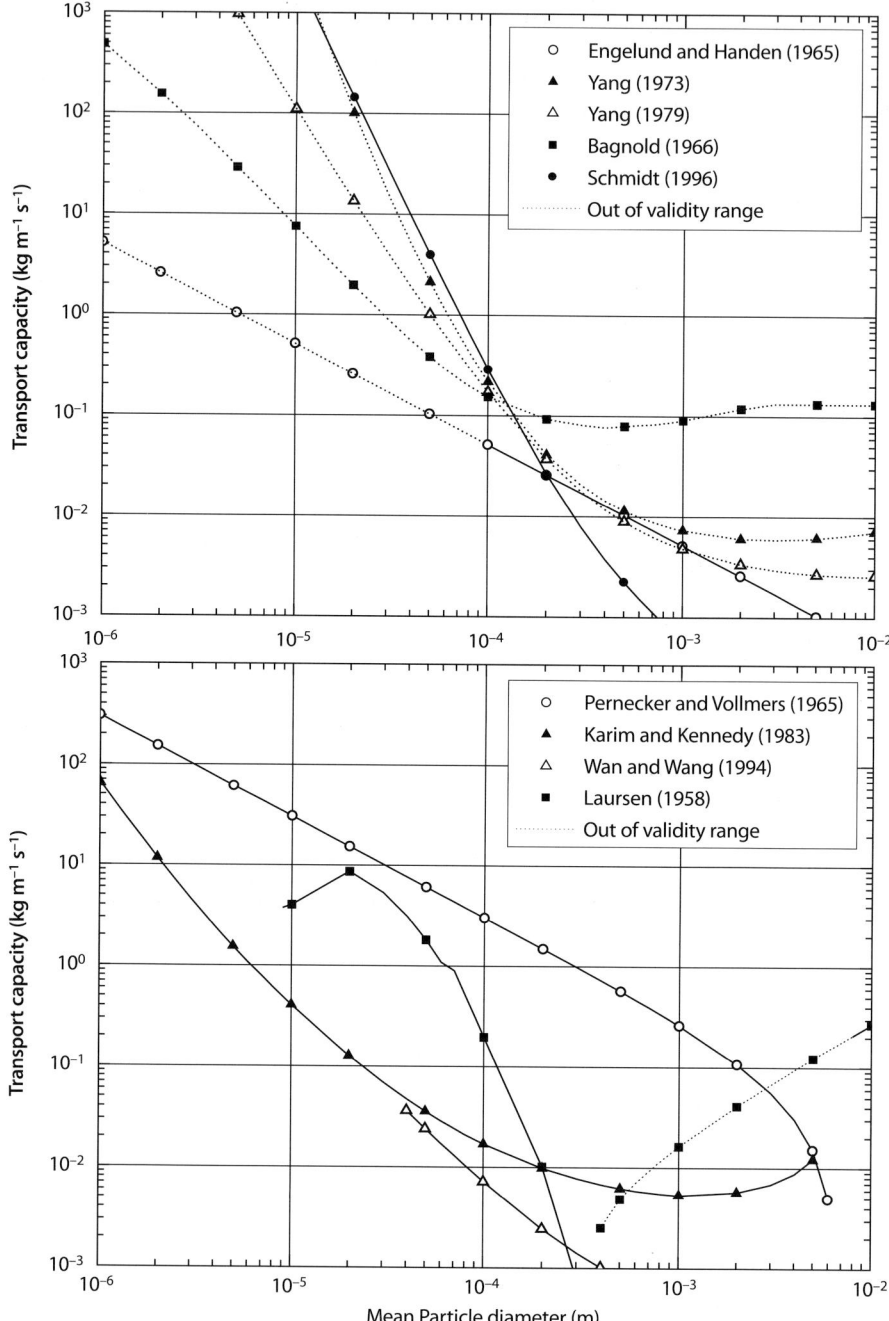

Fig. 11.2. Relationship between transport capacity and mean particle diameter as predicted by various formulas for steady sheet flow conditions. The values are calculated for a mean slope gradient of 0.05 and the hydraulic boundary conditions given in Fig. 11.3 (Schramm 1994)

The transport capacity equations of Yalin (1977), Govers (1990) and Schmidt (1996) were compared by Schramm (1994) to the approaches of Engelund and Hansen (1967), Yang (1973, 1979), Bagnold (1966), Pernecker and Vollmers (1965), Karim and Kennedy (1983), Wan and Wang (1994), and Laursen (1958). The relationship between the transport capacity TC and the mean particle diameter d is plotted in Fig. 11.2 and 11.3 for steady rill flow conditions. It can be seen that the Yalin equation underestimates the transport capacity for smaller particles in particular, whereas the Govers equation generally predicts rather low values ($TC = 0$ in Fig. 11.2). This characteristic of the Govers equation which is used in EUROSEM for rill flow conditions was the reason to develop the Everaert formula from which the transport capacity of the interrill sheet flow is estimated. Schmidt's transport capacity expression is the one which shows the highest sensitivity to changes in the particle diameter. The equations of Schmidt and Govers are the only ones included in this comparison which have been developed for the more sheet flow-like conditions of agricultural soil erosion; all others are based on uniform particle size distributions of non-cohesive sediments, and fully turbulent channel flow conditions. It can be seen from the Eq. 11.20, 11.30, 11.37, 11.42 and 11.44 that all transport capacity estimates depend on threshold or critical values beyond which the sediment particles are mobilised or held in suspension. Due to the varying flow depths and velocities these values are usually very difficult to determine for raindrop-impacted, shallow overland flow (Everaert 1991).

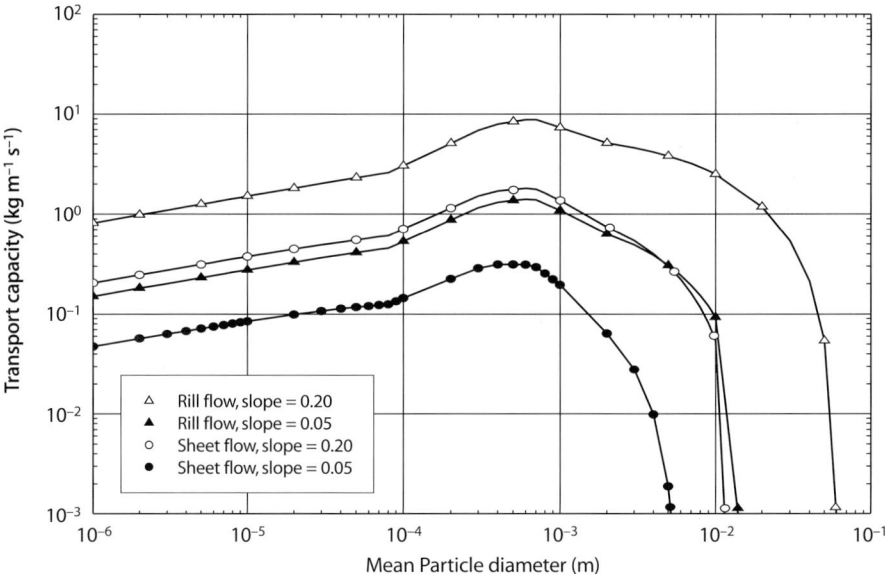

Fig. 11.3. Relationship between transport capacity and mean particle diameter as predicted by the Yalin (1977) formula. The equation is being applied to rill and sheet flow conditions. The values are calculated for a slope length of 200 m, a constant rainfall intensity of 30 mm h^{-1}, and a value of the Manning hydraulic roughness of 0.05 m s$^{-1/3}$. The mean slope gradient is given in the inset (Schramm 1994)

11.4
Sensitivity Analysis

Sensitivity analysis is a method for assessing the prediction errors of complex modeling systems. The soil erosion models investigated in this study are deterministic models in the sense that for a given set of identical input parameters identical results are produced regardless of the number of applications. In fact, the prediction error produced by complex modeling systems is usually influenced by all input and structural errors. Due to the intensive validation of the mathematical functions which are used in the models to describe the various processes of soil erosion it is assumed that the latter error is negligible. The prediction error may thus give hints to input errors which in our case may result mainly from estimation errors due to the inadequate representation of the spatial variability of the input parameter values (see Warrick and Nielsen 1980; Buchter et al. 1991; Immler and Zahn 1994 for a discussion of the variability of soil properties). The sensitivity analysis is used in this study as a procedure for estimating the change in the output with respect to changes in only one input parameter (McCuen 1973). With all other parameters held constant, this one-dimensional analysis assumes that the input parameters are independent of each other. This is usually not the case when considering the complex relationships in the soil-water system (e.g. among bulk density, soil texture and water content). However, this method allows the identification of those parameters which have a crucial impact on the model predictions, but it does not allow the identification of crucial parameter sets or constellations. For applicability and comparability reasons the parameter set of the sensitivity analysis was chosen from data for one Müncheberg (BW-5) and one Methau plot (71-T; see Sect. 1.5.1 for the location and numbering of the rainfall simulation plots).

11.4.1
Definition of Sensitivity Parameter

The sensitivity measure used in this study was proposed by Nearing et al. (1990). It is based on the deterministic sensitivity concept formulated by McCuen (1973). The measure is defined by the following expression:

$$S = \frac{O_2 - O_1}{O_{12}} \bigg/ \frac{I_2 - I_1}{I_{12}} \qquad (11.48)$$

where
- S = dimensionless sensitivity measure
- I_1 = least input value
- I_2 = greatest input value
- I_{12} = average of I_1 and I_2
- O_1 = output of argument I_1
- O_2 = output of argument I_2
- O_{12} = average of O_1 and O_2

S is a linear measure which represents the relative normalised change in the output to the normalised change in the related input (Fig. 11.4). The parameter is less than

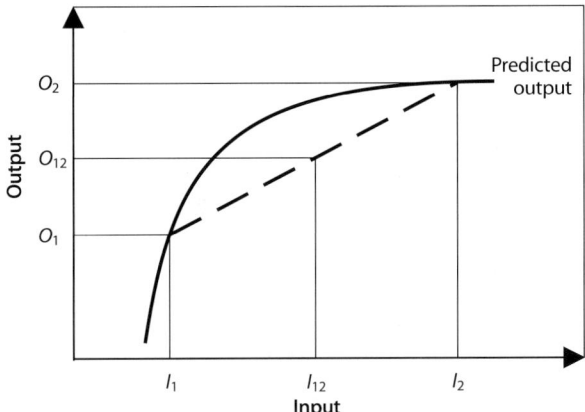

Fig. 11.4. Graphical representation of the sensitivity parameter S in Eq. 11.48. The output is more sensitive to variations in the input in the left than in the right region. S represents the average sensitivity as indicated by the dashed line

zero in case that an increase in input corresponds to a decrease in the output. As it is the primary purpose of WEPP, EUROSEM and E-2D to predict soil erosion, the total amount of soil loss, expressed in metric tons per hectare (1 t ha^{-1} = 1 000 kg/10 000 m^2), was chosen as the only output parameter from which the sensitivity measure was calculated. In most cases it was decided to calculate the sensitivity measure for the characteristic interval and not for the entire input-output range because of the apparent non-linear behaviour of the output curve, and in order to give an impression of the sensitivity for those ranges in which most of the input parameter estimates fall.

11.4.2
Results

The input parameters to which the predicted amounts of soil loss is most sensitive (i.e. which have a sensitivity value of $S \geq \pm 0.90$) are listed in Table 11.5. The compilation shows that most of these parameters influence both the runoff and soil loss predictions. Among these are those hydrological parameters that drive the model (i.e. rainfall intensity and duration), and those that determine the runoff generation

Table 11.5. Summary of the input parameters with a sensitivity value $S \geq \pm 0.90$. The parameters marked with an asterisk influence both runoff and soil loss (w/ = with, w/o = without, m_{sa} = sand content)

Rank	Parameter	Unit	Range	S
WEPP				
1	Rainfall intensity*	mm h^{-1}	25 – 60	1.92
2	Ridge roughness*	m	0.0 – 0.1	−1.87
3	Critical hydraulic shear ($m_{sa} \geq 30\%$, w/o rills)	Pa	1.5 – 5.0	−1.45
4	Baseline effective conductivity*	mm h^{-1}	3 – 15	−1.23

processes (i.e. initial soil water content, hydraulic conductivity, surface roughness). EUROSEM and E-2D are similar in that the initial soil water conditions have considerably higher sensitivity values than they have in WEPP. A characteristic of the WEPP model is the high number of input parameters with moderate to low sensitivity values ($S < \pm 0.90$). Seen in the context of the rather high number of input parameters (Table 11.1), it is likely that these parameters have a high impact on the model predictions as a whole. As expected those parameters which describe the soil erodibility in each model solely influence the predicted soil loss, but only in WEPP (hydraulic shear) and E-2D (erosion resistance) they cover high ranks. Soil properties have a rather high importance in WEPP due to the pedotransfer estimates of the critical hydraulic shear from which the rill erosion rate and – indirectly – the transport capacity of the rill flow are calculated (see Eq. 11.10, 11.20 and 11.31). Judged by the type of the input parameters with high sensitivity values, EUROSEM can be called the most hydrological model.

As it is the main objective of the sensitivity analysis to obtain information on model behaviour, it was decided to classify the input parameters accordingly to three main types of model reaction (Fig. 11.5, Table 11.6). The classification in Table 11.5 shows that the infiltration-related parameters such as initial soil moisture (WEPP, EUROSEM) and bulk density (E-2D) have a threshold-like impact on the runoff generation and thus soil erosion. This behaviour is in agreement with the findings of many long-term studies (cf. Boardman and Favis-Mortlock 1996) in which a certain threshold value has to be exceeded in order to produce a runoff. The input parameters which describe the soil erodibility have a high impact on the output over the entire input range (type 1) in WEPP and E-2D, but in EUROSEM they have a rather step-like influence (type 3). The plausibility of this model reactions is most difficult to assess because of the temporal changes of these parameters during a rainstorms due to aggregate destruction, soil surface sealing, microrelief lowering and the incision of concentrated flow paths (see Auerswald 1993 for description of the relevant processes).

An unexpected result of the sensitivity analysis is that certain input parameters have almost no impact (type 2) on the soil erosion predictions. Among these are the surface cover in E-2D, EUROSEM and WEPP, the side slope, depth and slope of rills, the

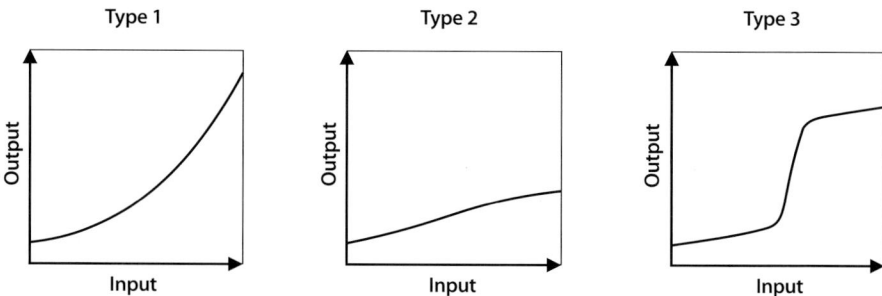

Fig. 11.5. The three main types of model output reactions: 1 = high sensitivity over the entire input range, 2 = moderate to low sensitivity over the entire input range, 3 = high sensitivity within a small input range

Table 11.6. Classification of the main input parameters (with sensitivity S) of WEPP, EUROSEM and E-2D according to the three types depicted in Fig. 11.5. The parameters marked with an asterisk influence both runoff and soil loss

Type 1	S	Type 2	S	Type 3	S
WEPP					
Rainfall intensity*	1.92	Organic matter content	−0.41	Ridge roughness*	−1.87
Critical hydraulic shear	−1.45	Interrill erodibility	0.28	Effective conductivity*	−1.23
Rainfall duration*	1.15	Cation exchange capacity*	0.25	Initial soil saturation level*	1.11
Slope steepness	1.07	Submerged residue mass	0.19	Slope length	1.09
Rill erodibility	0.83	Rock volume in topsoil	0.16	Rill width	0.88
		Interrill cover	−0.08	Rill spacing	−0.86
		Ridge height after tillage, dead root biomass	0.00	Rill cover	−0.81
		Bulk density	0	Cumulative rainfall since last tillage*	0.19
EUROSEM					
Maximum soil moisture content*	−4.52	Rock volume*	0.42	Initial soil moisture content*	2.79
Downslope roughness*	−1.37	Rill width*	−0.25	Rainfall intensity	2.32
Rainfall duration*	1.21	Manning rill hydraulic roughness	0.23	Slope steepness	1.14
Saturated hydraulic conductivity*	−1.06	Rill spacing	0.13	Manning interrill hydraulic roughness	−0.53
Net capillary drive*	−0.93	Impermeable surface cover	−0.06	Cohesion of soil-root matrix	−0.41
		Side slope of rills*	0.06	Soil detachability by rainfall	0.16
		Length of time step*	−0.01		
		Infiltration recession factor*	0.00		
		Soil porosity, rill depth, rill slope, interrill slope, across-slope roughness	0		
E-2D					
Rainfall intensity*	1.40	Slope length	0.67	Bulk density*	4.39
Rainfall duration*	1.07	Slope steepness	−0.55	Initial water content*	2.33
Erosion resistance	−1.00	Soil cover	−0.05		
Organic matter cont.*	−0.91				
Manning hydraulic roughness	−0.90				

profile length, the infiltration recession factor, the interrill slope towards the rills, the soil porosity and the across-slope roughness in EUROSEM, and the interrill erodibility, the ridge height, the bulk density, and all tillage-related parameters such as the ridge interval, the tillage depth and intensity in WEPP. This characteristic makes it difficult to use the event-based models for the assessment of the influences of different tillage and conservation practices on soil erosion.

Although the results of the sensitivity analysis depend on the input parameter set as a whole, the principal findings are in agreement with those of Nearing et al. (1990) for the WEPP model, of Albaladejo et al. (1994) for EUROSEM, and of Schmidt (1996) for E-2D. From a user's point of view the high sensitivity of the infiltration-related parameters is most difficult to deal with due to their high spatial and temporal variability even at the plot scale (Sisson and Wierenga 1981; Vieira et al. 1981; Alemi et al. 1988; Lauren et al. 1988), and the hitherto unknown algorithms for effective-parameter generation.

The reasons for the low sensitivity of many input parameters in WEPP and EUROSEM are unknown; it is likely that it results from an incomplete validation of the model algorithms. The findings sustain the suggestion of Troutman (1985) who recommends the use of sensitivity analysis not only for the testing of the basic model hypotheses, but also as a tool for identifying reasonable or illogical relationships among the variables used in a model. It cannot be concluded from the one-dimensional sensitivity analysis of this study whether the incorporation of a large number of input parameters with low sensitivity (WEPP) is superior to that of using a smaller number with higher sensitivities (EUROSEM, E-2D). Each sensitivity of the model parameters should be included in all model documents in order to provide the user with information on the model behaviour.

11.5
Model Application

The third step of the model assessment was the application of WEPP, EUROSEM and E-2D to simulated rainfall events in order to obtain a general impression of the accuracy of the model predictions and to identify possible sources of error in the process simulation. Another objective was to investigate to what extent the models are able to account for the impacts of different tillage and crop management systems on soil erosion. The problems associated with the use of a single-event model for such an assessment have been described above. In addition, the parameters which characterise common conservation practices such as continuous residue cover, reduced tillage intensity and depth, direct drilling and the growing of intermediate crops (Kahnt 1995; Buchner and Köller 1990) are very insensitive in WEPP, EUROSEM and E-2D. In nature, the described practices have the greatest impact on soil structure which – in the context of erosion – is mainly accounted for in each model by the soil erodibility parameters such as the erosion resistance in E-2D, the soil cohesion and rainsplash detachability in EUROSEM, and the critical shear, rill and interrill erodibility in WEPP. Because of these problems and in order to reduce the variability of the initial boundary conditions all plots were placed in seedbed condition with no residue cover before the start of the rainfall simulation.

WEPP, EUROSEM and E-2D were applied with no calibration to all dry runs of the rainfall experiments. The reasons for this approach were the missing dataset for calibration, missing information on which parameter to calibrate, the assumption that the amount of empirical information already contained in each model would result in reasonable predictions, and the assumption that – for a realistic simulation of natural rainfall events – the initial boundary conditions and governing processes as such would mainly determine the amount of soil loss. The input parameters determined from field samples were checked against the guide values or values calculated by the pedotransfer functions given in the user documents. As the models were developed for "every-day use" and in order to ensure realistic application conditions, it was tried to keep the expenditure for input parameter determination as low as possible.

11.5.1
Description of Test Plots

Eight rainfall simulations were carried out on the experimental sites of the Centre for Agricultural Landscape and Land Use Research (ZALF) in Müncheberg (ca. 50 km east of Berlin), and the Saxonian State Office for Agriculture (SLfL) in Methau (ca. 40 km south-east of Leipzig). These locations were chosen due to the different soil conditions and the nearby infrastructure which was required for the experiments.

The main characteristics of the experimental sites and plots are listed in Tables 11.7 and 11.8. General information on soil types and landscape characteristics can be found in Richter et al. (1970), Lieberoth (1982), Bork et al. (1995), Bork et al. (1997) for the Brandenburg, and in Lieberoth (1963), Haase (1978), and Mannsfeld and Richter (1995) for Saxony.

The rainfall simulators were provided by the ZALF and the Karlsruhe University (Schramm and Prinz 1993). Most input parameters were determined from samples taken before the start of the rainfall simulation. The rainfall, runoff and erosion rates were monitored during all experiments. The rill patterns and characteristics were mapped or measured after the cessation of the rainfall. The main results of the experiments are listed in Table 11.9.

11.5.2
Model Performance

The results of all simulation runs are given in Table 11.10. Because WEPP and EUROSEM allow the specification of rills, the rill geometry measured after the rainfall experiments was entered into the models in order to investigate the impact on the soil loss predictions. The (Govers or Everaert) transport capacity option in EUROSEM was also tested for the this reason. The bulk density measured after the experiments was used in E-2D in order to calculate runoff which was needed for the assessment of the soil erosion predictions. The accuracy of the model predictions was assessed for each experimental site by the efficiency coefficient R^2 suggested by Nash and Sutcliffe (1970) which was originally developped in order to assess the improvement in the model predictions due to parameter calibration. The dimensionless coefficient is defined as follows:

Table 11.7. Data summary for the Müncheberg and Methau experimental sites

Location	Müncheberg (Brandenburg)	Methau (Saxony)
Longitude, latitude	14°07'08" E, 52°31'40" N	12°52'17" E, 51°04'15" N
Mean annual precipitation	527 mm	693 mm
Mean temperature	8.2 °C	8.8 °C
Elevation about see level	ca. 55 m	ca. 280 m
Aspect of slope	North-west	South-east
Total slope length	Ca. 70 m	Ca. 160 m
Difference in elevation between ends	Ca. 9 m	Ca. 18 m
Slope steepness (top/middle/end)	3°/9°/5°	3°/10°/3°
Parent material	Fluvial sand over glacial till (Weichsel glaciation)	Eolian loess (Weichsel glaciation) over periglacial debris (Paleozoic porphyr)
Soil texture (USDA)	Sandy loam, loamy sand	Silt loam
Soil texture (KA 4, AG Boden 1994)	Slightly silty sand (Su2), medium loamy sand (Sl3)	Medium clayey silt (Ut3)
Main soil types (FAO 1994)	Cambic/Luvic Arenosol, Albic/Stagnic Luvisol	Haplic Luvisol, Gleyic Cambisol
Main soil types (AG Boden 1994)	Bänderparabraunerde, Pseudogley-Braunerde	Parabraunerde, Pseudogley-Parabraunerde
Main soil types (MMK, Lieberoth 1982)	Bändersand-Braunerde, Salmtieflehm-Fahlerde	Parabraunerde, Braunstaugley
Natural site unit (Lieberoth and Czwing 1989)	D3a-1	Lö5b-2
Soil score (Bodenschätzung)	S4D 31/33	L4Lö70/63
Type of primary tillage	Conventional (mouldboard plough, rotary plough)	Conservation (V-subsoiler, cultivator) and conventional (mouldboard plough)
Crops	Continuous fallow	Sugar beet-wheat-barley rotation

$$R^2 = \frac{\sum_{i=1}^{n}(x_i - \bar{x})^2 - \sum_{i=1}^{n}(x_i' - x_i)^2}{\sum_{i=1}^{n}(x_i - \bar{x})^2} \tag{11.49}$$

where
- R^2 = dimensionless efficiency measure
- i = event
- x_i = measured variable
- x_i' = simulated variable
- $\bar{x}$ = mean of all measured variables x_i from i to n

Table 11.8. Initial conditions of Müncheberg and Methau plots

Plot number	Unit	Müncheberg				Methau			
		BW-1	BW-3	BW-4	BW-5	69-T	71-T	73-T	75-T
Plot length	m	6	6	6	6	22	22	22	22
Plot width	m	2.5	2.5	2.5	2.5	2	2	2	2
Mean slope	°	9	5	10	6	10	10	10	10
Slope position	–	Middle	Top	Middle	Bottom	Middle	Middle	Middle	Middle
Texture (USDA)	%								
clay		11	5	9	3	15	15	15	15
silt		16	14	16	17	57	57	57	57
sand		73	81	75	80	28	28	28	28
Texture (KA 4)	%								
clay		11	5	9	3	15	15	15	15
silt		19	17	20	21	76	76	76	76
sand		70	78	71	76	9	9	9	9
Content of organic matter	g cm^{-3}	1.26	0.80	1.24	1.46	1.90	1.90	1.90	1.90
Initial bulk density	g cm^{-3}	1.24	1.20	1.21	1.21	1.32	1.19	1.14	1.37
Initial gravimetric moisture content	%	9.1	10.0	6.4	10.8	19.8	26.2	21.6	21.7
Volumetric rock content (≥20 mm)	%	0.1	0.1	0.1	0.1	0.1	0.0	0.0	0.0
Surface cover	%	0	0	0	0	0	0	0	0
Surface condition	–	Ploughed fallow				Seedbed with small ridges			
Depth of primary tillage layer	m	0.10	0.10	0.10	0.10	0.25	0.15	0.15	0.30
Days since last tillage	d	1	1	1	1	2	4	6	6

The coefficient increases up to 1 with improving model fit. If the R^2 is less than zero, the accuracy of the model predictions is less as if the mean of the measured values of x_i were used. The coefficient is very sensitive to the order of magnitude of the investigated events, i.e. R^2 approaches a value of 1 far earlier for the accurate simulation of large events than it does for small events. The efficiency values are summarised for the predicted runoff, soil loss and mean sediment concentrations in the Tables 11.12, 11.13 and 11.14. The results of the simulations are discussed in more detail in the following sections.

11.5.2.1
Runoff Predictions

The ranking of WEPP, EUROSEM and E-2D based on the Nash-Sutcliffe efficiency parameter is given in Table 11.11. Measured by the sign and amount of R^2 the general performance of all models is very low. The best runoff predictions are produced by

Table 11.9. Summary of the measured runoff and soil loss values for the Müncheberg and Methau plots. The accuracy of determination is about ±3% for the runoff volume and ±10% for the amount of soil loss (Müncheberg data reproduced with courtesy of ZALF, Methau data reproduced with courtesy of SLUG and SLfL)

Plot number	Unit	Müncheberg				Methau			
	–	BW-1	BW-3	BW-4	BW-5	69-T	71-T	73-T	75-T
Precipitation									
Volume	mm	15.3	16.3	18.3	19.9	36.1	35.0	36.7	34.9
Duration	min	21	22	25	28	60	60	60	60
Intensity	mm h^{-1}	43.8	44.4	43.8	42.6	36.1	35.0	36.7	34.9
Runoff									
Volume	mm	3.8	4.0	2.8	9.0	7.2	15.3	11.9	6.4
… as share of rainfall total	%	25	25	15	45	20	44	32	18
Time to runoff	min	9	4	10	4	26	18	21	26
Peak rate at end of dry run	mm h^{-1}	32	24	22	37	17	28	25	17
Infiltration									
Volume	mm	6.3	12.3	15.5	10.9	28.9	19.7	24.8	28.5
…as share of rainfall total	%	75	75	85	55	80	56	68	82
Final rate at end of dry run	mm h^{-1}	12	20	22	5	18	7	12	18
Final rate at end of wet run	mm h^{-1}	–	–	–	–	10	7	9	12
Soil loss									
Amount per unit area	t ha^{-1}	1.4	1.2	1.3	2.4	2.6	12.6	36.5	17.7
Approximate peak rate	g min^{-1}m^{-2}	22	12	23	19	13	44	195	80
Mean sediment concentration	g l^{-1}	36	30	45	26	36	83	308	276

EUROSEM for the Müncheberg plots, and by WEPP for the Methau experiments. The importance of the bulk density estimate for E-2D is emphasised by the first rank when using the consolidated value measured after the rainfall application. This behaviour is contradictory to the fact that freshly tilled fields are highly susceptible to soil erosion.

The figures in Table 11.11 show that all models have the tendency to underpredict runoff, with the only exception being the WEPP model for the silty soil of the Methau plots. Judged by sign and value of the mean difference between the measured and predicted runoff, this tendency is more clearly visible for the Müncheberg than the Methau

Table 11.10. Summary of measured and predicted runoff volumes, amounts of soil loss and mean sediment concentrations for the Müncheberg and Methau plots (w/ = with, w/o = without)

Plot number	Unit	Müncheberg				Methau			
	–	BW-1	BW-3	BW-4	BW-5	69-T	71-T	73-T	75-T
Measured values									
Precipitation volume	mm	15.3	16.3	18.3	19.9	36.1	35.0	36.7	34.9
Runoff volume	mm	3.8	4.0	2.8	9.0	7.2	15.3	11.9	6.4
Soil loss	t ha^{-1}	1.4	1.2	1.3	2.4	2.6	12.6	36.5	17.7
Mean sediment concentration	g l^{-1}	36	30	45	26	36	83	308	276
Predicted values									
WEPP									
Runoff volume	mm	2.6	2.1	3.9	4.1	13.0	15.2	15.3	16.0
Soil loss (w/o rills)	t ha^{-1}	2.2	0.8	2.8	1.7	42.9	51.1	46.4	45.3
Soil loss (w/ rills)	t ha^{-1}	2.9	0.9	2.2	1.8	7.9	47.2	45.0	41.1
Mean sediment concentration (w/o rills)	g l^{-1}	85	38	72	42	330	336	303	283
Mean sediment concentration (w/rills)	g l^{-1}	112	43	56	44	61	311	294	257
EUROSEM									
Runoff volume	mm	0.7	2.3	0.3	9.1	0	7.4	0.3	0
Everaert transport capacity equation									
Soil loss (w/o rills)	t ha^{-1}	3.4	2.7	1.5	46.2	0	37.1	1.5	0
Soil loss (w/ rills)	t ha^{-1}	0.6	0.4	0.2	6.7	0	8.9	0.7	0
Mean sediment concentration (w/o rills)	g l^{-1}	485	117	500	508	0	500	500	0
Mean sediment concentration (w/ rills)	g l^{-1}	86	17	67	74	0	120	233	0
Govers transport capacity equation									
Soil loss (w/o rills)	t ha^{-1}	0.2	0.02	0.05	3.7	0	12.2	0.3	0
Soil loss (w/ rills)	t ha^{-1}	0.6	0.4	0.2	6.7	0	8.9	0.7	0
Mean sediment concentration (w/o rills)	g l^{-1}	29	1	18	41	0	165	100	0
Mean sediment concentration (w/ rills)	g l^{-1}	86	17	67	74	0	120	233	0
E-2D									
Runoff volume (initial bulk density)	mm	0	0	0	0	0	0	0	0
Runoff volume (final bulk density)	mm	0	0	0	0	1.2	16.2	6.0	12.1
Soil loss (initial bulk density)	t ha^{-1}	0	0	0	0	0	0	0	0
Soil loss (final bulk density)	t ha^{-1}	0	0	0	0	0.07	1.8	1.4	1.8
Mean sediment concentration (initial bulk density)	g l^{-1}	–	–	–	–	–	–	–	–
Mean sediment concentration (final bulk density)	g l^{-1}	–	–	–	–	6	11	23	15

Table 11.11. Ranking of soil erosion models based on the Nash-Sutcliffe efficiency coefficient of the predicted runoff. The mean difference (%) between the normalised predicted and the normalised measured runoff (= 100% by definition), and the coefficient of variation (CV) are given for comparison. The rows in italics refer to the runoff value obtained when using the final bulk density as input for the E-2D calculations

Rank	Müncheberg				Methau			
	Model	R^2	Δ%	CV	Model	R^2	Δ%	CV
1	EUROSEM	0.19	−54	79	E-2D	0.37	−9	72
2	WEPP	−0.30	−24	49	WEPP	0.17	+65	35
3	E-2D	−4.13	−100	–	EUROSEM	−0.76	−88	21
4	*E-2D*	*−4.13*	*−100*	*–*	*E-2D*	*−1.84*	*−100*	*–*

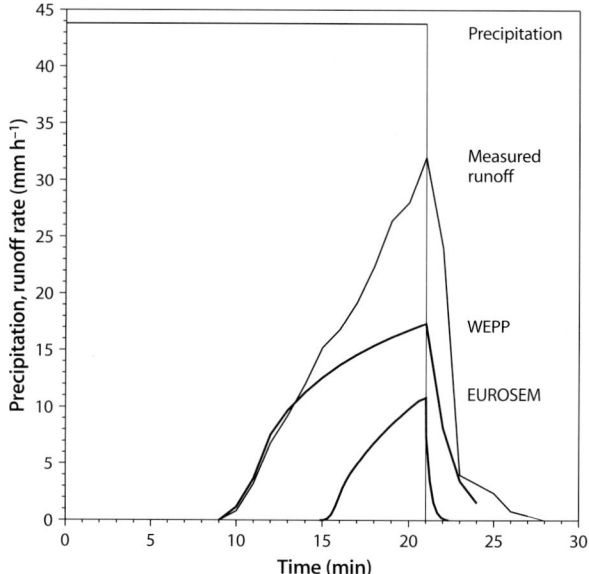

Fig. 11.6. Comparison of measured and predicted runoff hydrographs of the Müncheberg Plot BW-1

plots. Unfortunately this assessment is blurred by the circumstances that both soil texture *and* rainfall duration differed between the two locations. EUROSEM has a stronger tendency towards the underprediction of runoff than WEPP as it is indicated by the mean difference between the measured and simulated runoff. The zero runoff predictions of E-2D when using the initial plot conditions are somewhat surprising. The reasons for this model behaviour are the very low bulk density values of the uncompacted soils (J. Schmidt, pers. comm.) and the low initial moisture contents which result in an overestimation of the saturated hydraulic conductivity by a factor of about 12 for the Müncheberg plots, and a factor of 4 for the Methau plots. When entering the bulk density values measured after rainfall application, the predicted runoff for the Methau plots

Table 11.12. Predicted time to runoff and infiltration rate at the end of the rainfall simulation. All figures are expressed as percent of the measured values

	Müncheberg				Methau			
	BW-1	BW-3	BW-4	BW-5	69-T	71-T	73-T	75-T
WEPP								
Time to runoff	111	325	100	250	53	67	57	31
Final runoff rate	52	67	86	49	81	82	96	141
EUROSEM								
Time to runoff	156	225	160	125	–	122	238	–
Final runoff rate	83	80	23	560	–	242	33	–
E-2D								
Time to runoff	–	–	–	–	–	–	–	–
Final runoff rate	–	–	–	–	–	–	–	–

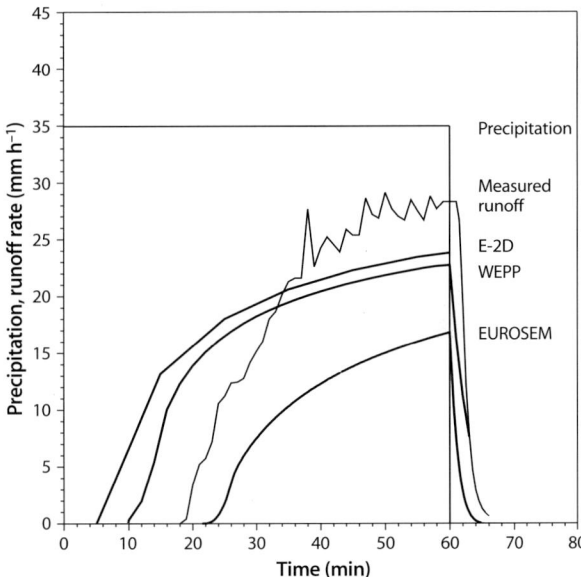

Fig. 11.7. Comparison of measured and predicted runoff hydrographs of the Methau Plot 71-T. Note the different scales of the time axes in Fig. 11.6 and 11.7.

almost equals the measured values. Table 11.11 shows that no increase in the accuracy of the runoff predictions for the Müncheberg plots is obtained by this method.

The reasons for the low accuracy of the runoff predictions are indicated by the plotting of the observed and simulated runoff hydrographs in Fig. 11.6 and 11.7, and the comparison between those parameters which mainly characterise the difference between the simulated and observed runoff curves (Table 11.12). A major difference between the predicted and measured curves is the more rectangular shape of the measured ones which consist of a rather steep initial section and a more flat limb when approaching saturation. The steep slope of the measured hydrograph of the Müncheberg plot indicates that the soil was still very much unsaturated at the end of the rainfall application. All simulated hydrograph are very similar in their initial slope and the following exponential decrease of the infiltration capacity. The generally low figures of the simulated final runoff rates in Table 11.12 indicate that steady-state runoff conditions are reached earlier in the experiments than in the simulations, or that the saturated hydraulic conductivity – and consequently the final infiltration rates – are overestimated by the models. The approach in EUROSEM which leaves the estimation of the saturated hydraulic conductivity to the user is less helpful for such cases as can be seen in the runoff predictions for the Methau Plots 69-T and 75-T: From the measured low runoff totals (Table 11.10) and the observed final infiltration rates of these plots it was concluded that the saturated conductivity would have been rather high. In fact, this estimate – together with the high sensitivity of the conductivity parameter and the low initial moisture – resulted in zero runoff predictions for these plots. The shape of the observed runoff hydrographs indicate that additional processes such as surface sealing or the saturation of large pores have a major impact on the runoff properties of the plots.

Except for WEPP (for the Methau plots) the models have the tendency to overpredict the time to runoff. This behaviour is mainly influenced by the low initial moisture content, the low bulk density and high surface roughness values due to the recent tillage, and the high impact of these parameters on the model predictions (see sensitivity values in Table 11.6). Because of the calculation of the available porosity mainly from the texture of the Methau silt loam, WEPP underpredicts the time to runoff for the these plots. Because soil porosity and initial moisture content have to be estimated directly for EUROSEM, deviations in the predicted time to runoff are most likely due to the initial moisture estimates (as the model is insensitive to changes in porosity; see Table 11.6). The magnitude of the deviations of the predicted runoff values is similar to that reported in Albaladejo et al. (1994).

11.5.2.2
Soil Loss Predictions

The ranking of the WEPP, EUROSEM and E-2D based on the Nash-Sutcliffe efficiency measure of the predicted soil loss is presented in Table 11.13. The decrease or increase in the accuracy of the model predictions due to the specification of rill (WEPP, EUROSEM) and bulk density (E-2D) information can also be assessed from this table. Despite of the close relationship between the runoff and soil loss predictions it can be seen that – judged by R^2 alone – the soil loss predictions are – with one exception – much better than the runoff simulations for the Müncheberg plots, and that they are – seen as a whole – worse than the runoff predictions for the Methau plots (see Table 11.11). However, when the models are assessed by the accuracy of the predicted soil loss only, the dependence of the predicted soil loss on the predicted runoff volume is ignored. It was therefore decided to use the mean sediment concentration (total soil loss normalised to runoff volume) as the criterion for the quality assessment of the erosion predictions. The possible non-linear relationship between runoff and soil loss

Table 11.13. Ranking of soil erosion models based on the Nash-Sutcliffe efficiency R^2 of the predicted soil loss. The mean difference (%) between the normalised predicted and the normalised measured soil loss (= 100% by definition), and the coefficient of variation (CV) are given for comparison. The rows in italics refer to the soil loss values obtained when using final bulk density as input for the E-2D calculations, and the rill characteristics measured after the rainfall for WEPP and EUROSEM (w/ = with, w/o = without)

Rank	Müncheberg Model	R^2	Δ%	CV	Methau Model	R^2	Δ%	CV
1	WEPP, w/ rills	0.92	+34	44	E-2D	−0.31	−93	70
2	WEPP, w/o rills	0.88	+27	47	EUROSEM, w/ rills, (E, G)	−0.32	−82	167
3	EUROSEM, w/o rills, (G)	0.82	−57	151	EUROSEM, w/o rills, (G)	−0.33	−75	171
4	E-2D	0.70	−100	–	E-2D	−0.48	−100	–
5	E-2D	0.70	−100	–	WEPP, w/ rills	−0.49	+165	35
6	EUROSEM, w/ rills, (E, G)	0.37	−6	115	EUROSEM, w/o rills, (E)	−0.75	−25	170
7	EUROSEM, w/o rills, (E)	−54.47	+529	119	WEPP, w/o rills	−0.92	+510	100

E   Everaert transport capacity equation;
G   Govers transport capacity equation.

is accepted because of the initial decision to renounce the runoff calibration before the carrying-out of the soil loss simulations due to missing information on the respective parameters, and the likely ignorance of mutual interdependencies among the variables of the soil-water system. The ranking of WEPP, EUROSEM and E-2D based on the efficiency measure of the sediment concentration is given in Table 11.14.

The ranking in Table 11.14 shows that the best predictions for the Müncheberg plots are produced by EUROSEM and WEPP when using the initial plot conditions. The high ranks for the Methau plots are also covered by these models, but only when the rill information obtained from the measurements after the rainfall experiments is used. The signs of the mean difference (%) between the measured and predicted sediment concentrations show that WEPP and EUROSEM have the tendency to overestimate the sediment concentration and thus the total amount of erosion. This tendency is also illustrated by those cases in Table 11.10 in which the good runoff predictions result in usually high overestimations of the soil loss (e.g. WEPP for 71-T, EUROSEM for BW-5). The tendency of the Govers formula to underestimate the transport capacity (see Sect. 11.3.2.5) is sustained by the sign of the mean difference in the sediment concentration predictions of EUROSEM for both experimental sites.

On the other hand, the Everaert transport capacity equation highly overestimates the sediment concentration for sheet flow conditions. This is clearly demonstrated by the high sediment concentration values in Table 11.10 which reach a maximum of about 500 g l^{-1} in all simulations (regardless of the differences in soil detachability), and by

Table 11.14. Ranking of soil erosion models based on the Nash-Sutcliffe efficiency R^2 of the predicted mean sediment concentration. The mean difference (%) between the normalised predicted and the normalised measured concentration (= 100% by definition), and the coefficient of variation (CV) are given for comparison. The rows in italics refer to the sediment concentration values obtained when using final bulk density as input for the E-2D calculations, and the rill characteristics measured after the rainfall for WEPP and EUROSEM (w/ = with, w/o = without)

Rank	Müncheberg					Methau			
	Model	R^2	Δ%	CV		Model	R^2	Δ%	CV
1	EUROSEM, w/o rills, (G)	0.48	−30	83		WEPP, w/ rills	0.69	+84	62
2	WEPP, w/o rills	0.06	+71	24		EUROSEM, w/ rills, (E, G)	0.51	+10	55
3	EUROSEM, w/ rills, (E, G)	−0.49	+83	48		EUROSEM, w/o rills, (G)	0.26	−24	65
4	WEPP, w/ rills	−0.75	+86	40		WEPP, w/o rills	0.13	+275	87
5	EUROSEM, w/o rills, (E)	−176.66	+1101	47		E-2D	0.10	−89	43
6	E-2D	–	–	–		EUROSEM, w/o rills, (E)	−0.67	+282	65

E Everaert transport capacity equation;
G Govers transport capacity equation.

the mean difference (+1101%, +282%) in Table 11.14. The high negative value of the Nash-Sutcliffe coefficient for the Müncheberg plots indicates the sensitivity of this measure to the overestimation of the smaller events when using the Everaert formula. The reason for the overproduction is that – for sheet flow conditions with no rills – the rate of soil detachment is solely controlled by rainfall intensity, which was in all cases sufficiently high to produce high detachment rates throughout the experiments. The predicted amount of soil loss was thus limited by the "available" runoff rate only. The strong impact of the transport capacity formulae on the erosion calculations is depicted in Fig. 11.8 for the Methau Plot 71-T. The finding indicates that the critical stream power in the Everaert transport capacity equation (Eq. 11.42) should be further investigated.

The amount of soil loss predicted with no rills is higher than that with rills regardless of the chosen transport capacity formulae. The reason for this behaviour is presumably the increased rill erosion rate which, in turn, leads to the exceedance of the runoff transport capacity after short flow travel lengths – due to the small size of the rills – and thus accalerated deposition.

The reasons for the low concentration predictions of E-2D are mainly the low runoff depth predictions, and the high sensitivity of the detachment calculations to the overestimated erosion resistance. Because only very few guidevalues were available for the latter parameter at the time of the simulation, the actual value was estimated from previous rainfall experiments which were carried out earlier on the same plots but on consolidated and crusted soil surfaces.

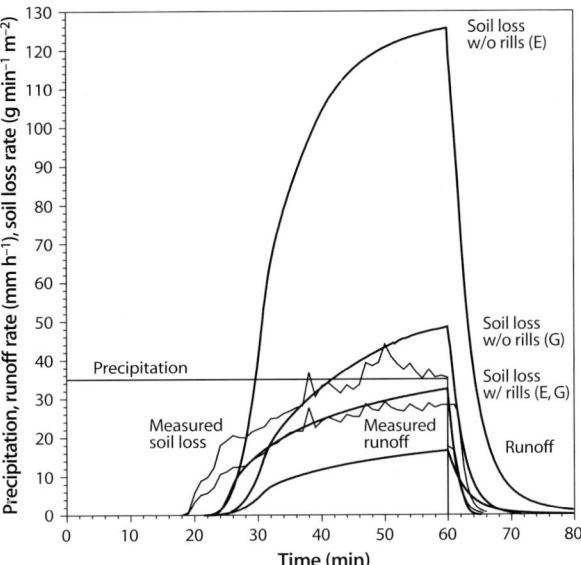

Fig. 11.8. Impact of Govers (*G*) and Everaert (*E*) transport capacity equations on the sediment discharge hy-drographs predicted by EUROSEM for the Methau Plot 71-T

The sediment and thus soil erosion predictions of EUROSEM and WEPP show a clear dependence on the soil texture (Table 11.10). WEPP produces rather accurate estimates for the two plough-tilled Methau plots (73-T, 75-T) which is mainly due to the reasonable runoff, erodibility and transport capacity calculations. For the Methau plots the rill, interrill erodibility and critical hydraulic shear values were calculated from the same soil texture, but the high overestimation of the mean sediment concentration for the subsoiler- (69-T) and cultivator-tilled plots (71-T) indicate that the erodibility parameters and their respective transfer functions are more valid for conventionally- than conservation-tilled soils (see compilation of WEPP erodibility experiments in Elliot et al. 1989). An interesting result is the decrease in the overestimation of the sediment concentration by WEPP for the Methau plots when entering the rill properties (Tables 11.10, 11.14). It is not clear whether this behaviour results from the default estimate of the rill width and spacing when no rills are specified, or from an increased deposition rate due to the increased erosion of the silty soil.

The spatial distribution of the net soil loss predicted by WEPP and E-2D is depicted in Fig. 11.9. For WEPP the increasing amount of soil loss per unit slope area indicates that the maximum transport capacity increases along the profile, whereas it is rather constant in E-2D due to the continuous balance of erosion and deposition. The sharp increase in the net erosion after a certain travelling length of the flow is predicted by both models, but the fluctuations in the lower section of the WEPP curve indicate abrupt changes from detachment- to transport-capacity limited regimes, and vice versa. The high net erosion estimate predicted by WEPP is the reason for the large difference in the predicted soil loss totals, which is 47.2 t ha^{-1} for WEPP and 1.8 t ha^{-1} for E-2D.

Because of their importance for the leaching of nutrients and pesticides, the sediment texture of the runoff samples was analysed and compared to those predicted by

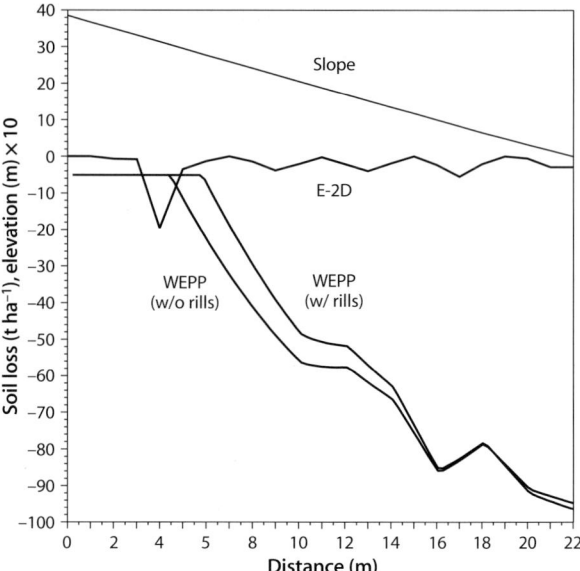

Fig. 11.9. Spatial distribution of net soil loss predicted by WEPP and E-2D for the Methau Plot 71-T. The runoff volumes are 15.2 and 16.2 mm. The calculations of E-2D are based on final bulk density

WEPP and E-2D (Table 11.15). The figures show that the enrichment of finer particles (i.e. clay) due to selective particle detachment and transport is not adequately simulated in WEPP because of the close similarity of the predicted and observed sediment compositions. The sediment enrichment ratio predicted by this model equals or is only slightly greater than 1. The composition of the eroded sediment is independent of whether rills are specified or not.

Despite the fact that E-2D ignores the sand content and overpredicts the silt content for the Müncheberg soils, the clay enrichment is predicted quite accurately for these soils. The high sediment enrichment potential especially of the Müncheberg loam is clearly demonstrated by the fact that – for all plots – the rainfall-detached soil (splash) has almost the same particle distribution as the parent soil. The actual sediment enrichment on the Müncheberg soil is caused by the selective flow detachment (and transport) which is facilitated by the low soil cohesion due to the high sand content. Similar results are reported in Foster et al. (1985) and Durnford and King (1993).

11.5.3
Assessment of Tillage Effects for the Methau Plots

Although there are certain limitations to the use of a single-event models for the assessment of conservation practices, the results of the Methau rainfall experiments indicate a close relationship among the type of tillage and the runoff and soil erosion characteristics of each plot. The comparison is facilitated by the very similar initial experimental conditions (i.e. slope, soil texture, rainfall intensity and duration, soil surface condition, see Table 11.8), and the long application (since 1992) of the different primary tillage practices to the Methau soil which ensures their influence on the soil properties. The first remarkable result is the low runoff volume of the subsoiler-tilled

Table 11.15. Comparison of the measured textural composition of the parent soil, the rainfall-detached sediment (splash), the sediment in runoff, and the predicted sediment for the Müncheberg and Methau plots. The E-2D simulations are based on final soil water content and bulk density in order to obtain runoff for the Müncheberg and Methau plots

	Müncheberg				Methau			
	BW-1	BW-3	BW-4	BW-5	69-T	71-T	73-T	75-T
E-2D (DIN clay/silt/sand)								
Parent soil	11	5	9	3	15	15	15	15
	19	17	20	21	76	76	76	76
	70	78	71	76	9	9	9	9
Measured rainfall-detached sediment composition	4	0	4	4	15	14	13	13
	21	11	18	15	79	80	83	84
	75	89	78	81	6	6	4	4
Measured runoff sediment composition	43	15	34	30	12	16	15	15
	48	47	43	48	82	79	80	82
	9	38	23	22	6	6	5	3
Predicted runoff sediment composition	35	21	30	13	15	15	17	17
	62	18	68	85	78	78	82	82
	3	1	2	2	7	7	4	4
WEPP (USDA clay/silt/sand)								
Parent soil	11	5	9	3	15	15	15	15
	16	14	16	17	57	57	57	57
	73	81	75	80	28	28	28	28
Measured rainfall-detached sediment composition	4	0	4	4	15	14	13	13
	19	10	16	13	70	71	74	74
	77	90	80	83	15	15	13	13
Measured runoff sediment composition	43	15	34	30	12	16	15	15
	46	45	41	45	82	09	80	82
	11	40	25	25	6	5	5	3
Predicted runoff sediment composition	11	5	9	3	15	15	15	15
	17	15	19	20	58	58	58	58
	72	80	72	77	27	27	27	27

(69-T) and deeply-ploughed plot (75-T) which have the highest initial bulk density values of all plots. The high unsaturated hydraulic conductivity is mainly a result of the high number of macropores which are produced by biologic activity and the loosening, non-inverting tool on Plot 69-T, and of the cracks produced by the plough in the soil matrix on Plot 75-T. The low runoff volume is only plausible with respect to the moisture storage capacity of the larger tillage depth although the total difference among the times to runoff on all Methau plots was less than seven minutes. The gravimetric or preferential flow in the macropores is not accounted for in the matrix infiltration approaches of all models, and it is usually left to the user to assess the impacts on the hydraulic conductivity. The tendency of all models to underpredict the runoff volume on all plots is mainly caused by this shortcoming.

The second surprising outcome of the Methau experiments are the large differences between the mean sediment concentration values of the conservation- (69-T, 71-T) and those of the conventionally-tilled plots (73-T, 75-T, see Table 11.10). The reason for the high values of the latter plots is the higher erodibility of the surface soil due to the disturbing impact of the turning plough on the soil structure. Although this soil characteristic is difficult to account for through a single parameter, the differences are fairly clearly reflected in the soil cohesion and rainfall erodibility indices of EUROSEM, and the erosion resistance of E-2D (Table 11.16). The stabilising effect of conservation tillage on soil structure due to the reduced tillage intensity is more difficult to account for by the critical hydraulic shear, rill and interrill erodibility parameters used in the WEPP model which are mainly calculated from soil texture. The guide values given in the model documents should be presented with more information on the boundary conditions under which they were determined for better assistance of the user in such cases.

The last finding which has to be emphasised here is the shape of the sedigraphs of the two plough-tilled Methau Plots 73-T and 75-T which clearly indicate detachment-limited erosion conditions. In all other experiments the positive slope of the sedigraph always closely follows that of the hydrograph which indicates the prevalence of transport capacity-limited erosion conditions. The reasons for the decrease in the soil loss rate which occurred on both plots approximately 20 minutes after the onset of runoff are the sealing of the soil surface due to aggregate breakdown, and the decreasing incision of the rills which had formed approximately 10 minutes earlier.

Table 11.16. Summary of soil erodibility parameters. The erosion resistance of E-2D was calculated from model calibration. The EUROSEM parameters were measured in the field with splash cups and a torvane. The WEPP erodibility and shear values are based on the pedotransfer functions suggested in the user documents. The guide values for EUROSEM are from Morgan et al. (1993), and from Flanagan (1994) for WEPP

	Unit	Müncheberg				Methau			
		BW-1	BW-3	BW-4	BW-5	69-T	71-T	73-T	75-T
E-2D									
Erosion resistance	10^{-4} N m^{-2}	4.10	0.27	0.80	0.67	1.70	1.10	0.45	0.30
EUROSEM									
Soil erodibility to raindrop impact	g J^{-1}	1.96	3.80	4.18	2.61	0.48	0.86	1.15	0.98
Guide values	g J^{-1}	1.7–3.1	1.9–4.0	1.7–3.1	1.9–4.0	0.8–2.3			
Soil cohesion	kPa	4.21	2.81	4.86	3.99	5.29	5.72	3.99	2.91
Guide values	kPa	2–4	2–3	2–4	2–3	2–5			
WEPP									
Interrill erodibility	10^6 kg s m^{-4}	4.73	4.65	4.68	4.73	5.22			
Guide values	10^6 kg s m^{-4}	3.41–4.98				5.08			
Rill erodibility	10^{-3} s m^{-1}	9.0	13.8	9.0	7.8	13.4			
Guide values	10^{-3} s m^{-1}	5.3–10.2				12.1			
Critical hydraulic shear	Pa	2.72	2.40	2.65	2.27	3.50			
Guide values	Pa	2.5–3.2				3.50			

11.6
Model Assessment

An assessment of each model based on selected criteria is presented in Table 11.17. The given criteria were found to be the most important ones among others such as model plausibility or appropriateness of the fundamental model equations. Each ranking is briefly discussed in the following.

The user and technical manuals are probably the most important documents for "coming to grips" with a computer model. The first rank in this category was assigned to E-2D because of the detailed description of the parameter estimation procedures for various boundary conditions, and the documentation of the basic model equations (Schmidt et al. 1996). The last ranks of WEPP (user manual, Flanagan and Livingston 1995) and EUROSEM (technical documentation, Morgan et al. 1992; Morgan 1994) are due to the insufficient description of the application of the single event mode in WEPP, and the insufficient description of the fundamental model equations in the EUROSEM technical documents. Most of this information was obtained from B. Diekkrüger (personal communication).

The first rank in model handling is ascribed to WEPP because of its user interface programme which allows the entry, retrieval and editing of all data via text editors, file builders and on-screen help features from within the model. Only slightly less user support is provided by E-2D, although the model has to be quit before assessing the input and output files. EUROSEM has almost no interface, and all input and output files are stored in the same directory with the core module.

Table 11.17. Assessment of WEPP, EUROSEM and E-2D based on selected criteria and comparative ranking

	Rank		
	WEPP	EUROSEM	E-2D
User manual	3	2	1
Technical model documentation	2	3	1
Model and data handling	1	3	2
Number of input parameters	3	2	1
Number of input parameters with high sensitivity ($S \geq \pm 0.90$)[a]	3	2	1
Accuracy of Müncheberg runoff predictions[b]	2	1	3
Accuracy of Methau runoff predictions[b]	1	2	3
Accuracy of Müncheberg sediment concentration predictions[b]	2	1	3
Accuracy of Methau sediment concentration predictions[b]	2	1	3
Representation of tillage systems[c]	1	1	1
Mean	2.0	1.8	1.9

[a] Sensitivity S with respect to impact of input changes on soil loss (Eq. 11.48);
[b] Ranking based on Nash-Sutcliffe coefficient (Eq. 11.49);
[c] Ranking of WEPP is based on the single-event version.

The number of input parameters is a fundamental criterion which has a high influence on the decision about whether a model is considered for application or not. Despite of the fact that many parameters are not needed for the WEPP event simulation, one cannot ignore or delete these parameters because they are used for internal checks of the consistency of the whole input parameter set. The model applications also suggest that a high number of parameters does not necessarily result in a higher accuracy of the model predictions due to the better representation of the physical processes or boundary conditions. With respect to the numbers of input parameters, E-2D has a clear first rank in this row.

The number of input parameters with high sensitivity measures ($S \geq 0.90$) determines to what extent particular expenses for parameter surveying should be considered. It should be noted that the highest sensitivity measure of the WEPP input parameters is about half of that found for EUROSEM and E-2D. Very sensitive parameters and their possible impact on the model predictions should be indicated and discussed in the model documents. As the predictions of all models are very sensitive to changes in the infiltration-related parameters, users may be confronted with particular difficulties when estimating and assessing the hydrological properties of the modelled soil-water system. Because of the low number of very sensitive parameters, E-2D was ascribed the first rank in this category.

The accuracy of the model predictions is a fundamental criterion for assessing the relevance of the simulated processes, and thus the performance of each model. However, the low number and the high variability of the model predictions on which the ranking is based in this study are of only limited value when the behaviour or the plausibility of a simulation system are to be assessed. When considering the dependence of the model predictions on the initial conditions (e.g. with or without rill specification), and the decisions left to the user (e.g. transport capacity formula), which had a considerable impact on the ranks of WEPP and EUROSEM for the Müncheberg and Methau plots, it is very difficult to use assess the models on the basis this criterion. The main conclusion to be drawn from this part of the study is the need for a large number of model predictions in order to arrive at a more sound assessment of erosion, but this approach is contradictory the time- and cost-demanding procedures required for the determination of the input parameters for these models.

Due to the limited number of parameters by which the *specific* characteristics of tillage systems are accounted for, the single-event models are not recommended for assessing the effects of different tillage systems on soil loss. The soil properties (i.e. soil structure, types of soil pores) responsible for the considerable differences among the observed hydro- and sedigraphs on the Methau plots are only indirectly represented by the input parameters of each model. Because of these difficulties all models were given the same rank in Table 11.17. These results also show the clear need for more detailed experimental investigations of the behaviour of the soil-water systems especially with respect to the long-term effects of these implements.

The averages of the individual rank positions in Table 11.17 show that the various model properties are almost being balanced out against each other. The individual criteria were not given any weighting for it was felt that all of them are of equal importance when the application of a model is considered. However, the distribution of the high and low ranks shows the different strengths and weaknesses may require different approaches to "come to grips" with each model.

11.7
Summary and Conclusions

The basic problem with which this study is concerned is how and with what accuracy natural processes can be simulated by physically-based modeling systems. The soil erosion models which were compared and assessed in this study are the single-event and profile versions of the USDA-Water Erosion Prediction Project model (WEPP), the European Soil Erosion Model (EUROSEM), and EROSION-2D (E-2D).

The approach of the study can be divided into three parts. The first step is the comparison and discussion of the individual model concepts and fundamental equations. The conceptual representation of erosion as a sequence of infiltration, Horton runoff generation, particle detachment, transportation and deposition processes as the basis of modern, event-based soil erosion modeling is emphasised. However, all models use different approaches in the way to describe these processes. An example are the infiltration parameters which are calculated mainly from pedotransfer functions in WEPP and E-2D but which have to be directly estimated by the user in EUROSEM, or the different approaches to the description of the runoff movement (kinematic wave) along the hillslope. Considerable differences exist among the number of the required input parameters for each model which is mainly a result of whether the model was developed for uncomplicated use (E-2D), the comprehensive representation of the main processes of erosion (EUROSEM), or the assessment of the long-term effects of different tillage and conservation systems (WEPP). Despite of the differences in the individual model components (e.g. rainfall erosivity, infiltration, particle detachment, transport capacity) the theoretical analysis showed the similar dependence of these approaches on fundamental assumptions such as the raindrop diameter and velocity distributions, the soil flow regime, and a threshold value above which the soil particles are detached by overland flow.

The most important finding of the one-dimensional sensitivity analysis which was performed as the second step of the investigation is the high sensitivity with which the runoff and soil-loss predictions depend on the infiltration-related parameters. This behaviour is of particular importance because of the close relationship between these two parameters. Other crucial parameters which have rather high impact on the accuracy of the predicted amount of soil loss are rainfall intensity and duration for all models, erosion resistance and surface hydraulic roughness in E-2D, across-slope roughness for EUROSEM, slope steepness in EUROSEM and WEPP, and ridge roughness and hydraulic shear in WEPP. Generally, the sensitivity of WEPP and EUROSEM to the parameters describing rill geometry is surprisingly low which is presumably due to their small rill widths and depths.

The third step in the study was the application of the (uncalibrated) models to simulated rainfall events in sandy and silty plots near Berlin (Müncheberg) and Leipzig (Methau). Judged by the Nash-Sutcliffe efficiency parameter, the most accurate runoff predictions are produced by WEPP for the Methau silt loam, and by EUROSEM for the Müncheberg sandy loam.

Judged by the mean difference between the observed and predicted values, all models have the moderate (WEPP) to strong tendency (EUROSEM, E-2D) to underpredict the runoff volume. E-2D predicts no runoff for all plots when using initial bulk density which leads to a high overprediction of the saturated conductivity rate. The strong influence of the infiltration-related parameters on the runoff predictions is demonstrated by the fact that E-2D produces the most accurate runoff predictions for the Methau plots when using final bulk density.

Due to the large differences in the measured and predicted runoff volumes, the predicted amounts of soil loss cannot be directly compared to the measured ones. For comparison, the soil loss totals were divided by the respective runoff totals, thus obtaining the mean sediment concentration of the runoff as a rough measure of the erosive "efficiency" of runoff. This approach assumes that the runoff-soil loss ratio remains constant for different runoff volumes. Except from the plough-tilled Methau plots (73-T and 75-T), the sediment concentrations predicted by WEPP are about two times higher than the measured ones (average of all plots). The predicted concentrations were lower if rills were specified. The sediment concentrations predicted by EUROSEM show a strong dependence on the chosen transport capacity equation. For a non-rilled soil surface the concentrations predicted by the Everaert transport capacity equation are about 2 to 20 times higher than the measured concentrations. The Govers transport capacity equation predicted more realistic values for all plots. The latter equation is used by default when rills are specified. The sediment concentrations predicted by E-2D were considerably lower than those predicted by WEPP and EUROSEM. Almost all overpredictions of the amount of soil loss are due to the overprediction of sediment transport capacity because runoff was usually under-predicted. The two exceptions are E-2D, and EUROSEM when using the Govers transport capacity equation. Judged by the Nash-Sutcliffe efficiency coefficient, the accuracy of the WEPP soil loss predictions increases only for the Methau plots if rills are considered in the simulation runs. The best soil loss estimates of EUROSEM are produced for the Müncheberg plots when not specifying the rill geometry but using the Govers transport capacity equation (which was originally developed for rills). The fluvial sorting of grains which resulted in the relative enrichment of finer particles in the sediment of the Müncheberg plots was more accurately predicted by E-2D than by WEPP.

An increase in the accuracy of the runoff and soil loss predictions either due to specification of rills or the application of pedotransfer functions for predicting basic infiltration parameters was not observed. The high degree of variation in the individual model predictions indicate that the accuracy of the prediction is influenced far more by other factors than these. The rather large numerical differences among the measured and predicted runoff and soil loss values, and among the individual model predictions are the most surprising findings of this study.

The following conclusions can be drawn from the findings of this study. First, the results of both the sensitivity analysis and the model applications indicate that most of the problems associated with the accurate simulation of soil erosion are related to the mathematical and processual treatment of unsaturated flow through soils. Judged by the differences in the observed and predicted runoff hydrographs and totals, the various factors which control the formation of runoff are less well represented in each model. The second conclusion is the fact that despite it is the objective of WEPP, EUROSEM and E-2D to predict soil erosion an agricultural land, the key relationships among those parameters which describe the structure of soil (e.g. density, porosity, erodibility, content of organic matter) are only poorly accounted in the input parameter estimation procedures or guide values. Especially the somewhat contradictory relationship between the (high) bulk density and infiltration capacity of conservation-tilled soils deserves further research. The third conclusion is that sensitivity analysis should be more applied during model development in order to identify and assess the individual contribution of each input parameter or functional relationship to the accuracy of the predicted variables. The last conclusion to be drawn from the experimental part of this study is the need for more systematic investigations of those (few?) crucial parameters which determine the efficiency of integrated soil erosion protection measures.

Acknowledgements

The Methau rainfall experiments were jointly conducted and funded by the Saxonian State Office of Environment and Geology (SLUG) at Freiberg, and the Saxonian State Office of Agriculture (SLfL) at Leipzig. The Müncheberg experiments were conducted and funded by the Centre for Agricultural Landscape and Land-Use Research (ZALF) in Müncheberg. The carrying-out of this study would not have been possible without the support of Walther Schmidt and Detlef Deumlich.

References

AG Boden (1994) Bodenkundliche Kartieranleitung, 4th edn. Schweizerbart'sche Verlagsbuchhandlung, Stuttgart
Albaladejo J, Castillo V, Martinez-Mena M (1994) EUROSEM: Preliminary validation on non-agricultural soils. In: Rickson RJ (ed) Conserving soil resources: European perspectives. CAB, Wallingford, pp 314–325
Alemi MH, Shariari MR, Nielsen D (1988) Kriging and cokriging of soil water properties. Soil Technology 1:117–132
Auerswald K (1993) Bodeneigenschaften und Bodenerosion. Relief, Boden, Paläoklima 8. Bornträger, Berlin, 208 p.
Bagnold RA (1966) An approach to the sediment transport problem from general physics. US Geological Survey Professional Paper 422–I. US Government Printing Office, Washington DC
Beasley RP, Huggins LF, Monke EJ (1980) ANSWERS, a model for watershed planning. Transactions of the American Association of Agricultural Engineers 23:938–944
Bennett JP (1974) Concepts of mathematical modeling of sediment yield. Water Resources Research 10:485–492
Boardman J, Favis-Mortlock D (eds) (1998) Modelling soil erosion by water. NATO ASI Series I: Global Environmental Change, Vol. 55. Springer, Berlin
Bork H-R, Dalchow C, Kächele H, Piorr H-P, Wenkel K-O (1995) Agrarlandschaftswandel in Nordost-Deutschland unter veränderten Rahmenbedingungen: ökologische und ökonomische Konsequenzen. Ernst and Sohn, Berlin
Bork H-R, Dalchow C, Schatz T, Frielinghaus M, Höhn A, Schmidt R (1997) The soil and sediment profile Bäckerweg in the natural reserve "Märkische Schweiz", East-Brandenburg, Germany. Mitteilungen der Deutschen Bodenkundlichen Gesellschaft 84:327–330
Brandt CJ (1989) The size distribution of throughfall drops under vegetation canopies. Catena 16:507–524
Brandt CJ (1990) Simulation of the size distribution and erosivity of raindrops and throughfall drops. Earth Surface Processes and Landforms 15:687–698
Buchner W, Köller K (1990) Integrierte Bodenbearbeitung. Ulmer, Stuttgart
Buchter B, Aina PO, Azari AS, Nielsen D (1991) Soil spatial variability along transects. Soil Technology 4:297–314
Campbell GS (1985) Soil physics with BASIC. Elsevier, Amsterdam
Chu S-T (1978) Infiltration during an unsteady rain. Water Resources Research 14:461–466
Dunne T (1978) Field studies of hillslope flow processes. In: Kirkby MJ (ed) Hillslope hydrology. Wiley, Chichester, pp 227–293
Durnford D, King PK (1993) Experimental study of processes and particle-size distributions of eroded soils. Journal of Irrigation and Drainage Engineering 119:383–398
Eagleson PS (1970) Dynamic hydrology. McGraw-Hill, New York
Elliot WJ, Liebenow AM, Laflen JL, Kohl KD (1989) A compendium of soil erodibility data from WEPP cropland soil field erodibility experiments 1987/88. National Soil Erosion Research Laboratory Rep 3). US Department of Agriculture-Agricultural Research Service, West Lafayette

Engelund F, Hansen E (1967) A monograph on sediment transport in alluvial streams. Teckniks Verlag, Copenhagen

Everaert W (1991) Empirical relations of the sediment transport capacity of interrill flow. Earth Surface Processes and Landforms 16:513–532

Everaert W (1992) Processes of interrill erosion: laboratory experiments. Unpublished Ph.D. thesis. Katholic University, Leuven (in Flemish)

FAO (ed) (1994) FAO-Unesco soil map of the world, revised legend reprinted with corrections. World Resources Report 60. FAO, Rome

Ferreira VA, Smith RE (1992) OPUS, an integrated simulation model for transport of nonpoint-source pollutants at the field scale, vol. II, user manual. U.S. Department of Agriculture-Agricultural Research Service 90. US Depart Agriculture-Agricultural Research Service, Washington DC

Finkner SC, Nearing MA, Foster GR, Gilley JE (1989) A simplified equation for modeling sediment transport capacity. Transactions American Association of Agricultural Engineers 32:1545–1550

Flanagan DC (ed) (1994) Water Erosion Prediction Project (WEPP), erosion prediction model version 94.7 user summary. National Soil Erosion Research Laboratory Report 9. US Department of Agriculture-Agricultural Research Service, West Lafayette

Flanagan DC, Livingston SJ (eds) (1995) US Department of Agriculture Water Erosion Prediction Project (WEPP) version 95.7, user summary. National Soil Erosion Research Laboratory Report 11. US Department of Agriculture-Agricultural Research Service, West Lafayette

Flanagan DC, Nearing MA (eds) (1995) US Department of Agriculture Water Erosion Prediction Project (WEPP) version 95.7, hillslope profile and watershed model documentation. National Soil Erosion Research Laboratory Report 10. US Department of Agriculture-Agricultural Research Service, West Lafayette

Foster GR, Lane LJ (1987) US Department of Agriculture Water Erosion Prediction Project (WEPP), user requirements. National Soil Erosion Research Laboratory Report 1. US Department of Agriculture-Agricultural Research Service, West Lafayette

Foster GR, Meyer DL (1972) A closed-form soil erosion equation for upland areas. In: Shen HW (ed) Sedimentation. Colorado State University, Fort Collins, 12.1–12.19

Foster GR, Meyer LD (1975) Mathematical simulation of upland erosion by fundamental soil erosion mechanics. Agricultural Research Service S-40. US Department of Agriculture-Sedimentary Laboratory, Oxford

Foster GR, Young RA, Neibling WH (1985) Sediment composition for nonpoint-source pollution analyses. Transactions of the American Association of Agricultural Engineers 28:133–139

Fread DL (1993) Flow routing. In: Maidment DR (ed) Handbook of hydrology. McGraw-Hill, New York, 10.1–10.36

Ghidey F, Alberts EE (1994) Interrill erodibility affected by cropping systems and initial soil water content. Transactions of the American Association of Agricultural Engineers 37:1809–1815

Govers G (1990) Empirical relationships for the transport capacity of overland flow. In: Walling DE, Yair A, Berkowicz S (eds) Erosion, transport and deposition processes. International Association of Hydrological Sciences Publication 189. Internat Assoc Hydrol Sc Press, Wallingford, 45–63

Govers G (1991) Spatial and temporal variations in splash detachment: a field study. In: Okuda S, Netto A, Slaymaker O (eds) Extreme landforming events. Z Geomorph, N.F., Suppl 46. Bornträger, Berlin, 15–24

Green WH, Ampt GA (1911) Studies on soil physics I: The flow of air and water through soils. Journal of Agricultural Science 4:1–24

Gunn R, Kinzer GD (1949) The terminal velocity of fall for water droplets in stagnant air. Journal of Meteorology 6:243–248

Haase G (1978) Leitlinien der bodengeographischen Gliederung Sachsens. Beiträge zur Geographie 29:7–79

Horton RE (1933) The role of infiltration in the hydrologic cycle. Transactions of the American Geophysical Union 14:446–460

Immler LG, Zahn MT (1994) Die flächenhafte Variabilität bodenphysikalischer Parameter und des C_{org}-Gehaltes in den Pflugsohlen je eines Ton-, Sand- und Lößstandortes. Zeitschrift für Pflanzenernährung und Bodenkunde 157:251–257

Julien PY, Simons DB (1985) Sediment transport capacity of overland flow. In: Transactions of the American Society of Agricultural Engineers 28:555-761

Kahnt G (1995) Minimalbodenbearbeitung. Ulmer, Stuttgart

Karim MF, Kennedy JF (1983) Computer-based predictors for sediment discharges and friction factors for alluvial streams. Iowa Institute of Hydraulic Research Report 242. University of Iowa, Ankeny

Kinnell PIA (1991) The effect of flow depth on sediment transport induced by raindrops impacting shallow flow. Transactions of the American Association of Agricultural Engineers 34:161–168

Knisel WG (ed) (1980) CREAMS: a field scale model for chemicals, runoff, and erosion from agricultural management systems. US Department of Agriculture Conservation Research Report 26. US Department of Agriculture-Science and Education Administration, Washington DC
Lauren JG, Wagenet RJ, Bouma J, Wosten JHM (1988) Variability of saturated hydraulic conductivity in a Glossaric Hapludalf with macropores. Soil Science 145:20–28
Laursen EM (1958) The total sediment load of streams. Journal of the Hydraulics Division, Proceedings of the American Society of Civil Engineers 84
Laws JD (1941) Measurements of the fall velocity of water drops and raindrops. Transactions of the American Geophysical Union 21:709–721
Laws JD, Parsons DA (1943) The relation of raindrop size to intensity. Transactions of the American Geophysical Union 24:452–460
Lieberoth I (1963) Lößsedimentation und Bodenbildung während des Pleistozäns in Sachsen. Geologie 12:149–187
Lieberoth I (1982) Bodenkunde, 3rd edn. Deutscher Landwirtschaftsverlag, Berlin
Lieberoth I, Czwing E (1989) Weiterentwickelte natürliche Standorteinheiten auf der Grundlage der MMK: Inhalt, Gliederung und Ausgrenzung der Natürlichen Standorteinheiten (NStE neu). Agriculture Exhibition of the GDR, Markkleeberg
Mannsfeld K, Richter H (eds) (1995) Naturräume im Sachsen. Forschungen zur Deutschen Landeskunde 238). Zentralausschuß für deutsche Landeskunde, Trier
Marshall JS, Palmer WM (1948) The distribution of raindrops with size. Journal of Meteorology 5:165–166
McCuen RH (1973) The role of sensitivity analysis in hydraulic modeling. Journal of Hydrology 18:37–53
Mein RG, Larson CL (1973) Modeling infiltration during a steady rain. Water Resources Research 9:384–394
Merriam RA (1973) Fog drip from artificial leaves in a fog wind tunnel. Water Resources Research 9:1591–1598
Meyer LD, Wischmeier WH (1969) Mathematical simulation of the processes of soil erosion by water. In: Transactions of the American Society of Agricultural Engineers 12:754-758, 762
Morgan RPC (1994) The European Soil Erosion Model: an update on its structure and research base. In: Rickson RJ (ed) Conserving soil resources: European perspectives. CAB, Wallingford, 286–299
Morgan RPC, Rickson RJ (1988) Erosion assessment and modeling. Commission of the European Communities Report EUR 10860 EN. Office for Official Publications of the European Communities, Luxembourg
Morgan RPC, Quinton JN, Rickson RJ (1992) EUROSEM documentation manual, version 1. Silsoe College, Silsoe
Morgan RPC, Quinton JN, Rickson RJ (1993) EUROSEM, a user guide, version 2. Silsoe College, Silsoe
Nash JE, Sutcliffe V (1970) River flow forecasting through conceptual models I, a discussion of principles. Journal of Hydrology 10:282–290
Nearing MA, Deer-Ascough L, Laflen JM (1990) Sensitivity of the WEPP hillslope version soil erosion model. Transactions of the American Association of Agricultural Engineers 33:839–849
Onstad CA (1984) Depressional storage on tilled soil surfaces. Transactions of the American Association of Agricultural Engineers 27:729–736
Pernecker L, Vollmers HJ (1965) Neue Betrachtungsmöglichkeit des Feststofftransportes in offenen Gerinnen. Die Wasserwirtschaft 55
Poesen J (1985) An improved splash transport model. Z Geomorph N.F. 29:193–211
Prendergast AG (ed) (1983) Soil erosion. Commission of the European Communities Report EUR 8427 EN. Office for Official Publications of the European Communities, Luxembourg
Rawls WJ, Brakensiek DL (1983) A Procedure to Predict Green and Ampt Infiltration Parameters. In: ASAE Conference on Advances in Infiltration. Am Soc Agricultural Engineers, Chicago 102–112
Rawls WJ, Stone JJ, Brakensiek DL (1989) Infiltration. In: Lane LJ, Nearing MA (eds) US Department of Agriculture Water Erosion Prediction Project (WEPP) hillslope profile version, model documentation. National Soil Erosion Research Laboratory Report 2. US Department of Agriculture-Agricultural Research Service, West Lafayette, 4.1–4.12
Renard KG, Foster GR, Weesies GA, McCool DK, Yoder DC (1997) Predicting soil erosion by water: a guide to conservation planning with the Revised Universal Soil Loss Equation (RUSLE). US Department of Agriculture Handbook 703. US Department of Agriculture-Agricultural Research Service, Washington DC
Richter H, Haase G, Lieberoth I, Ruske R (1970) Periglazial-Löß-Paläolithikum im Jungpleistozän der Deutschen Demokratischen Republik. Petermanns Geogr Mitt Ergänzungsheft 274. Haack, Gotha
Rose CW, Williams JR, Sander GC, Barry DA (1983) A mathematical model of soil erosion and deposition processes I, theory for a plane land element. Soil Sc Soc Am J 47:991–995

Schmidt J (1988) Wasserhaushalt und Feststofftransport an geneigten, landwirtschaftlich bearbeiteten Nutzflächen. Unpublished Ph.D. thesis. Department of Geography of the Free University, Berlin

Schmidt J (1996) Entwicklung und Anwendung eines physikalisch begründeten Simulationsmodells für die Erosion geneigter, landwirtschaftlicher Nutzflächen. Berliner Geographische Abhandlungen 61. Department of Geography of the Free University, Berlin

Schmidt J, Werner M von, Michael A (1996) EROSION-2D: ein Computermodell zur Simulation der Bodenerosion durch Wasser. Sächsische Landesanstalt für Landwirtschaft, Sächsisches Landesamt für Umwelt und Geologie, Dresden Freiberg

Schramm M (1994) Ein Erosionsmodell mit räumlich und zeitlich veränderlicher Rillenmorphologie. Institut für Wasserbau und Kulturtechnik Mitteilungen, 190. Technische Hochschule, Karlsruhe

Schramm M, Prinz D (1993) Rainfall simulation tests for parameter determination of a soil erosion model. In: Wicherek S (ed) Farm land erosion. Elsevier, Amsterdam, 373–387

Sharpley AN, Williams JR (eds) (1990a) EPIC: Erosion/Productivity Impact Calculator 1, model documentation. US Department of Agriculture Technical Bulletin 1768. US Department of Agriculture-Agricultural Research Service, Washington DC

Sharpley AN, Williams JR (eds) (1990b) EPIC: Erosion/Productivity Impact Calculator 2, user manual. US Department of Agriculture Technical Bulletin 1768. US Department of Agriculture-Agricultural Research Service, Washington DC

Sisson JB, Wierenga PJ (1981) Spatial variability of steady-state infiltration rates. Soil Science Society of America Journal 45:699–704

Smith RE (1992) OPUS, an integrated simulation model for transport of nonpoint-source pollutants at the field scale I, documentation. US Department of Agriculture-Agricultural Research Service 98. US Department of Agriculture-Agricultural Research Service, Washington DC

Smith RE, Parlange J-Y (1978) A parameter-efficient hydrologic infiltration model. Water Resources Research 14:533–538

Stone JJ, Lane LJ, Shirley ED (1992) Infiltration and runoff simulation on a plane. Transactions of the American Association of Agricultural Engineers 35:161–170

Styczen M, Nielsen SA (1989) A view of soil erosion theory, process-research and model building: possible interactions and future developments. Quademi die Scienza del Suolo 2. Firenze

Troutman BM (1985) Errors in parameter estimation in precipitation-runoff modeling I, theory. Water Resources Research 21:1195–1213

Van Genuchten MT (1980) A closed-form equation for predicting the hydraulic conductivity of unsaturated soils. Soil Science Society of America Journal 44:892–898

Vereecken H, Maes J, Feyen J, Darius P (1989) Estimating the soil moisture retention characteristics from texture, bulk density and carbon content. Soil Science 148:389–403

Vieira SR, Nielsen DR, Biggar JW (1981) Spatial variability of field-measured infiltration rate. Soil Science Society of America Journal 45:1040–1048

Werner M von (1995) GIS-orientierte Methoden der digitalen Reliefanalyse zur Modellierung von Bodenerosion in kleinen Einzugsgebieten. Unpubl Ph.D. thesis. Depart Geogr Free Univ Berlin

Wan Z, Wang Z (1994) Hyperconcentrated flow. Balkema, Rotterdam

Warrick AW, Nielsen DR (1980) Spatial variability of soil physical properties in the field. In: Hillel D (ed) Applications of Soil Physics. Academic Press, New York, 319–344

Wischmeier WH, Smith DD (1965) Predicting rainfall erosion losses from cropland east of the Rocky Mountains. Agriculture Handbook 282. US Department of Agriculture, Washington DC

Wischmeier WH, Smith DD (1978) Predicting rainfall erosion losses – a guide to conservation planning. Agriculture Handbook 537. US Department of Agriculture-Science and Education Administration, Washington DC

Woolhiser DA, Liggett JA (1967) Unsteady, one-dimensional flow over a plane – the rising hydrograph. Water Resources Research 3:753–771

Woolhiser DA, Smith RE, Goodrich DC (1990) KINEROS, a kinematic runoff and erosion model, documentation and user manual. US Department of Agriculture-Agricultural Research Service 77. US Department of Agriculture-Agricultural Research Service, Washington DC

Yalin MS (1977) Mechanics of sediment transport, 3rd edn. Pergamon Press, Oxford

Yang CT (1973) Incipient motion and sediment transport. Journal of Hydrology 99:1679–1704

Yang CT (1979) Unit stream power equations for total load. Journal of Hydrology 49:123–138

Young RA, Onstad CA, Bosch DD, Anderson WP (1987) AGNPS, Agricultural Nonpoint Source Pollution Model, a watershed analysis tool. Agricultural Research Service Conservation Research Report 35. US Department of Agriculture-Agricultural Research Service, Washington DC

Chapter 12

Simulating Hydrological and Erosional Processes Using the PEPP-HILLFLOW Model – Parameter Determination and Model Application

K. Gerlinger

12.1 Introduction

The aim of the multidisciplinary research project 'Weiherbach' is the development of an operational, physically based numerical model for describing transport processes of water, eroded soil, fertilizer and other substances in a small rural catchment (Plate 1992). The implementation of an erosion component is required due to the damage that soil erosion causes on agricultural land. Outside the fields, eroded sediment can be a major pollutant and a carrier of polluting chemicals, such as pesticides and plant nutrients (e.g. phosphorus).

Existing soil erosion models are either empirically based (e.g. USLE; Wischmeier and Smith 1978) or require a large set of input parameters (e.g. WEPP; Flanagan and Nearing 1995). Therefore, an operational model with high spatial and temporal resolution was required for the transport processes of eroded soil following individual storm events in small rural catchments. Furthermore, intensive field measurements had to be carried out to give a reliable data base for the model. Since the reliability of the erosion model results depends strongly on the quality of the hydrological simulation, the combination of the erosion model with a sophisticated hydrological model system is required.

12.2 PEPP-HILLFLOW Model

In a first investigation step, the slope erosion model PEPP (Process orientated Erosion Prediction Program) was developed (Schramm 1994; Gerlinger 1997). In the model, runoff, erosion and deposition are calculated either for rill or for sheet flow. The processes, which provide a continuous, time and space dependent transition between a surface with and without rills, can not, as of yet, be simulated.

Therefore, it must be known beforehand if sheet flow or rill flow, e.g. between planting rows or in other fixed linear flow path, is predominant in the field. If the overland flow is specified as concentrated flow in rills, the rill geometry is variable and can be computed for deposition areas according to discharge and sediment load.

12.2.1 Surface Runoff

For modeling the surface runoff, the kinematic wave approach is applied to account for unsteady flow processes. To solve the momentum equation, the energy losses are determined by the Manning-Strickler-formula and the friction slope is assumed to be equal to the bed slope:

$$I = v^2 n^2 \frac{1}{R^{4/3}}$$ (12.1)

where
- I = bed slope ( )
- v = flow velocity (L/T)
- n = roughness coefficient (Manning's n) (T/L$^{1/3}$)
- R = hydraulic radius (area by wetted perimeter) (L)

The conservation form of the continuity equation for one dimensional flow is:

$$\frac{\partial Q}{\partial x} + \frac{\partial A}{\partial t} = q(x,t)$$ (12.2)

where
- Q = discharge (L^3 T^{-1})
- A = cross-sectional area (L^2)
- q = lateral inflow (L^2 T^{-1})

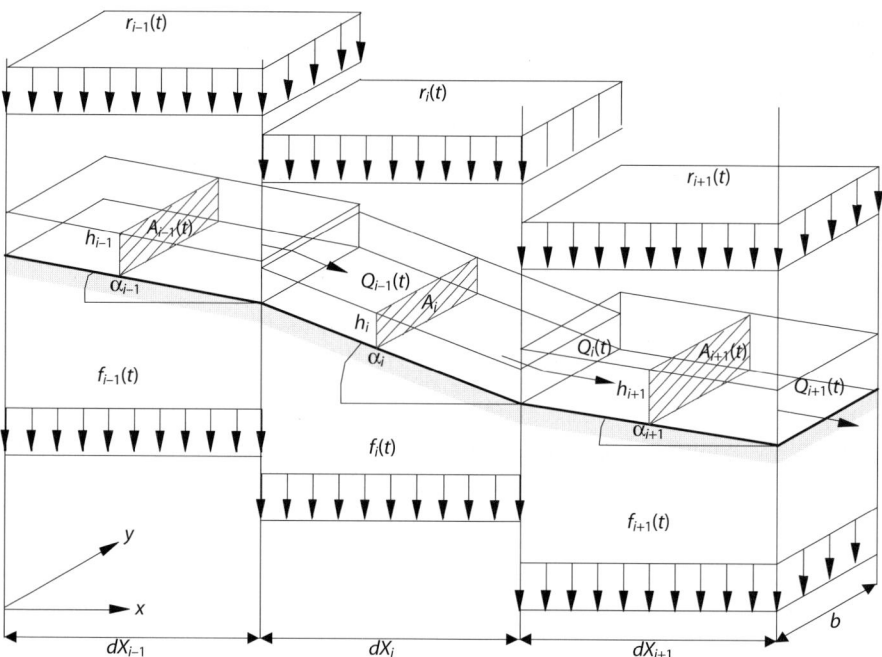

Fig. 12.1. Calculation of overland flow by dividing the slope in discrete straight elements (Gomer 1995)

The lateral inflow corresponds to the rate of surface flow resulting from the difference of rainfall and infiltration:

$$q(x,t) = [r(t) - f(x,t)]b \cos \alpha(x) \tag{12.3}$$

where
- r = rainfall intensity $(L\,T^{-1})$
- f = infiltration rate $(L\,T^{-1})$
- b = width of the flow (L)
- α = slope angle ( )

To avoid numerical instabilities a simplified explicit numerical solution of the kinematic wave equation was implemented. The solution of the continuity equation is carried out in discrete time and space steps by considering the slope segments as storages of the length Δx with constant flow area A. The flow area is calculated using the effective rainfall and the difference between the flow $Q_{i-1}(t)$ in and the flow $Q_i(t)$ out of the segment (Fig. 12.1). A comparison of this simplified solution with a finite-difference-solution of the kinematic wave equation demonstrated similar results.

12.2.2
Transport Capacity

Since there is no universal equation for determining transport capacity, four different methods have been implemented in the program: Engelund and Hansen 1967; Yalin 1977; Yang 1979; Schmidt 1996. Figure 12.2 shows the transport capacity calculated with

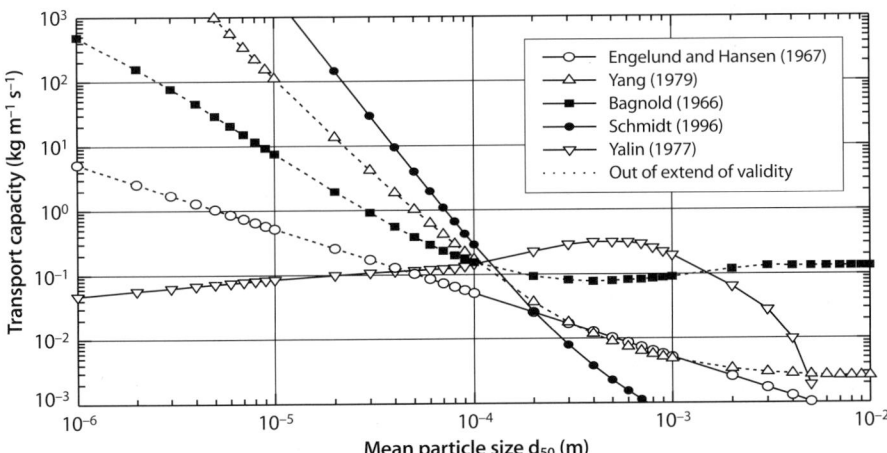

Fig. 12.2. Comparison of the methods for calculating transport capacity dependent on particle size (results of a test simulation with stationary sheet flow, slope length 200 m, slope gradient 5%, Manning's n 0.05 s m$^{-1/3}$; based on Schramm 1994)

these four methods (including the approach of Bagnold 1966) versus the mean particle size diameter for a test simulation with stationary sheet flow. The method of Yalin (1977) leads to a reduction in the transport capacity for small particle sizes (<0.06 mm) instead of the expected increase. The method of Schmidt (1996) is based on the calculation of the sinking velocity under no flow conditions and is very sensitive to the particle size. The method of the Bagnold (1966) calculates an almost constant transport capacity value for particle sizes greater 1 mm. The methods of Engelund and Hansen (1967) and Yang (1979) showed the most reliable results over a range of hydraulic conditions.

For all methods, the enrichment of the fine particle fraction in the flow due to selective deposition of coarser particles is computed. The transport capacity is calculated for each class of the particle size distribution, and the respective deficit or surplus is then redistributed.

12.2.3
Detachment Capacity

The determination of the potential erosion rate follows the basic concept of Schmidt (1991) by calculating the external forces acting on the soil particles. Detachment occurs if the resistance of the soil to erosion caused by internal friction, cohesion, and gravity is overcome.

a   The forces of the rainfall can be characterized by the momentum flux of rainfall m_r (Fig. 12.3). It is calculated by:

$$m_r = \rho r \cos\alpha v_f (1 - C) \tag{12.4}$$

where
- m_r = momentum flux of rainfall (M L^{-1} T^{-2})
- ρ = fluid density (M L^{-3})
- v_f = fall velocity of raindrops (L T^{-1}); can be approximated by a function of the rainfall intensity: $v_f = 4.459 + 0.613\ln(r\cos\alpha)$
- C = soil cover ( )

It is assumed that the momentum of the raindrops augments the momentum flux of the runoff. To describe the impact of the raindrops precisely, the microscopic transformation processes of the drops, falling on the soil surface and on the overland flow, must be considered. Their momentum is split partially straight down in the soil, which is dependent on bulk density, linking forces of the particles, aggregate breakdown and the resulting flow depth, while the other part of the momentum is transformed in a lateral shear strength. The exact share of increase in the momentum flux of overland flow by the raindrops is not calculable. Therefore, only the component of the momentum flux of rainfall pointing downslope is considered, by multiplying m_r with $\sin\alpha$, because the vectors of both rainfall and runoff then point in the same direction. This is a crude approximation of the natural processes.

b   The forces of the overland flow are described by the momentum flux of overland flow m_q per unit area (Fig. 12.4):

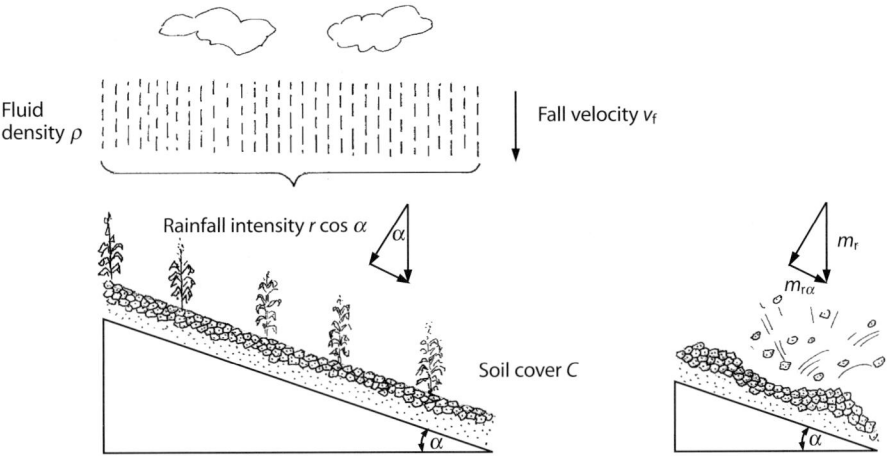

Fig. 12.3. Representation of an upland profile for calculation of momentum flux of rainfall

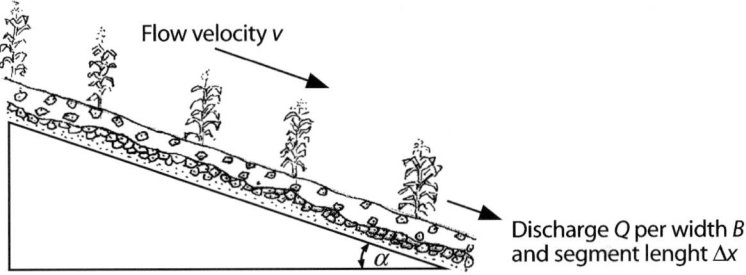

Fig. 12.4. Representation of an upland profile for calculation of momentum flux of overland flow

$$m_q = \frac{\rho v Q}{B \Delta x} \tag{12.5}$$

where
- m_q = momentum flux of overland flow (M L^{-1} T^{-2})
- B = distance between rills (1 m for sheet flow) (L)
- Δx = length of unit segment (L)

Schmidt (1996) observed by laboratory experiments with a rainfall simulator using loess soil an empirical relationship between sediment load q_s and the momentum fluxes:

$$q_s = 1.75 \times 10^{-4} \left(\frac{m_q + m_r \sin \alpha}{m_{\text{crit}}} - 1 \right) \tag{12.6}$$

where
- q_s = sediment load (M L^{-1} T^{-1})
- m_{crit} = critical momentum flux of the soil (M L^{-1} T^{-2})

The critical momentum flux m_{crit} corresponds to an *erosion resistance* of the soil. It is a soil specific parameter which has to be determined by measuring the values of m_r, m_q and q_s of rainfall experiments, solving the equation for m_{crit} and inserting the measured values.

The detachment capacity d_c of the flow per segment is calculated by division of the sediment load by flow and/or rainfall of the respective segment and the sediment load from the immediate upslope segment:

$$\frac{q_{s,i}^{j+1} - q_{s,i-1}^{j+1}}{\Delta x_i} = d_{c,i}^{j+1} \tag{12.7}$$

where
- d_c = detachment capacity (M L^{-2} T^{-1})

The empirical relationship for determining the sediment load q_s has the disadvantage of not being transferable to conditions other than the ones which were applied for the evaluation of the equation. In addition, a more physically sound equation should respect the *change* of the momentum fluxes in space and time instead of their absolute value. This is, however, a suitable operational concept which was already applied with success in the models of von Werner (1995) and Gomer (1995).

12.2.4
Erosion and Deposition

For the calculation of the actual erosion or deposition rate, two different concepts can be applied. Both concepts are included in the PEPP model, and the model user may decide which concept corresponds better with his/her specific interest:

The *transport/erosion rate approach* of Meyer and Wischmeier (1969) is based on the assumption that the surface flow picks up soil particles dependent on the detachment capacity until the transport capacity limit is reached. At locations where the transport capacity limits sediment load, sediment load is set equal to transport capacity. A further augmentation of the sediment load leads to deposition of the surplus particles. This concept is implemented in the pepp_1 program version.

The *transport deficit approach* of Foster and Meyer (1975) and Foster (1982) is based on the assumption that the detachment/deposition by flow d_f is linearly proportional to the difference between transport capacity and sediment load:

$$d_f = C_e(TC - q_s) \tag{12.8}$$

where
- d_f = detachment/deposition rate (M L^{-2} T^{-1})
- C_e = erosion/deposition coefficient (1/L)
- TC = transport capacity (M L^{-1} T^{-1})

A negative difference indicates deposition. The idea is that as sediment load fills transport capacity, less energy is available for flow to detach sediment. Therefore, the detachment capacity d_c is proportional to the transport capacity and their ratio equals the erosion/deposition coefficient C_e. This concept is implemented in the PEPP_2 program version.

The transport deficit approach leads to a smooth augmentation of the sediment load along the slope if the transport capacity increases continuously as well. However, the influence of the method for determining the transport capacity on the result is considerable since there exists no general applicable transport capacity equation for sheet flow on cohesive soils. Dependent on the method which is used to calculate transport capacity, the results for the sediment load differ substantially. The erodibility of the soil has less of an importance. Therefore, it is recommended to use the PEPP program version pepp_1 with the transport/erosion rate approach which is more soil-related. The following text refers always to this model version.

The transport/erosion rate approach can lead to a discontinuity in the course of calculated sediment load along a slope due to a sharp transition between the segments where sediment load is over and the segments where it is below transport capacity. All sediment load which passes the transport capacity will be deposited immediately in the segment. This may lead to unrealistic deposition rates, as the sinking velocities of the small sized particles are very slow. Therefore, deposition is reduced in both PEPP versions by calculating the deposition coefficient dependent mainly on the settling velocity of the sediment particles. Foster (1982) proposed for the deposition coefficient:

$$C_e = 0.5 \frac{V_s B}{Q} \tag{12.9}$$

where
- V_s = settling velocity of the sediment particles (LT^{-1})

Woolhiser et al. (1990) included the difference between sediment load and transport capacity:

$$C_e = \frac{V_s B}{Q}\left(1 - \frac{TC}{q_s}\right) \tag{12.10}$$

In the PEPP model the deposition coefficient is calculated corresponding to the settling of the particles. For deposition within a segment, a sinking particle has to reach the soil (Fig. 12.5). The particle which is at the beginning of the segment near the water surface has the longest distance to travel. For the distance Δx within the segment follows:

$$\Delta y = \frac{V_s}{V}\Delta x \tag{12.11}$$

For $\Delta y \geq$ flow depth h all particles will deposit, while for $h \geq \Delta y$ within the time step Δt the particle deposit is dependent on the ratio of Δy to h. Therefore:

Fig. 12.5. Definition sketch for the calculation of the deposition coefficient C_e (Schramm 1994)

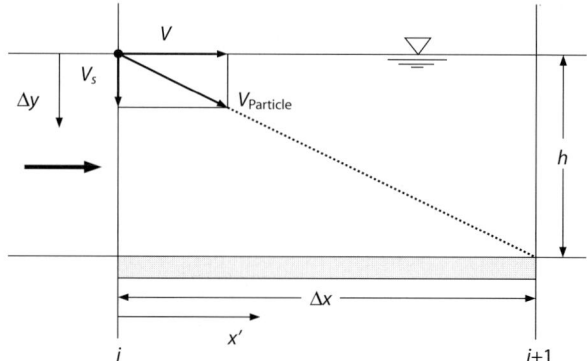

$$C_{e,1} = \frac{\Delta y}{h} = \frac{V_s}{V}\frac{\Delta x}{h} \qquad (12.12)$$

For the calculation of the settling velocity of the particles the influence of flow turbulence is not considered. This may lead to high deposition rates. The deposition coefficient may therefore be reduced dependent on the transport capacity according to Woolhiser et al. (1990):

$$C_{e,2} = \frac{V_s}{V}\frac{\Delta x}{h}\left(1 - \frac{TC}{q_s}\right) \qquad (12.13)$$

Both deposition coefficient $C_{e,1}$ and $C_{e,2}$ are included in the PEPP model and the model user may choose.

12.2.5
Sensitivity Analysis PEPP

A sensitivity analysis of a quantitative model is used to examine the effects of variations in input parameter values upon the model behaviour and output. To quantify this influence, the sensitivity analysis of the PEPP model was performed by defining a sensitivity parameter SP (McCuen and Snyder 1986):

$$SP = \frac{O_2 - O_1}{O_{12}}\frac{I_{12}}{I_2 - I_1} \qquad (12.14)$$

where
- SP = sensitivity parameter
- I_1, I_2 = highest, smallest input value
- O_1, O_2 = respective output values
- I_{12}, O_{12} = mean value of the input respectively output values

A high amount of SP reflects a high influence on the model result. The parameter is less than zero if an increase in input corresponds to a decrease in output, and vice

Table 12.1. Input parameter for the standard slope (with rills)

Model parameter standard slope	Input value	Model parameter standard slope	Input value
Slope form linear, length	120 m	Rainfall intensity	40 mm h^{-1}
Slope gradient	15%	Duration of rainfall	40 min
Rill spacing	1 m	Infiltration capacity	40→10 mm h^{-1}
Width-depth-ratio of the rill	1.5	Transport capacity model	Engelund and Hansen
Rill geometry	variabel	Median particle size (uniform)	0.02 mm
Time discretization	2 s	Mean particle density	2650 kg m^{-3}
Space discretization	5 m	Erosion resistance m_{crit}	0.001 N m^{-2}
Soil cover	0%	Manning's n	0.05 s m$^{-1/3}$

Table 12.2. Range of the input values and resulting values of *SP* in relation to the soil loss (in order of sensitivity)

Model parameter	Range			Unit	Sensitivity *SP*
Effective rainfall[a]	1381	–	2444	l	+1.142
Erosion resistance m_{crit}	10^{-4}	–	10^{-2}	N m^{-2}	−1.000
Manning's n	0.03	–	0.3	s m$^{-1/3}$	−0.873
Slope gradient	1	–	50	%	+0.782
Rill spacing	0.25	–	2.0	m	+0.334
Slope length	20	–	995	m	+0.269
Width-depth-ratio of the rill >2[b]	2	–	6		−0.130
Rainfall intensity	15	–	80	mm h^{-1}	+0.090
Width-depth-ratio of the rill <2[b]	1	–	2		+0.086
Space discretization	1	–	120	m	+0.027
Time discretization	1	–	10	s	+0.023
Soil cover	0	–	100	%	−0.014

[a] By constant duration of simulation (40 min);
[b] The sensitivity parameter *SP* is calculated for the width-depth-ratio (*wdr*) of the rills twice as there is no linear relation between *wdr* and soil loss. The *wdr* leads to a maximum of soil loss for *wdr* = 2 when the discharge in the triangular shaped rills is hydraulically favourable.

versa. The equation for the calculation of *SP* is based on the assumption of linearity between the input and output values and of independence of the parameters. To conduct the sensitivity analysis a standard slope was defined (Table 12.1) and the values of each input parameter were modified within a certain range (Table 12.2). The model behaviour may change when it is applied to a complete different situation. However, the resulting order of the sensitivity parameter *SP* for this standard situation reveals the importance of a precise determination of the effective rainfall per time, the erosion resistance and the Manning's *n* for the model results of soil loss.

12.2.6
Infiltration

The sensitivity analysis of the PEPP model shows that one of the main influencing parameters is the effective rainfall. Since the model does not implicitly calculate the infiltration, the effective rainfall has to be known beforehand. Alternatively, the model can be applied in direct combination with the HILLFLOW-2D model for hillslopes (Bronstert 1994), which was developed in the same research project. This distributed and physically based modeling system simulates all relevant hydrological processes, such as interception, evapotranspiration, infiltration, soil-moisture movement, surface runoff, and subsurface stormflow. Aside from the two-dimensional version, one- and three-dimensional versions of the model are available. The discretization of the model allows the natural spatial and temporal variability of the event and state variables (e.g. topography, vegetation, soil data, precipitation, climatic data) to be taken into account and enables the model user to produce the simulation results in the desired resolution. The variable time discretization is controlled by the different process velocities.

The infiltration component of HILLFLOW is based of the assumption of a dual porous soil within the upper soil horizon and includes the calculation of infiltration rates into the soil matrix (micropores) and into the macropores. The distinction between these two processes is necessary to explain correctly the high infiltration rates observed in nature. Since the distribution of the pore geometries covers a wide range of lengths and widths, the term macropore implies here that the effect of capillary

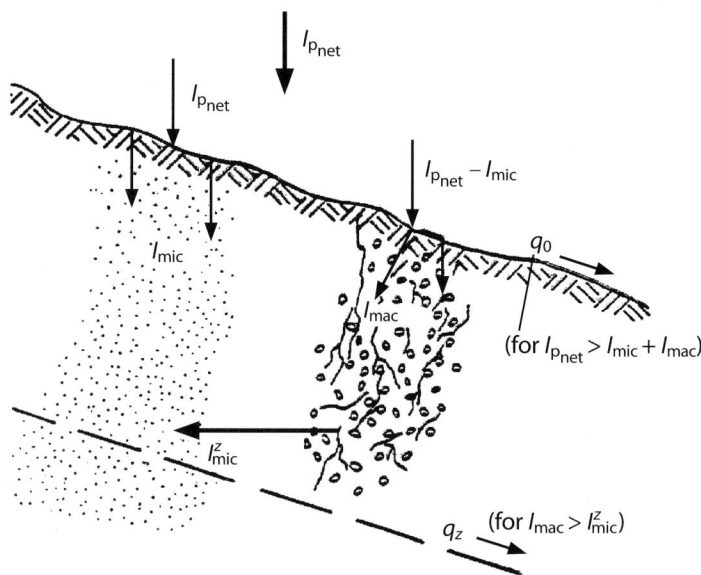

Fig. 12.6. Scheme of the infiltration component (Bronstert 1994). I_{pnet} = Intensity of net rainfall; I_{mic} = Infiltration into the soil matrix (micropores); I_{mac} = Infiltration into the macropores; I_{mic}^z = Infiltration into the matrix from macropores; q_o = Surface runoff; q_z = Interflow

potential within the macropore is negligible and gravity is the only driving force for water flow. In Fig. 12.6 a schematic presentation of the infiltration component is shown. To ease understanding the two pore systems are drawn apart. Although the portion of macropore volume is small compared to total pore volume, macropores are capable of conducting a significant water flux due to their high transport capacity.

A disadvantage of the HILLFLOW model is the limitation of the macroporous soil layer to the upper soil horizon, especially in agricultural soils where regular tillage dominates, the macropores are destroyed in the plowing horizon. Field investigations have shown that the depth of the macropores can reach up to 2 m. Also, the water flow in the macropores may already occur at low infiltration rates when the soil matrix is not saturated with water. Nevertheless, the HILLFLOW model includes at least the preferential flow in the macropores which represents a big advantage compared to other infiltration models without this important component.

For modeling the matrix water flow the Richards equation is applied which combines the unsaturated Darcy law and the law of conservation of mass. Instead of a numerical solution a simplified model has been developed to calculate the water matrix dynamics by means of a fuzzy logic approach using fuzzy rules to calculate the unsaturated flow between two adjacent soil elements. Alternatively, the approach of Van Genuchten (1980) can be used to determine the relation of the soil moisture content to the soil suction head as well as to the hydraulic conductivity. These relations are stored in so called look-up-tables to speed up the calculation.

12.2.7
Sensitivity Analysis HILLFLOW

A complete sensitivity analysis of a model system like HILLFLOW, with it's huge sets of input parameters, is very complex. But to estimate at least the influence of the infiltration related parameters, a short sensitivity analysis was conducted by varying the parameters Manning's n, free volume of the macropores, depth of the macroporous soil layer and saturated hydraulic conductivity. A standard slope was defined, which equals a natural slope in the experimental basin Weiherbach (length 70 m, mean slope angle 12%, silty loam) and a rainfall event with a maximum intensity of 60 mm h^{-1} was simulated with different input values (Table 12.3, for more details see Koch 1995).

Regarding the calculated volume of discharge for the different input values, the sensitivity of the model to the saturated hydraulic conductivity is evident. Due to the

Table 12.3. Calculated discharge volume and sensitivity parameter for the sensitivity analysis of same HILLFLOW input parameters

	Standard slope	Manning's n (s m$^{-1/3}$)		Free volume of macropores (%)		Depth of macroporous soil layer (cm)		Saturated hydraulic conductivity (mm h^{-1})	
Range of parameters		0.200	0.067	0.1	1	20	40	4.5	13.1
Discharge volume (l)	139	63	188	310	5	176	102	847	0
Sensitivity parameter SP		−1.18		0.99		−0.79		−2.05	

high natural variability a precise determination of this parameter is difficult (see Carsel and Parish 1988). Also the model parameters free volume of macropores and depth of the macroporous soil layer are very sensitive but hard to measure. Higher values of these parameters leads to higher infiltration rates as they augment the complete pore volume. Manning's n influences mainly the surface flow, but if the discharge remains longer on the soil due to a hydraulic rough surface, the infiltration increases as well.

12.3
Determination of the Model Parameters for PEPP

The sensitivity analysis of PEPP revealed the influence of roughness coefficient n and erosion resistance m_{crit}. The application of the PEPP model for single slopes as well as in catchments requires, therefore, a precise prediction of these parameters in space and time. A spatial and temporal analysis of the surface roughness and especially of the erosion resistance was performed to provide the user with information about the required model input parameters.

The main investigation area is the 6.3 km² agricultural Weiherbach Catchment in the hilly Kraichgau region (Southwest Germany), which is, for the most part, loess covered. In order to obtain sufficient data, a transportable rainfall simulator (12 m × 2 m) was incorporated into the study. Rainfall intensity is regulated by varying the velocity of the oscillating nozzles (Veejet 80/100). Usually, a rainfall intensity of approximately 60 mm h^{-1} was applied to the plots until steady state runoff conditions had been established for a certain time.

Manning's n was determined first by measuring the mean flow velocity at steady state conditions with a tracer. A slug of dye was injected into the main rill and the time required for the concentration peak to pass a downstream point was measured. This method only shows the velocity in the main rills which are not representative of the mean velocity on the surface. Therefore, the roughness coefficients were finally estimated by fitting the recessing limb of the simulated model hydrographs to the observed hydrographs of the rainfall simulations (see Engman 1986). This leads to a mean roughness coefficient composed of the surface areas with and without rills.

12.3.1
Temporal Variability of the Erosion Resistance

12.3.1.1
Temporal Variability throughout a Year

Before the erosion model can be applied, one must know whether the erosion resistance, as a soil specific model parameter, is time invariant. A temporal variability throughout a year of other soil erodibility indices was found by several authors (e.g. Coote et al. 1988; Nearing et al. 1988; Imeson and Vis 1984).

For this reason, rainfall experiments were carried out during the growing season on sugar beet and maize fields. Fields with these crops are susceptible to erosion due to the late leaf cover. In each experiment, the rainfall simulator was moved across the slope and assembled bordering on the previous plot to guarantee comparable but undisturbed soil conditions. The similarity of rainfall experiments conducted in adjacent plots has been proven by several comparative tests.

During the growing seasons of 1993 and 1994, 15 rainfall simulations were carried out, from the time the crops were seeded until complete leaf cover. The results of the 9 simulations from 1994 are presented as an example. Table 12.4 gives an overview of the varying initial conditions of the experiments and shows the results for soil loss and runoff. In all simulations it was observed that erosion susceptibility remains high between planted rows during vegetation growth. This can lead to high sediment loads caused by early summer storms (Fig. 12.7).

The calculated erosion resistance values during the rainfall experiments at the different dates throughout the year are also depicted in Fig. 12.7. At the beginning of the rainfall experiments, the erosion resistances vary considerably; however, since steady state conditions *at the end* of the simulations were obtained, the erosion resistances become constant as well. The variation of the erosion resistance values *at the beginning* of the experiments can be explained by the varying initial and unsteady state conditions throughout the year.

The different amounts of rainfall before the experiments, for example, lead to different conditions of the crust, but for the relatively uniform erosion resistance *at the end* of the simulations, the process of the first formation of a crust after the aggregation processes following the tillage seems to be important. The soil moisture content at the first rainstorm during the vegetation period which leads to a stable crust influences the resistance of the crust for the following rainstorms. This is based on an intense aggregate breakdown of dry aggregates due to slaking, creating a dense and smooth crust, whereas wet aggregates would resist the rain drop impact causing a rougher and more stable surface.

Table 12.4. Initial conditions of the experiments maize/sugar beet 1994 and the results of runoff and soil loss after 30 mm of rainfall and at steady state conditions. Rainfall intensity amounts to 34.4 mm h^{-1} (maize) and 62.6 mm h^{-1} (sugar beet)

	Date 1994	Days after tillage	Soil cover (%)	Initial soil moisture (Vol.-%)	Rainfall to start runoff (mm)	Runoff (mm) 30 mm^a	Soil loss (t ha^{-1}) 30 mm^a	Runoff rate^a (mm h^{-1})	Soil loss rate^a (g m^{-2} min^{-1})
Maize	20.05.	24	0	29.5	1.1	19.1	2.3	30.5	15.8
12.9% clay	01.06.	36	10	14.3	4.6	5.9	1.3	18.6	7.9
81.6% silt	16.06.	51	40	20.4	1.7	12.8	2.9	20.0	7.6
5.5% sand	01.07.	66	90	22.2	0.6	15.9	2.5	24.1	6.8
Gradient 10%									
Sugar beet	03.04.	4	0	21.8	3.1	17.1	23.8	58.9	94.1
15.2% clay	12.04.	13	0	25.6	2.1	15.9	15.2	56.2	57.2
78.4% silt	27.04.	28	5	21.6	2.0	23.1	22.4	62.1	98.8
6.4% sand	30.05.	61	30	14.7	7.3	18.6	44.0	58.2	138.1
Gradient 16%	17.06.	79	50	18.3	4.2	13.2	28.6	50.7	96.0

^{a} Runoff and soil loss are for 30 mm of simulated rainfall; runoff rate and soil loss rate are at the equilibrium flow.

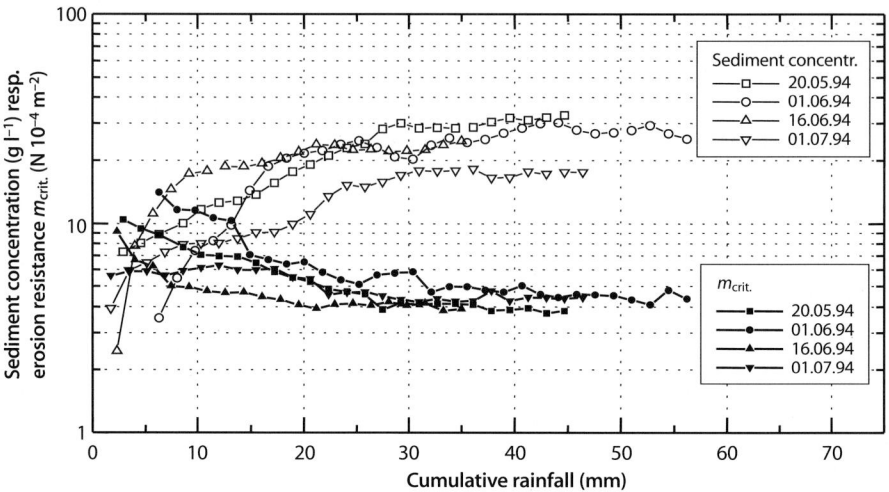

Fig. 12.7. Values of the measured sediment load and the calculated erosion resistance from the rainfall experiments during the vegetation growth on a maize field

During the simulation at different dates throughout the year, however, the once formed crust is broken up and after some time comparable conditions prevail. In general, it can be concluded that the erosion resistances for each field at the different dates are relatively uniform. There seems to be only a small temporal variability during the growing season of maize and sugar beet.

12.3.1.2
Temporal Variability between the Years

Apart from the temporal variability during the growing season, the between year variation of the erosion resistance was also determined. Changes in erodibility on bare, recently tilled soils from year to year have been found, for example, by Alberts et al. (1987). To investigate if this assumption is also applicable for the erosion resistance, rainfall experiments were carried out twice on identical plots. In August 1993, 21 rainfall experiments were carried out before the seeding of the intercrop. In April 1994, 8 rainfall experiments were repeated after the seedbed preparation for maize. In both cases, the soil was tilled one week before the experiments with a rotary hoe. The initial conditions and the results of two rainfall experiments on the bottom and in the middle of a slope at different times are shown in Table 12.5 as an example. It can be concluded that a relatively constant erosion resistance is achieved at the same location and following the same tillage, independent of the season. Only the antecedent soil moisture content may influence the results.

The investigations of the temporal variability of the erosion resistance reveal a relatively small variation throughout the year. There are variations in sediment load and runoff which indicate a temporal variability. These variations are due to changes in

Table 12.5. Initial conditions of the experiments 31 August 1993 (dry conditions) and 20 April 1994 (wet conditions). Results of runoff and soil loss after 30 mm of rainfall and at steady state conditions at the bottom and in the middle of the slope

	Clay (%)	Orgn. matter (%)	Initial soil moisture (Vol.-%)	Rainfall to start runoff (mm)	Runoff (mm) 30 mm[a]	Soil loss (t ha^{-1}) 30 mm[a]	Runoff rate[a] (mm h^{-1})	Soil loss rate[a] (g m^{-2} min^{-1})	Erosion resistance (N 10^{-4} m^{-2})
Bottom '93	19.8	1.6	12.8	16.6	1.5	0.6	40.1	40.8	8.36
Bottom '94	23.1	2.2	27.7	13.2	6.6	4.6	56.7	43.6	10.22
Middle '93	14.7	1.5	7.9	13.8	3.5	2.8	50.2	154.1	2.87
Middle '94	17.9	1.7	24.3	8.3	16.1	32.5	61.1	125.7	4.72

[a] Runoff and soil loss are for 30 mm of simulated rainfall; runoff rate and soil loss rate are at the equilibrium flow; slope gradient: for the bottom experiments: 16.2%, for the middle experiments: 18.0%; rainfall intensity 62.2 mm h^{-1}.

conditions, such as soil cover or surface roughness. Since these values are included in the calculation of the erosion resistance, this model parameter is a better indicator for erodibility than the bare comparison of soil losses.

12.3.2
Spatial Variability of the Erosion Resistance

To investigate the spatial variability of erosion resistance, rainfall experiments were carried out on different soil types. In the Weiherbach Catchment, the only soil types are those which developed on loess and on a Mesozoic formation (Keuper). To extend the investigation, rainfall experiments were also carried out in different regions in Saxony (East Germany) on erodible soils which developed on loess and different Paleozoic formations. In each region different tillage and management practices were investigated (Schmidt and Michael 1999).

The variability of erosion resistance within the investigated areas and the influencing parameters can not be determined by these investigations. They allow for an estimation of the erosion resistance of different regions and an application of the erosion model outside the Weiherbach Catchment, however, in order to simulate erosion in a catchment it is necessary to determine the small scale variability of the erosion resistance. By doing this, detailed information of the influencing factors of the required parameter is obtained.

Therefore, selected slopes in the erodible loess covered regions of the Weiherbach Catchment were divided into strips, and the rainfall simulator was moved on every strip from the bottom of the slope to the top. The soil moisture contents of the rainfall experiments have been different: low water content in summer 1993 and high in spring 1994. Within these two periods the rainfall experiments were carried out over a short time so that initial conditions of the respective experiments would be comparable. To estimate the erosion resistance without time and cost consuming rainfall experiments, a detailed investigation of the influencing soil properties in the examined region was carried out. Since the erosion resistances are not randomly distributed but show a spatial dependency, a determination of the erosion resistance using soil parameters was sought.

12.3.2.1
Determination of the Erosion Resistance using Soil Parameters

In addition to the rainfall experiments, investigations of particle and aggregate size distribution, content of organic matter, soil moisture content, aggregate stability, plasticity limits and shear strength were carried out to determine the suitability of these soil properties for the estimation of the erosion resistance.

The aggregate stability was determined by the percolation stability method (see Kainz 1981; Murschel 1991), in which water is percolated through a small tube filled with 1–2 mm aggregates. A constant percolating rate is achieved by the breaking down the aggregates. This rate is then used as a measure for the aggregate stability.

A second method which was used is the rainfall stability (see Roth 1992; Le Bissonnais et al. 1989; Henk 1989), in which aggregates were placed into sieves of different size classes containing a small layer of gravel. Rainfall simulation was then carried out. An intensity of 60 mm h^{-1} was applied over 10 min., and the amount of soil washed through the sieve was used as a measure for the aggregate stability. Shear strength of the soil was determined using a pocket shear meter (see Torri 1987; Brunori et al. 1989) and a more elaborate laboratory vane apparatus.

The results indicate that the values of the pocket shear meter and of the percolation stability method do not correspond with the erosion resistance. Also the sediment concentration, by achieving steady state conditions at the end of the rainfall experiments, shows no correlation to these soil parameters. The variability of these measurements seems to be higher than the natural variation within the investigated soil texture class.

The other parameters are suited to reveal at least the *relative* differences on the slopes. The mean aggregate size and the rainfall stability of the aggregates, particularly, show a relation to the erosion resistance because smaller aggregates are more unstable due to slaking (Henk 1989; Gäth 1995). In addition, higher values of the plasticity limit correspond to higher values of erosion resistance which results in a significant correlation between these parameters.

Soil parameters which a more easily available also showed significant relations to the erosion resistance. The clay content, the amount of organic matter and the initial soil moisture content seem to be the main influencing factors.

On one slope the clay content of the soil was suitable to estimate the erosion resistance because the erosion resistance augmented from the top to the bottom (see Table 12.5) as well as to the eastern part of the slope. This was caused by a respective augmentation of the clay content on this slope which corresponded with an augmentation of the amount of organic matter and antecedent soil moisture.

The amount of organic matter is a function of other factors beside the clay content. On two slopes the amount of organic matter was significantly higher since these slopes lied fallow before tillage. This resulted in the highest erosion resistance and lowest sediment concentration of the investigation (Table 12.6a).

Table 12.6. Initial conditions and results of some experiments: *a* with a higher amount of organic matter and initial soil moisture, *b* with different clay content and comparable soil moisture and *c* with comparable clay content and different soil moisture. Results of runoff and soil loss after 30 mm of rainfall and at steady state conditions

Plot name	Clay (%)	Orgn. matter (%)	Initial soil moisture (Vol.-%)	Rainfall to start runoff (mm)	Runoff (mm) 30 mm[a]	Soil loss (t ha^{-1}) 30 mm[a]	Runoff rate[a] (mm h^{-1})	Soil loss rate[a] (g m^{-2} min^{-1})	Erosion resistance (N 10^{-4} m^{-2})
a High amount of organic matter									
Altenberg	28.1	3.2	28.7	6.2	5.4	0.48	23.2	5.4	27.8
Weiherbächle	40.8	3.4	34.1	27.0	0.8	0	16.6	0.3	209.5
b Different clay content, comparable soil moisture									
Neuenbürg 11	22.5	1.8	20.2	7.3	12.1	11.9	53.1	66.1	6.1
Neuenbürg 12	27.6	1.8	19.1	7.3	10.4	14.2	44.4	52.8	7.8
Neuenbürg 22	31.4	1.7	18.6	9.4	5.8	6.1	39.8	54.0	5.3
Neuenbürg 31	21.3	1.8	19.8	5.2	13.8	12.7	52.6	67.4	6.1
c Comparable clay content, different soil moisture									
Feldmägen 11	16.7	1.7	26.8	35.4	0	0	41.4	29.5	9.5
Feldmägen 21	17.7	1.6	24.4	35.3	0	0	43.4	28.2	8.6
Feldmägen 12	15.6	1.3	21.8	6.8	18.4	22.4	62.1	142.9	3.4
Feldmägen 22	16.6	1.1	20.1	4.2	13.6	21.3	53.4	101.6	4.5

[a] Runoff and soil loss are for 30 mm of simulated rainfall; runoff rate and soil loss rate are at the equilibrium flow; range of slope gradients: 12.1%–17.2%; rainfall intensity 62.6 mm h^{-1}.

Apart from clay content and amount of organic matter, the antecedent soil moisture influences the infiltration, erosion resistance and sediment concentration in a complex way. On the one hand, rain on a soil with a high moisture content can lead to an early saturation of the soil, creating saturation overland flow. Compared to low initial soil moisture conditions, the amount of rainfall which provokes runoff is less if wet conditions in general prevail (Table 12.5, experiments 1993 compared to 1994). On the other hand, wet aggregates resist longer to aggregate breakdown and crusting which prevents Hortonian (infiltration-excess) overland flow. Since crusting is the main cause of overland flow on tilled loess soil following individual thunderstorms, a wet soil is able to maintain high infiltration rates while dry areas produce higher discharge and sediment concentration. High moisture content in aggregated soils leads to high erosion resistance. Therefore, the amount of rainfall to produce runoff is higher on the plots with relatively wet conditions compared to the other plots (Table 12.5, experiments bottom compared to middle).

Similar to the amount of organic matter the antecedent soil moisture depends not only on the clay content and the respective soil suction, but it is also influenced by external factors like heavy rainfall, stagnating water or a groundwater level close to the surface. One slope, with a relatively wide range of clay content over the field, had comparable contents of soil moisture due to the rainfall characteristics before the investigation (Table 12.6b). Therefore, the erosion resistance and the sediment concentration also showed comparable results. Another slope, with comparable contents of clay and organic matter over the investigated area, showed higher erosion resistance and lower sediment concentration at the bottom of the slope (Table 12.6c). Due to the small distance to the groundwater and the capillary fringe, the soil moisture was higher at the bottom of the slope, which resulted in a higher aggregate stability and less crusting. The amount of rainfall before surface runoff started was significantly higher and the soil loss stayed behind the upper part of the slope.

12.3.2.2
Flowchart for the Determination of the Erosion Resistance

To ease the application of the PEPP model, equations were sought to determine the erosion resistance by it's main influencing factors water content, humus content and clay content. Multiple step-wise regressions were conducted with these factors, and the correlation coefficients were compared. A flowchart was established to estimate the erosion resistance dependent on the informations which are available for the model user (Fig. 12.8). Already with two of the three factors the erosion resistance can be estimated and the validity of the chosen equation can be seen by the respective correlation coefficient. The rainfall experiments on the loess soil in the Weiherbach catchment revealed a different reaction of the soil to erosion, dependent on the general soil moisture conditions in the investigated area. This is equal to the susceptibility of the soil to crusting. If the general weather conditions are such that the upper soil layer is dry and that aggregated soil particles are available, the soil will be prone to crusting. On the contrary, if the soil is humid and the aggregate stability high, no crusting will occur and the erosion resistance will be higher. If the model user has no information about the susceptibility of the regarded soil to crusting, the path in the flowchart with 'crusting unknown' should be chosen. The term soil moisture content refers here always to the mass water content before rainfall.

12.3.3
Estimation of Manning's *n*

The *geometric* roughness of the soil surface can be measured in detail for example by laser. An approximation method is the measurement of the difference between a straight line distance connecting two points and the actual distance measured over all microtopographic irregularaties. To do this a 1 m long chain is laid on the ground and the length of the chain compared to a stable 1 m rod is measured. However, the Manning's *n* is a parameter for the *hydraulic* roughness of the surface which differs from the *geometric* roughness, since Manning's *n* for example depends on the flow depth. Nevertheless, a correlation between the Manning's *n* of the rainfall experiments and the soil parameters was sought. The mean aggregate diameter was suitable to estimate Manning's *n* since greater aggregates resist slaking and keep a rough surface. Also, the shortening of the chain can be used to identify the roughness. The chain must be laid in the direction of the slope before the rainfall. However, the calculated correlation coefficients were quite low, and therefore, no equation can be recommended for

CHAPTER 12 · **Simulating Hydrological and Erosional Processes** 269

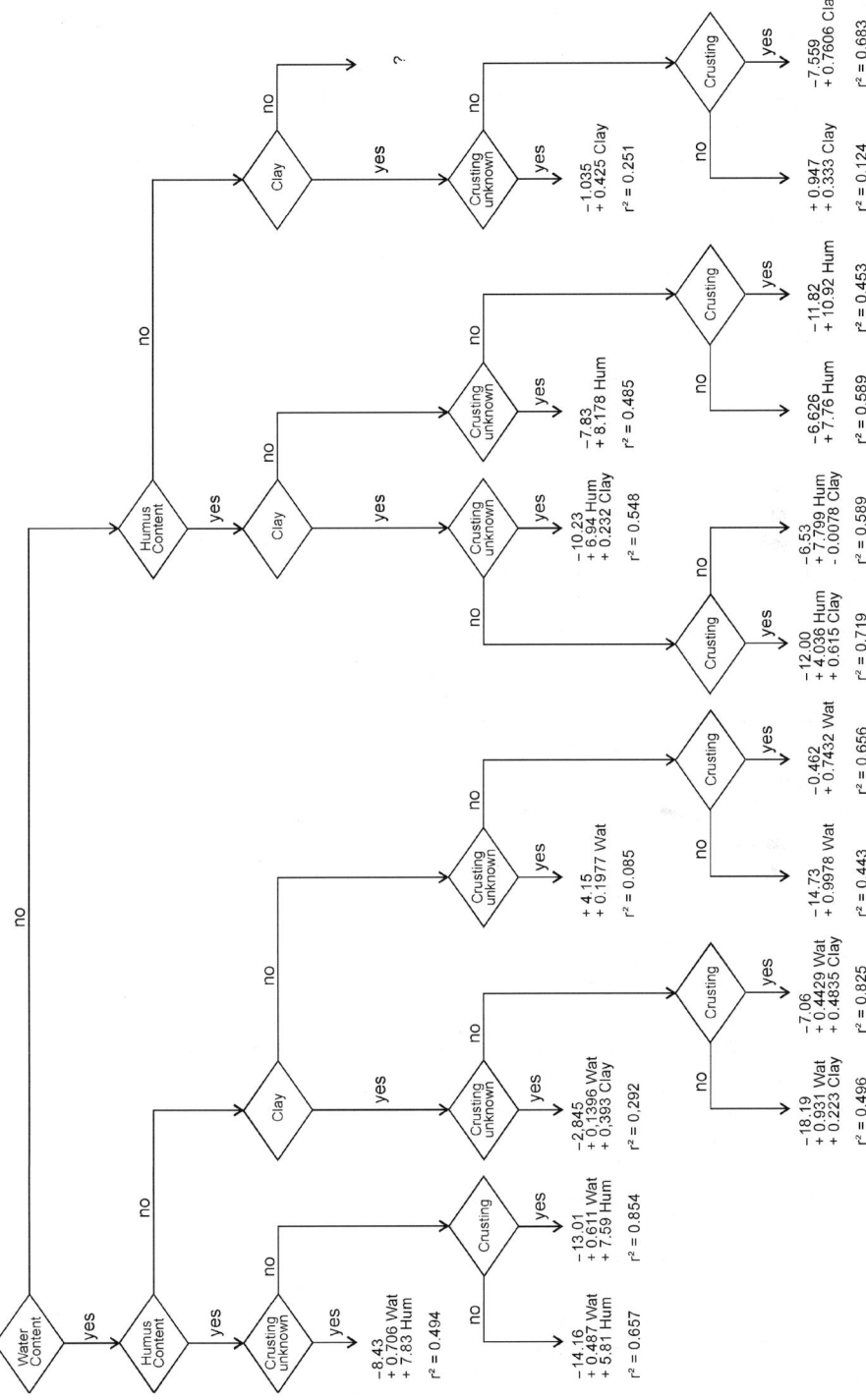

Fig. 12.8. Flow chart for the estimation of the erosion resistance by soil properties dependent on the informations available for the user ("Crusting unknown" means that the model user does not know if crusting of the soil happened. "Crusting" means that the aggregated soil was dry when the rainfall started, then a aggregate breakdown occurred, followed by crusting of the soil surface)

Table 12.7. Estimated values for the roughness coefficient (Manning's n), based on the results of the rainfall experiments

Land use or soil cover		Manning's n (m s$^{-1/3}$)		
		Low	Mean	High
Corn (seed bed to maturity)		0.015	0.042	0.145
Sugar beet (seed bed to maturity)		0.019	0.036	0.123
Freshly tilled soil (harrowed), crusted		0.015	0.037	0.074
Cereals (height up to 10 cm), crusted		0.010	0.026	0.050
Freshly tilled soil, chain shortening downslope:	<4 cm	0.010	0.030	0.067
	4 – 8 cm	0.012	0.036	0.123
	8 – 12 cm	0.020	0.059	0.190

the determination of Manning's n. A table for different crops was established with ranges of Manning's n (Table 12.7). The model user must decide if the soil is dominated by clayey or wet conditions which result in a higher surface roughness. For dry conditions or mainly small aggregates the lower value of Manning's n should be applied. In addition, the chain shortening of a 1 m chain can be helpful to estimate Manning's n.

12.4
Application of the PEPP-HILLFLOW Model

12.4.1
Simulation of Rainfall Experiments

For an application of the model on a small scale, the rainfall experiments of Schramm (1994) were simulated which were not included in the determination of the above regression equations. Bronstert (1994) simulated the hydrologic part of these rainfall experiments with the HILLFLOW model. Thus, the parameter for HILLFLOW were already available for this simulation.

As an example for the comparison of measured and simulated results, two consecutive rainfall experiments (plot size 22 m × 4 m) are depicted in Fig. 12.9. The first rainfall intensity used was 44 mm h^{-1}, on 19 October 1990, until steady state conditions were achieved. On 24 October 1990, a rainfall intensity of 33 mm h^{-1} was applied until surface runoff started and then, after a two hour break, the second measurement started. The erosion resistance for the model simulation of 7.6×10^{-4} N m^{-2} was determined by the flow chart (Fig. 12.8) using the content of clay (22.2%) and of organic matter (1.8%). It was equal for both rainfall experiments. Figure 12.9 shows the correlation of the simulated and measured results. The influence of the soil moisture on the erosion resistance is limited to the conditions when the first heavy rainfall starts. The saturation which is achieved after the first experiment and before the second experiment does not change the erodibility of the soil. The kind of crusting of the first high intensity rainfall determines the detachment capacity for the following events so that the

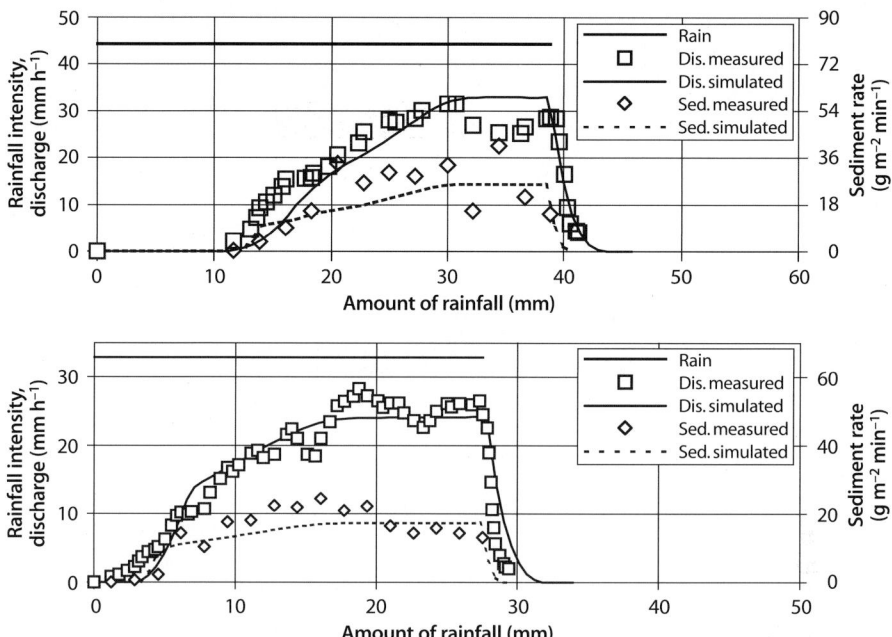

Fig. 12.9. Rainfall, discharge and sediment rate of the rainfall experiments Neuenbürg on 19 October 1990 (above) and 24 October 1990 (below). Comparison measured and simulated results from PEPP-HILLFLOW

erosion resistance must only be chosen once. This corresponds of the experiences made by the investigations about the temporal variability of the erosion resistance.

12.4.2
Simulation of Natural Rainfall Events on Slopes

As an example for the application of the PEPP-HILLFLOW model on slopes, the simulation of the slope *Leierfass* in the Weiherbach catchment is presented. At this slope a test plot with a field trough was installed to sample the surface runoff and the sediment load of a 4 m wide and 69 m long stripe. A cross-section of this slope is depicted in Fig. 12.10b (mean gradient 12.8%, maximal gradient 16.5%). During the growing season 1994, sugar beet was planted in the direction of slope. Within this period, two erosive rainfalls led to measurable soil loss (Table 12.8).

For the model application the spatial distribution of the soil types on the slope has to be known. The loess soils in the Weiherbach catchment can be classified in three groups according to the soil forming processes. Soil type 1 corresponds to the colluvial soils, soil type 2 to the lower part of the slopes and soil type 3 to the upper parts (Table 12.9). The soil hydrological variables and the Van Genuchten parameter were determined using the pedo-transfer functions of Carsel and Parrish (1988) based on the mean soil texture data of the respective soil class. In addition, measured values from field soil samples were used.

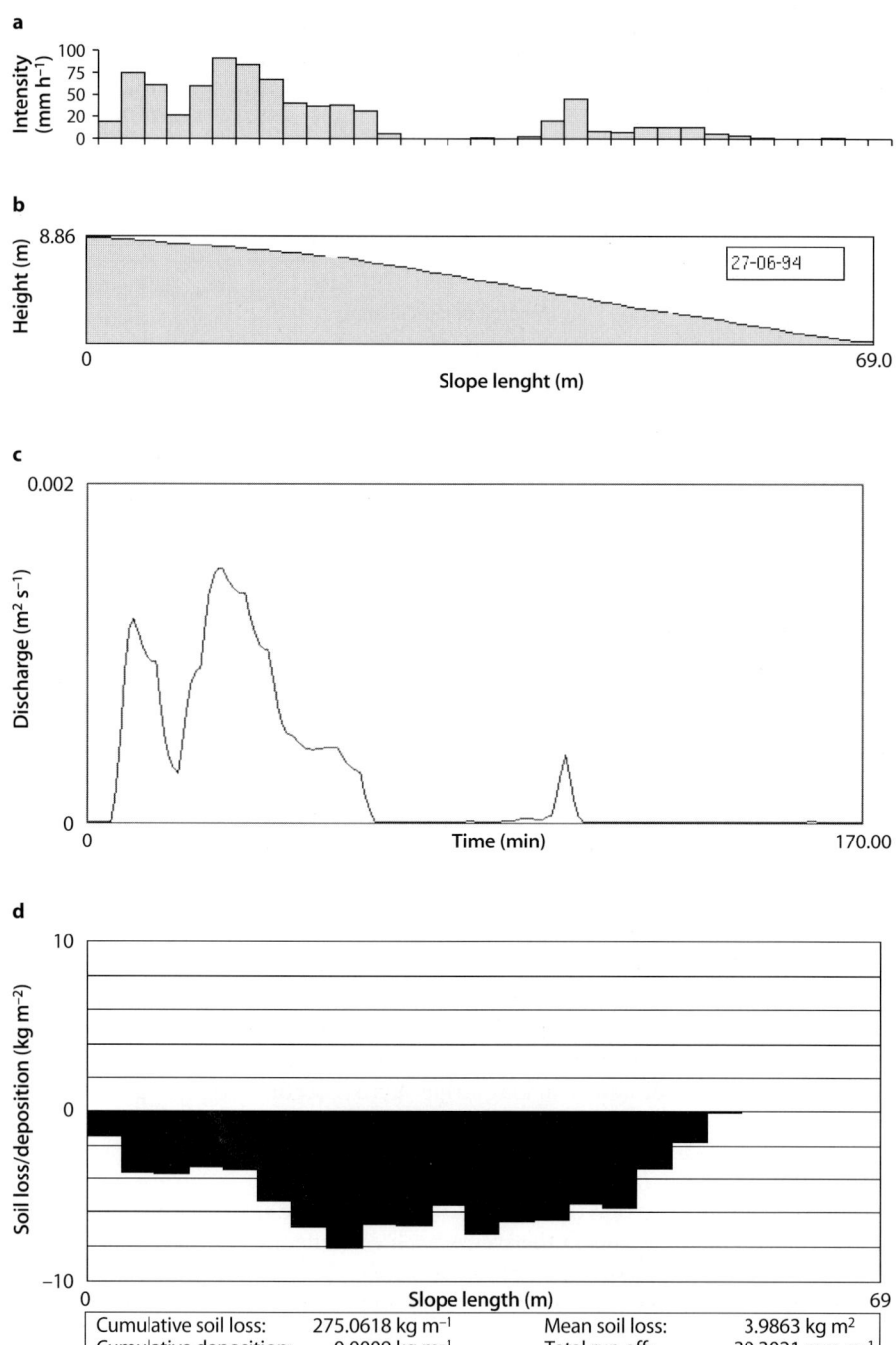

Fig. 12.10. a Rainfall intensity of the event of 27 June 1994; **b** Slope profile; **c** Calculated discharge of the event (model result of HILLFLOW); **d** Result of the erosion simulation (screen shot from PEPP)

On the simulated slope *Leierfass* only the soil types 2 and 3 occur. Their spatial distribution on the slope was determined using aerial photographs. To keep the price low, these photographs were taken by a small aircraft with an ordinary camera. They were processed on a photo-CD and can be geometrically corrected. The classification can be done according to spectral differences and in the comparison of determined soil texture analysis, since the areas with lower content of clay and humus show a brighter color. By the same means, the spatial distribution of the erosion resistance was estimated using Fig. 12.8. Combining the respective clay content of the soil types with measurements of the content of organic matter leads to an erosion resistance of 7.7×10^{-4} N m^{-2} (soil type 2) and 4.1×10^{-4} N m^{-2} (soil type 3). The transition between the two values of the erosion resistance along the slope was done continuously to avoid numerical instabilities.

For the HILLFLOW model the variables were chosen according to Bronstert (1994). The conductivity of the macroporous soil layer where the interflow occurs was estimated to 0.002 m s^{-1}. It's depth was set equal to the depth of the plowing horizon (0.3 m), and the free volume of the macropores was estimated to 0.5%. The initial soil moisture content was determined by field measurements with TDR, which where interpolated on the slope by the means of a specific kriging method (Bayes–Markov–Kriging, Lehmann 1995).

The roughness coefficient was chosen dependent on the development of the sugar beets. At the first erosive event in 1994 the field was under seed bed preparation. Therefore, a Manning's n of 0.04 s m$^{-1/3}$ was estimated according to Table 12.7. At the second event, the sugar beets covered already 90% of the surface and only in the gap between the planting rows (every 0.5 m) surface runoff could occur. Thus, the first event was simulated under the assumption of sheet flow while for the second event, rill flow with a rill distance of 0.5 m was assumed. To count for the higher roughness due to the plants at the second event the roughness coefficient was set to 0.06 s m$^{-1/3}$.

In Fig. 12.10c and 12.10c the simulated discharge and soil loss for the second erosive event are depicted as screen shots of the PEPP-HILLFLOW model. There is no soil loss in the lower part of the slope due to the higher erosion resistance in this area but the transport capacity calculated by the approach of Engelund and Hansen (1967) is sufficient to transport the above eroded material to the end of the slope. The simulated mean soil loss of 3.9 kg m^{-2} (equal to 39 t ha^{-1}) corresponds well to the observed soil loss of 42.5 t ha^{-1}. Also for the first erosive event, the simulated value (5.2 t ha^{-1}) is comparable to the measured soil loss (5.17 t ha^{-1}). This shows the suitability of the PEPP-HILLFLOW model to simulate erosive rainfall on slopes.

12.4.3
Simulation of Natural Rainfall Events on Small Catchments

The PEPP-HILLFLOW model is a slope model but it can also be used to simulate small catchments by calculating individual slopes and combining them with a channel network. The watershed is thus represented by a cascade of planes and channels. This procedure is also applied in other models, such as KINEROS (Woolhiser et al. 1990) or IHDM4 (Beven et al. 1987).

The catchment has to be divided in single slopes according to the land use and the topography. Field observations have to be included, because single structures like furrows, walls and farming roads strongly influence the direction of the surface runoff flow path. A digital elevation model is in general not dense enough to count for these features. For each slope the main flow path has to be determined. Its length and gradient can be taken from a Geographical Information System. As the PEPP-HILLFLOW model only simulates slopes of 1 m width, the selected slopes should have a rectangular form. The model results may then be multiplied by the respective width of the slope to obtain the total discharge and soil loss. For irregularly shaped slopes, by division of the field area by the length of the flow path, a theoretical width of a rectangular slope may be calculated.

By dividing the watershed into a branching system of channels with plane elements contributing lateral flow to the channels or to the upper end of first order channels, runoff is routed over each plane and through the channel system. For hydrological flow routing in the channel the Kalinin-Miljukov method was applied (Plate et al. 1977).

As examples, the results of the simulation of two rainfall events in a sub-catchment of the Weiherbach area will be presented. The spatial distribution of the three soil types (Table 12.8) in the sub-catchment was determined by soil survey data, topography and aerial photographs. Single point measurements of the initial soil moisture content were interpolated for areas of the watershed using Bayes—Markov—Kriging. The change of the soil moisture distribution from the last measuring day to the start of the erosive rainfall event was calculated with HILLFLOW.

The moisture data, together with the clay content of the soil, were used to determine the erosion resistance using Fig. 12.8. On 12 August 1994 for example, the soil was prone to crusting due to the low water content in the soil. Therefore, a soil moisture content of 8.8% for soil type 3 corresponds to an erosion resistance of 3.6×10^{-4} N m^{-2}, of 10.7% (soil type 2) to 8.3×10^{-4} N m^{-2} and of 14.0% (soil type 3) to 13.7×10^{-4} N m^{-2}.

The hydraulic roughness coefficient was chosen according to Table 12.7 dependent on the respective land use and soil cover. Manning's n numbers of 0.16 s m$^{-1/3}$, 0.08 s m$^{-1/3}$ and 0.03 s m$^{-1/3}$ were selected for cereals, corn and sugar beet, and vegetables (cucumber), respectively. For the infiltration related parameters, Table 12.9 was used. The conductivity of the macroporous soil layer where the interflow occurs was estimated to 0.0015 m s^{-1} and it's depth was set equal to the depth of the plowing horizon (0.35 m). The free volume of the macropores was estimated to 1.5%. All three of these parameters influence considerably the amount of infiltration in the upper soil layer, especially for heavy rainfall events when the flow in the macropores is important. As there presently exist no detailed information about the macropore distribution in the soil, the determination of the macropore volume is linked with uncertainty. Further research in determining the influencing factors and the spatial distribution of this parameter is currently proceeding in the research project.

Table 12.8. Data of the natural erosive rainfall events on the slope Leierfass 1994

Date	Total rain (mm)	Rain duration	Mean intensity (mm h^{-1})	Max. 5 min int. (mm h^{-1})	Soil loss (kg)	Meas. soil loss (t ha^{-1})
25 April 1994	22.2	165	8.1	73.2	142.7	5.2
27 June 1994	64.4	160	24.2	91.2	1 172.2	42.5

Table 12.9. Classified soil types of the loess soils in the Weiherbach catchment and respective soil hydraulic parameters

	Abbrev.[a]	Abbrev.[b]	Clay (%) <2 μm	Silt (%) 2–63 μm	Sand (%) 63–2000 μm
Soil type 1: stark schluffiger Ton[a]	Tu4	SiCL	30	67	3
Soil type 2: stark toniger Schluff[a]	Ut4	SiL	22	74	4
Soil type 3: mittel toniger Schluff[a]	Ut3	SiL	14	80	6

Parameter	Unit	Soil type 1	Soil type 2	Soil type 3
Saturated soil moisture content Θ_S	$cm^2\ cm^{-2}$	0.42	0.43	0.46
Residual soil moisture content Θ_R	$cm^2\ cm^{-2}$	0.096	0.088	0.067
Saturated hydraulic conductivity k_S	$mm\ h^{-1}$	3.00	3.90	4.90
Van-Genuchten-parameter α	(1/m)	1.50	1.75	2.00
Van-Genuchten-parameter n	(–)	1.25	1.45	1.41

[a] German name and abbreviation according to AG Boden (1994: 134);
[b] Name according to US Dept. of Agriculture (SiCl: silty clay loam, SiL: silty loam).

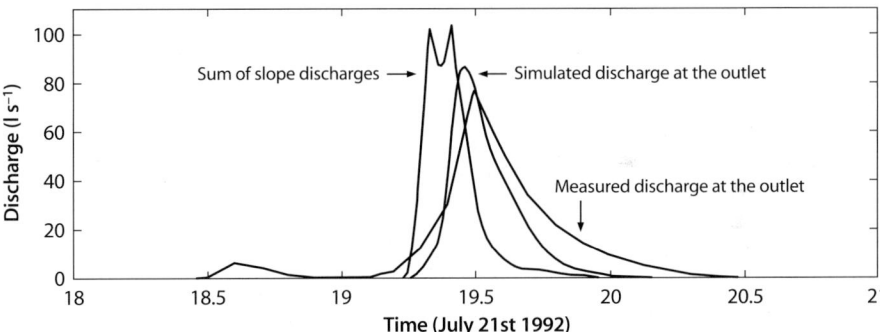

Fig. 12.11. Comparison of measured and simulated discharge (as sum of the calculated discharges of the 16 isolated slopes and after the flood routing) for the event on 21 July 1992

In Fig. 12.11, the results of the simulated and measured discharge for the sub-catchment on 21 July 1992 are depicted. A positive correlation is evident, however, the sensitivity of the HILLFLOW model to the macropore volume is very high, because a macropore volume of 0.5% leads to a significantly over-estimated discharge.

For the validation of catchment models, the comparison of the results with the discharge measurement on the outlet is insufficient. The comparison of the interpolated, spatial distributed soil moisture contents with simulated values allows a distributed evaluation, but the interpolated values may also include errors. In addition, erosive rainfall events are related to water contents close to saturation in the complete catch-

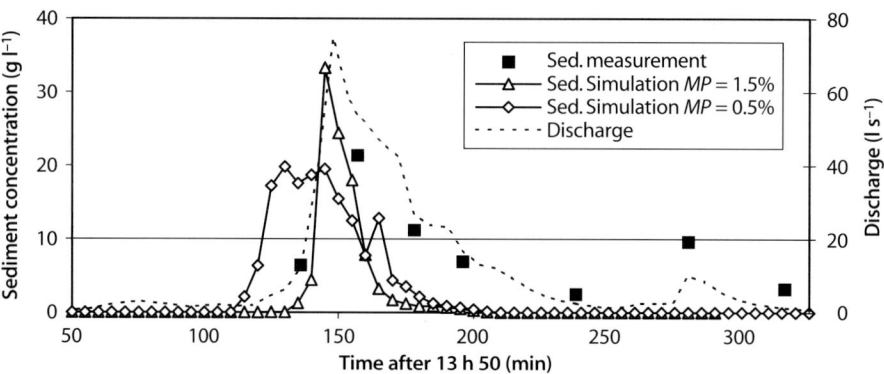

Fig. 12.12. Measured and simulated sediment concentration at the outlet of the sub-catchment for the event of 12 August 1994 (*MP* = free volume of macropores)

ment so that differences between simulated and interpolated values are minimal. As long as distributed models are validated only with outlet data their advantage compared to lumped models is not evident.

This applies also for the erosion part of the model. In addition, the erosion processes are superimposed on hydrologic variability which may lead to an error of propagation.

In Fig. 12.12, the results of the simulated sediment concentration at the outlet, with a macropore's free volume of 0.5%, are represented with the measured values for the event on 12 August 1994. The comparison with measured values is reasonable.

The results of the simulated sediment concentration must be considered with precaution. A similar reasonable sediment concentration curve can be achieved with a range of combinations of input parameters, for example with a free volume of macropores of 1.5% (Fig. 12.12).

Since the sediment concentration is not unique, detailed measurements of the spatial distribution of the soil loss on the slopes in the catchment have to be carried out. The level of Caesium-137 in the soil can be used to map soil erosion patterns in a landscape, but the uncertainty of this method is considerably high (De Roo 1993). Mapping of erosion damages or the use of marks (e.g. pins) are time consuming and difficult to quantify (e.g. for sheet erosion).

Further improving of the models should be connected to an improvement of the data collecting methods. Due to the development of the computer technique, not the calculation but the validation and verification of the model results with measured data are the main problem.

12.5
Summary and Outlook

The aim of the multidisciplinary research project 'Weiherbach' at Karlsruhe University is the development of an operational physically based numerical model for predicting transport of water, eroded soil, phosphorus, nitrogen and pesticides in a small rural catchment. An operational, event-orientated model for soil erosion called *PEPP* was developed. Since there is no universal equation to determine transport capacity, four different methods have been implemented in the *PEPP* model. The increase of the fine particle fraction in the flow due to selective deposition of coarser particles is computed.

The determination of the potential erosion rate is based on the calculation of the momentum flux acting on the soil particle produced by raindrops and by overland flow. The soil particle is detached if these external forces overcome the erosion resistance of the soil. The effective detachment capacity can be determined by two different methods. Also different deposition coefficients can be chosen by the model user.

To determine the model parameter (especially the erosion resistance and the hydraulic roughness) a transportable rainfall simulator (12 m × 2 m) was incorporated into the study. The main investigation area was the 6.3 km^2 Weiherbach catchment (SW-Germany) which is for the most part loess-covered.

Through the vegetation period the erosion resistance is influenced by rainfall characteristics, farmer's practices and the development of the soil structure. To estimate this temporal variability, 16 rainfall experiments were carried out on maize and sugar beet in the Weiherbach catchment over a specific time spacing. The results showed that once a crust on the soil is established a relatively small temporal variability throughout the vegetation period occurred.

To apply the erosion model to a catchment the spatial distribution of the erosion resistance has to be known. With the data of more than 60 rainfall experiments in different areas, the erosion resistances of the different soil types can be estimated. In addition, the rainfall experiments demonstrated that the main influencing parameters of soil loss are clay content, amount of organic matter and antecedent soil moisture content. A flow chart was developed to estimate the erosion resistance dependent on the information available for the model user. For a precise determination of the erosion resistance the variability can be determined from surveys of characteristic soil properties like aggregate stability or plasticity indices.

The soil erosion model can be applied in direct combination with the advanced hydrological model system HILLFLOW for hillslopes (Bronstert 1994). This distributed and physically based model simulates all relevant hydrological processes, such as interception, evapotranspiration, soil-moisture movement, surface runoff, subsurface stormflow, and especially infiltration in micro- and macropores. It was developed in the same research project. The application of the combined PEPP-HILLFLOW model demonstrated a simularity between measured and simulated results. Further investigations should also focus on a detailed parameter determination, beside the erosion resistance.

Currently, an intensive study with a small rainfall simulator is going on to analyse the macropore distribution (number, depth, width) in the soil and its relevance to infiltration and pesticide transport. In addition, the implementation of the PEPP model in the new hydrological catchment model CATFLOW (Maurer 1997) is being investigated. This will ease the calculation of the soil erosion in catchments. To account for diffusive nutrient transport the PEPP model will be extended to calculate the transport and enrichment of phosphorus (Scherer 2000).

References

AG Boden (1994) Bodenkundliche Kartieranleitung. 4. Aufl. Schweizerbart'sche Verl., Stuttgart

Alberts EE, Laflen JM, Spomer RG (1987) Between year variation in soil erodibility determined by rainfall simulation. Trans Am Soc Agric Eng 30:982–987

Bagnold RA (1966) An approach to the sediment transport problem from general physics. US Geological Survey Profess Paper 422-I

Beven KJ, Calver A, Morris EM (1987) The Institute of Hydrology distributed model. Institute of Hydrology, Wallingford, Report No. 98

Bronstert A (1994) Modellierung der Abflußbildung und der Bodenwasserdynamik von Hängen. Mitteilungen Inst f Hydrologie und Wasserwirtschaft, Heft 45, Univ Karlsruhe

Brunori F, Penzo M, Torri D (1989) Soil shear strength: its measurement and soil detachability. Catena 16:59–71

Carsel RF, Parrish RS (1988) Developing joint probability distributions of soil water retention characteristics. Water Resources Research 24:755–769

Coote D, Malcolm-McGovern C, Wall G, Dickinson W, Rudra R (1988) Seasonal variation of erodibility indices based on shear strength and aggregate stability in some Ontario soils. Can J Soil Sci 68:405–416

De Roo APJ (1993) Modeling surface runoff and soil erosion in catchments using Geographical Information Systems. Nederlandse Geografische Studies No. 157, Utrecht

Engelund F, Hansen E (1967) A monograph on sediment transport in alluvial streams. Tekniks Verlag, Copenhagen

Engman ET (1986) Roughness coefficients for routing surface runoff. J Irr Drain Eng 112(1/2):39–53

Flanagan DC, Nearing MA (ed) (1995) USDA-Water Erosion Prediction Project: hillslope profile and watershed model documentation. NSERL Report No. 10, USDA-ARS, West Lafayette, Indiana

Foster GR (1982) Modeling the erosion process. In: Haan CT, Johnson HP, Brakensiek DL (eds) Hydrological Modeling of Small Watersheds. ASAE Monograph No. 5, St. Joseph, MI: 297–380

Foster GR, Meyer LD (1975) Mathematical simulation of upland erosion by fundamental erosion mechanics. Proc. of Sediment Yield Workshop, ARS-S-40, USDA Sediment Lab, Oxford, MI:190–207

Gäth S (1995) Ursachen der Luftsprengung. Bodenökologie und Bodengenese, H 15, TU Berlin

Gerlinger K (1997) Erosionsprozesse auf Lößböden: Experimente und Modellierung. Mitt Inst Wasserbau und Kulturtechnik, Heft 194, Univ Karlsruhe

Gomer D (1995) Oberflächenabfluß und Bodenerosion in Kleineinzugsgebieten mit Mergelböden unter einem semiariden mediterranen Klima. Mitt Inst Wasserbau und Kulturtechnik, Heft 191, Univ Karlsruhe

Henk U (1989) Untersuchungen zur Regentropfenerosion und Stabilität von Bodenaggregaten. Landschaftsgenese und -ökologie, Heft 15, TU Braunschweig

Imeson AC, Vis M (1984) Seasonal variations in soil erodibility under different land-use types in Luxembourg. J Soil Sc 35:323–331

Kainz M (1981) Auswirkung einer langjährigen Stallmistdüngung auf das Bodengefüge. Diplomarbeit, Lehrst. f. Bodenkunde, TU München

Koch J (1995) Modellierung des Feststoffaustrags landwirtschaftlich genutzter Flächen durch Kombination eines Erosionsmodells mit einem hydrologischen Modell. Vertieferarbeit Inst. f. Wasserbau und Kulturtechnik, Univ Karlsruhe

Le Bissonnais Y, Bruand A, Jamagne M (1989) Laboratory experimental study of soil crusting: relation between aggregate breakdown mechanisms and crust structure. Catena 16:377–392

Lehmann W (1995) Anwendung geostatistischer Verfahren auf die Bodenfeuchte in ländlichen Einzugsgebieten. Mitt Inst Hydrologie und Wasserwirtschaft, Heft 52, Univ Karlsruhe

Maurer T (1997) Physikalisch begründete, zeitkontinuierliche Modellierung des Wassertransports in kleinen ländlichen Einzugsgebieten. Mitteilungen Inst. f. Hydrologie und Wasserwirtschaft, Heft 61, Univ. Karlsruhe

McCuen RH, Snyder WM (1986) Hydrologic modeling: statistical methods and applications. Prentice-Hall, Englewood Cliffs, New Jersey

Meyer LD, Wischmeier WH (1969) Mathematical simulation of the process of soil erosion by water. Trans Am Soc Agric Eng 12(6)

Murschel B (1991) Die Entwicklung eines Informationssystems zur Reduzierung der Erosion und des Stoffaustrages am Beispiel von Ackerböden im Kraichgau. Dissertation, Agrarwiss I, Univ Hohenheim

Nearing MA, West LT, Brown LC (1988) A consolidation model for estimating changes in rill erodibility. Trans Am Soc Agric Eng 31:696–700

Plate EJ (ed) (1992) Prognosemodell für die Gewässerbelastung durch Stofftransport aus einem kleinen ländlichen Einzugsgebiet. Schlußbericht zur 1. Phase des BMFT-Verbundprojektes. Mitteilungen Inst für Hydrologie und Wasserwirtschaft, Heft 41, Univ Karlsruhe

Plate EJ, Schultz GA, Seus GJ, Wittenberg H (1977) Ablauf von Hochwasserwellen in Gerinnen. Verl. P. Parey, Schriftenreihe des DWK

Scherer U (2000) Modelling phosphorus transport processes in an small southern German rural catchment. Proceedings of the IAHS-Symposium: The role of erosion and sediment transport in nutrient and contaminant transfer, 10.–14. Juli, 2000, Waterloo, Canada (in press)

Schmidt J (1991) A mathematical model to simulate rainfall erosion. In: Bork HR, De Ploey J, Schick AP (eds) Erosion, transport and deposition processes – theory and models. Catena Supplement 19:101–109

Schmidt J (1996) Entwicklung und Anwendung eines physikalisch begründeten Simulationsmodells für die Erosion geneigter, landwirtschaftlicher Nutzflächen. Berliner Geogr Abh 61, Inst Geogr, FU Berlin

Schmidt W, Michael A (1999) Bodenabtrag und Wasserinfiltration auf Einzelflächen und in Einzugsgebieten in Sachsen bei Bodenbearbeitung mit und ohne Pflug. Mitt Dtsch Bodenkdl Gesellschaft, Band 91, Heft 1:79–82

Schramm M (1994) Ein Erosionsmodell mit zeitlich und räumlich veränderlicher Rillengeometrie. Mitt Inst Wasserbau und Kulturtechnik, Heft 190, Univ Karlsruhe

Roth C (1992) Die Bedeutung der Oberflächenverschlämmung für die Auslösung von Abfluß und Abtrag. Bodenökologie und Bodengenese, Heft 8, TU Berlin

Torri D (1987) A theoretical study of soil detachability. Catena supplement 10:15–20

Van Genuchten MT (1980) A closed-form equation for prediction the hydraulic conductivity of saturated soils. Soil Sci Soc Am J 44:892–898

Werner M von (1995) GIS-orientierte Methoden der digitalen Reliefanalyse zur Modellierung von Bodenerosion in kleinen Einzugsgebieten. Dissertation, Inst Geogr, FU Berlin

Wischmeier WH, Smith DD (1978) Predicting rainfall erosion losses. A guide to conservation planning. US Dept of Agriculture, Agriculture Handbook, No. 537

Woolhiser DA, Smith RE, Goodrich DC (1990) KINEROS, a kinematic runoff and erosion model: documentation and user manual. US Dept of Agriculture, Agric Res Service, ARS-77, March 1990

Yalin MS (1977) Mechanics of sediment transport. 2nd edn. Pergamon Press Ltd., Oxford, GB

Yang CT (1979) Unit stream power equation for total load. J of Hydrology 40:123–138

Part III
Recent Model Developments

Chapter 13

Dynamics and Scale in Simulating Erosion by Water

R.E. Smith · J.N. Quinton

13.1
Introduction

Most models that simulate water-based erosion in use today are simple, lumped models, and most are empirical. There are good reasons for simplicity: for most engineering problems where an erosion estimate is needed, there is neither data, nor time or money to generate the data needed for more extensive analysis. However, engineering approaches have often limited the research directions, with relatively little effort expended to learn more about the complex process of soil movement by runoff water. In this chapter we will argue for the value of distributed models, either as better descriptors of the unsteady, nonuniform process of water erosion, or as means to improve the more simple models which will continue to be used in engineering applications.

Dynamic models are by definition distributed in time, and those we describe below are also distributed in space. Models lumped either in space or time ignore variations in those dimensions. Plot and field-scale lumped models such as the USLE (Wischmeier and Smith 1978) and RUSLE (Renard et al. 1994) are empirical regression models which are based on a specific set of experimental data covering differences in slope, plot length, soil, plant cover, and rainfall. The model WEPP (Lane and Nearing 1989) treats spatial variations in flow rate and slope, for example, but WEPP uses a single steady water flow profile to represent local flow conditions during the storm. This is time-lumping. The AGNPS watershed model (Young et al. 1987) employs a complex treatment of distributions in space using raster or grid divisions of a catchment, but uses lumped methods in each element. Application of lumped erosion regression models at each raster of a large area has little support in hydrologic science. Lumped models embody a particular set of implied assumptions, and such applications usually involve inappropriate assumptions regarding the material actually carried through a given element.

In this regard, scale is also important. Transport capacity estimates based on local hydraulic conditions, often assuming steady conditions, may be inappropriate when applied at the scale of a 100 m^2 element, within which much variation in flow conditions exist. But such scale extrapolations are commonly done in sediment modeling.

The two dynamic, distributed models which we describe below are not presented as ideal examples, but rather as representative of some of the features which should be part of erosion simulation models. There is a considerable challenge in incorporating natural heterogeneities into tractable numerical models, and much remains to be understood in the mechanics of entrainment of soil particles. We use dynamic models here to demonstrate their advantages over lumped models, and how their use may improve those more simple models of erosion.

13.2
Basic Features of KINEROS2

KINEROS2 ('K2') is a dynamic, unsteady, distributed model simulating runoff and sediment transport varying in time and space across a catchment. The erosion and runoff features of this model are similar to the models EUROSEM (and OPUS, Smith 1992), and K2 will be used to represent the simulation of erosion as a dynamic transport process.

K2 requires the catchment to be approximated by a treelike network of cascading elements. Elements may be rectangular surfaces ("overland flow") or channels, or ponds. Figure 13.1 illustrates the basic elements used. Surfaces may be completely pervious or be partially impervious, such as an urban source area with house roofs and streets. Channels may be pervious as well, with simple or compound trapezoidal geometry. For catchment simulation, a number of raingages scattered around the catchment may be used, with space-time interpolation to find the expected rain pattern on any element. The features of K2 have been summarized in more detail by Smith et al. 1995a.

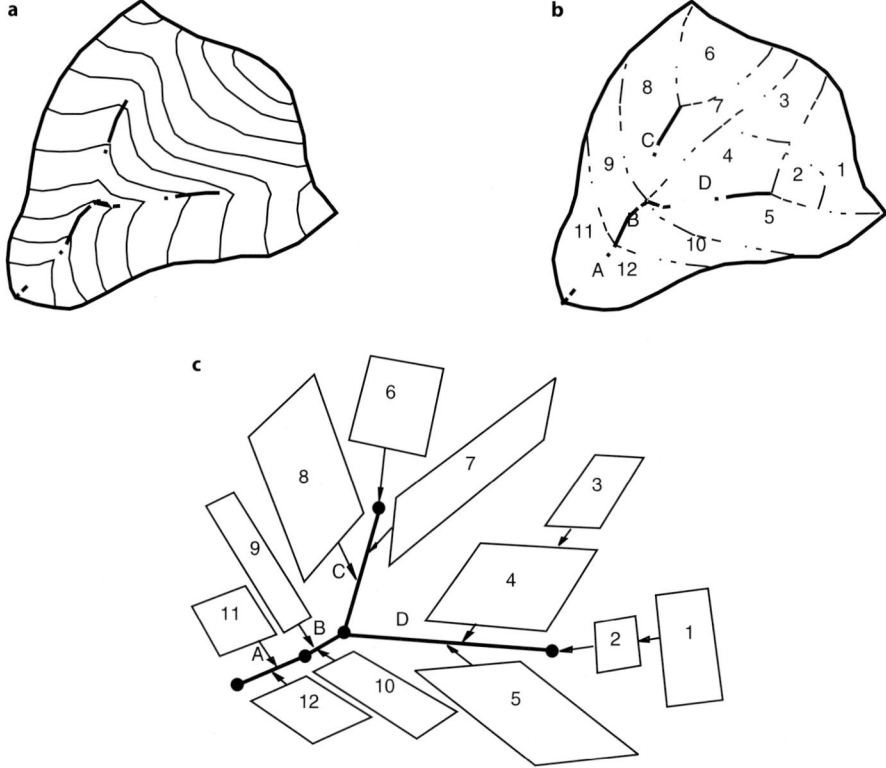

Fig. 13.1. Schematic illustration of the abstraction of a small catchment into the KINEROS2 network of planes and channels which represents catchment flow paths

13.2.1
Runoff

Runoff on surfaces is assumed to move parallel to one axis, and is thus treated with a one-dimensional dynamic continuity relation applicable to both surface runoff and channel flow:

$$\frac{\partial A}{\partial t} + \frac{\partial Q}{\partial x} = q_r(x,t) \qquad (13.1)$$

in which A is local flow area, Q is local discharge ($L^2\,T^{-1}$), and q_r is the local rate of production of rainfall excess. The excess rate is 0 prior to ponding and filling of surface detention, and is rainrate less infiltration rate, $r(t) - f(x,t)$, after ponding and while any water remains on the surface. Infiltration rate f is found using the Smith-Parlange infiltration equation (Smith and Parlange 1978), based on unsaturated soil hydraulic properties. Infiltration calculations are made interactively with runoff calculations, to simulate infiltration losses during recession flow, when rainfall may have ceased, or to simulate runoff advancing down an ephemeral channel.

The kinematic wave equation is used for surface and channel flow routing, by combining Eq. 13.1 with a Manning normal flow relation for local discharge $Q(A)$ or velocity $v(h)$. Solution of Eq. 13.1 is obtained by an implicit finite difference method (Woolhiser et al. 1990). In this manner the model obtains values of surface water velocity, unit discharge, depth, and flow width at each time and each location on the catchment for use in simulating erosion and sediment transport. Microtopography is considered in runoff simulation, through the $Q(A)$ relation, and also plays a role in interdependencies of runoff and infiltration (Woolhiser et al. 1996).

13.2.2
Sediment Transport

KINEROS2 treats sediment movement as an unsteady, one-dimensional convective transport problem, using a continuity equation similar to that for water:

$$\frac{\partial(AC)}{\partial t} + \frac{\partial(QC)}{\partial x} = e_s + q_s C_s - w f_C \qquad (13.2)$$

in which C (= $C(x,t)$) is concentration of sediment; e_s is sediment entrainment rate; q_s is local addition of water; w is flow width; and C_s is the sediment concentration in inflow q_s, such as for lateral inflow to a channel. Eqation 13.2 is applicable to either runoff surfaces or channels. It is also applicable to the concentration of any representative particle size, and K2 treats a soil as made up of as many as five different representative sizes. A negative value of e_s represents local deposition. The last term f_C represents deposition induced by infiltration flux f. Sediment entrainment e_s is composed, potentially, of two independent sources; rainsplash detachment e_r and hydraulic erosion e_n:

$$e_s = e_r + e_n \qquad (13.3)$$

Hydraulic erosion may be positive or negative, and is treated as the net difference between continuous processes of soil detachment q_e, and deposition q_d:

$$e_n = q_e - q_d \tag{13.4}$$

where q is a volume rate of material lost or gained per unit bed area. Moreover, deposition rates for a particle size class i are assumed to be directly estimable from particle settling velocity v_s and concentration:

$$q_{d_i} = (wv_sC)_i \tag{13.5}$$

We further reason (Smith et al. 1995b) that "transport capacity" concentration, C_{mx}, at a given flow condition represents a dynamic equilibrium, where rate of bed erosion just matches the rate of deposition, and thus a relation for transport capacity and erosion rate emerges. This assumes a bed of loose material, i.e., a cohesionless soil. The modification for crusted or cohesive soil will be treated subsequently. On this premise, for $C_i = C_{mxi}$, we can set

$$q_{e_i} = q_{d_i} \tag{13.6}$$

since in this case e_n is 0 in Eq. 13.4.

Transport capacity concentration C_{mx} is a function of local slope, runoff velocity and depth, and particle properties. Numerous relations have been proposed for transport capacity, several of which indicate a threshold condition, below which (flow or hydraulic shear, or stream power, for example) C_{mx} is zero. Based on Eq. 13.5 and 13.6, we may write an expression for erosion rate, based on any transport capacity relation, for any particle class i,

$$q_e = wv_sC_{mx} \tag{13.7}$$

understanding that C_{mx} can be found for any flow condition. Combining Eq. 13.7 and 13.4 we form an expression for net erosion rate, assuming reversible processes at the bed:

$$e_{n_i} = bwv_{si}(C_{mx} - C)_i \tag{13.8}$$

Coefficient b ($0 < b < 1$) is introduced to account for cohesive resistance to erosion, and relative abundance of material of particle class size i. It is 1 for deposition conditions ($C > C_{mx}$) or cohesionless material, but is much smaller for erosion conditions ($C < C_{mx}$) on cohesive or crusted soil.

The other major source of surface soil entrainment is rain splash energy e_r in Eq. 13.3. The value of e_r is usually assumed a function of the energy or momentum of raindrops. The EUROSEM model also treats secondary energy, which results from drops forming on elevated vegetation and subsequently falling to the ground. In any case, knowledge of cover extent and height can be used to modify the total erosive energy of the rain underneath cover of whatever type.

K2 assumes, from (Meyer and Wischmeier 1969), that rainfall induced entrainment rate is proportional to the square of the rainfall rate:

Fig. 13.2. Comparison of the rainfall splash detachment functions of KINEROS2 and EUROSEM. Changing parameter values can make the two functions similar in any desired range, but assumptions regarding rainfall energy are different

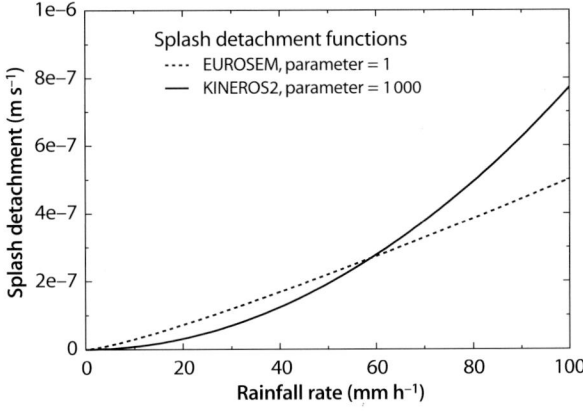

$$e_r = c_f r^2 \exp(-hc_h) \tag{13.9}$$

in which c_f is a soil condition-based coefficient, and the exponential factor represents the dampening of splash detachment due to mean water depth h. Parameter c_h is suggested to be 2 for h in mm, providing significant damping for depth h on the order of raindrop size or larger.

EUROSEM assumes a logarithmic relation (Morgan et al. 1997) for splash erosion, and the two assumptions are similar though different, as illustrated in Fig. 13.2. Both are based on observations of rainfall or rainfall energy spectrum. As can be appreciated from this figure, changes in the coefficient can make the two curves similar in any range of rainfall rate, but not in all ranges simultaneously.

13.2.3
Initial and Boundary Conditions

Solution of Eq. 13.2 is facilitated numerically by the determination through Eq. 13.1 of values of $Q(x,t)$ and $A(x,t)$ at each increment of time, dt. In addition, one may assume, for upland reaches, the boundary condition $Q = A = 0$ at $x = 0$. An initial condition applies at the instant that local runoff begins, t_p (ponding time), not at the beginning of rainfall. Since rainfall detachment occurs at this point, Eq. 13.2 may be solved for the conditions $A = 0$ and $dQ/dx = 0$, and $f = r$, with $b = 1$, and deposition rates equal to rain detachment rates, giving an initial condition for C_o of:

$$C_o(t = t_p) = \frac{c_f r^2}{w(r + v_{si})} \tag{13.10}$$

13.2.4
Solution Characteristics

Equation 13.2, as a concentration-based transport equation, has noteworthy features, some more obvious than others. During the rising hydrograph, flow depths are increasing, and there is tendency for dilution of the flow as well as particle entrainment. Thus a steady or limited source of sediment can show reduced sediment concentrations as flow increases. In addition, during the hydrograph recession, while rainsplash entrainment is reduced or eliminated, and sediment transportability is decreasing, there may be sustained transport since concentrations must be above transport capacity concentration for net deposition to occur. In addition, the linear relation of Eq. 13.8, for small settling velocities and thus a long characteristic time, may cause concentrations to be increasing after the flow peak as long as C_{mx} exceeds C. The result is that we expect concentrations to be elevated during recessions compared to the earlier part of a hydrograph. Infiltration of water with concentration C during recession flow is assumed to deposit a proportional amount of sediment at the soil surface, rather than concentrating the sediment into the smaller depth of surface water.

13.2.5
Treating Soil with Distributed Particle Sizes

There are several reasons for explicitly considering the range of particle sizes found in soil. First of all, existing transport capacity relations were based on experiments with uniform particle sizes in flumes, but soils are rarely so constituted. The relation of particle size to transport capacity may be nonlinear, and thus the transport capacity of the mean particle size d_{50} may not represent the net transport capacity of the ensemble of particle sizes. Further, the transport processes during a storm on a land slope will usually result in enrichment: smaller particle sizes may be transported and larger ones deposited. This sorting and enrichment may be very important in water quality studies related to soil-adsorbed chemicals, but it cannot properly be simulated using a single average soil d_{50}. Particle erosion is assumed to be limited, in part, by the abundance and transportability of the least transportable particle class size. Otherwise, soil surfaces would become "paved" by larger particles left behind by removal of the more erodible classes.

Questions of the distribution of flow transporting capacity among different particle sizes in a soil are as yet theoretically or experimentally unresolved. There are, however, some reasonable assumptions about the transport of an ensemble of particle sizes that can be made, allowing us to treat them appropriately:

a Conservation of mass equations apply to each particle size.
b Particle class erosion rates must be proportional to the relative availability of that class in the surface soil. This applies to both rain and flow detachment.
c Hydraulic erosion from a soil surface is controlled by the transportability of the largest and least transportable particle size. On the converse assumption, an "erosion pavement" would quickly occur, shutting off erosion.
d Deposition rates of different particle sizes are mutually independent.

KINEROS2 allows soil particle size distributions to be different on each hydrologic element, so that, for example, a channel bed may be composed of loose sand, whereas an upland surface may be mostly silt or loam. The soil of any element must be described, however, by proportions from the same 5 particle size classes. EUROSEM uses a representative particle size. For most soil particle sizes this appears reasonable, given the evidence that, in the silt to sand size range, particle size is not a significant factor in transport capacity (Govers 1990).

13.3
Application Examples

13.3.1
Splash vs. Erosion Limiting Conditions

Erosion patterns are affected by the distribution of erosive causes between rainfall energy and entrainment by flowing water. This division of erosion sources is traditionally labeled "interrill" and "rill" erosion. However, soil particles may be entrained by overland flow where there is no evidence of rills, and rainsplash energy can also act within rills. Those traditional labels will therefore not be used here.

Splash erosion is often dominant in the early stages of a storm. One reason for this is that rainfall kinetic energy can be transmitted directly to the soil in the absence of a protective cover of water. As surface water accumulates, of some of the kinetic energy of raindrops is dissipated (Hartley and Alonso 1991). Another reason for this is that runoff often occurs in the latter parts of storms, since earlier high intensities may occur prior to inception of runoff. The higher intensities creating runoff will tend to be towards the beginning of runoff periods. Time lumped simulation models, by treating the entire storm with a steady water flow profile along the runoff path, will not be able to make an accurate distinction in these processes. Such models will rather have splash erosion dominate the upper reaches and rill erosion dominate the lower reaches.

When splash energy suspends soil particles in overland flows that have little or no transport capacity, the asymptotic concentration in the moving water is that required for deposition rate to balance the splash detachment rate. This results in an expression for C similar to Eq. 13.10:

$$C = C_{mx} + \frac{c_f r^2}{v_s w} \tag{13.11}$$

Movement of particles in the absence of transport capacity has been termed "rain flow transportation" (Moss et al. 1979). For high rainfall energies this concentration may be many times higher than C_{mx}, and may carry particles until a concentrated flow area is reached which can transport the sediment out of the catchment.

Figure 13.3 illustrates a condition in which splash suspension is the dominant overland erosion process. The catchment is a 4.4 ha watershed in Walnut Gulch experimental watershed in Arizona. The example is rather extreme, insofar as the catchment surface is quite resistant to flow erosion, but produces significant sediment. In this storm simulation, b (Eq. 13.8) is set small (0.0005) to account for flow erosion resistance, but rain intensities are high and c_f (Eq. 13.9) is quite large (800 to 80 000). Sampled data are consistent with this dominance of splash erosion, as illustrated here. Simulation of even small amounts of hydraulic erosion would have C increase somewhat with the flow rate, and have much higher concentrations later in the event, soon after the peak flow. It should be noted that sediment transport capacity remains important when splash detachment dominates, as indicated in Eq. 13.11, but for small values of C_{mx}, it is the runoff rate which limits sediment discharge from the catchment.

Figure 13.4 illustrates the temporal dynamics of erosion components during a storm. The example here is a runoff event on a plot at Woburn, Bedfordshire England. This is a

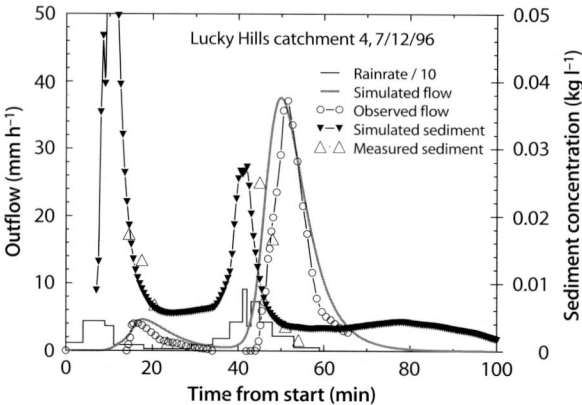

Fig. 13.3. KINEROS2 simulation of runoff and sediment concentration on Lucky Hills catchment 104, Tombstone Arizona. Because of the dominance of splash detachment, concentration peaks fall well before the runoff peak

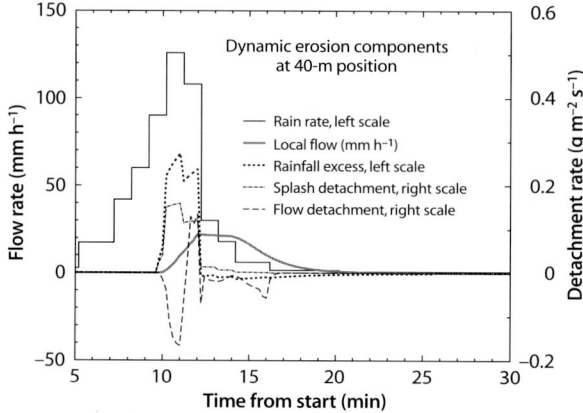

Fig. 13.4. Changes in simulated flow, rainfall excess, and detachment components at the lower end of a 40-m plot on the Woburn catchment, Bedfordshire, England

furrowed plot, and furrows are represented as regularly spaced rills. Changes in runoff and sediment production 40 m down the slope are illustrated. Negative values of flow detachment represent net deposition rates. In the earliest runoff, splash erosion is producing far more suspended matter than transport capacity, and a dynamic balance exists between suspension and deposition. Then as furrow flow increases, flow detachment increases until the runoff peak. During recession, the transport capacity drops, but first the excess load previously supported by splash detachment is deposited, and the remaining deposition is slower as concentration C closely tracks the declining transport capacity C_{mx}. Deposition rate follows C_{mx}, which abruptly declines at about 16 minutes as flow falls below the transport threshold stream power (vS) value of 0.004 m s^{-1}.

13.3.2
Temporal Changes

Implicit in the coefficient c_f is a factor that represents the exposure of bare soil to rainfall. In KINEROS and EUROSEM this is directly reflected by a cover factor (0–1). Complete cover will eliminate splash detachment, but such cases usually also create significant hydraulic resistance, so that surface water velocity (and transport capacity as a consequence) is considerably reduced. Most of these important erosion controls change throughout the year, requiring significant temporal changes in splash parameter c_f (Smith et al. 1997). Much more research is needed to evaluate temporal changes in the susceptibility of soil to rainfall energy. It is well known that rainsplash energy creates surface soil crusts on most soils, which increase runoff rates (Moore 1981), but may also decrease soil detachability.

13.3.3
Steady Flow and Length Effects

A weakness of simulating erosion by assuming a steady flow profile is that most erosive storms are short and intense, and a steady flow condition is never reached. While assuming a steady flow profile may on occasion provide a reasonable estimate of the storm sediment discharge, there is often little resemblance between steady flows and actual storm flows along a slope.

Equation 13.2 may be reduced to the steady flow erosion equation by eliminating the time variable terms. Flow, in that case, is steady but nonuniform. Assuming no lateral sediment inflow, as for a runoff surface, and ignoring the last term in Eq. 13.2 which is particular to recessions, the resulting ordinary differential equation is

$$Q\frac{dC}{dx} + qC = e_r + bwv_s(C_{mx} - C) \tag{13.12}$$

This equation is straightforward to solve along a runoff surface where $h(x)$, $Q(x)$, and $(q = dQ/dx)$ are known.

Figure 13.5 illustrates the water and sediment profiles $h(x)$ and $C(x)$ on the Woburn plot slope, at the time of peak flow, as compared with that for steady flow at the same runoff rate. Steady flow is the WEPP model assumption. Because of differences in upslope conditions and associated transport capacities, erosion rates and sediment concentration profiles are significantly different, despite the similarity at the downslope end. For the steady assumption, sediment concentrations rise steeply only after a certain distance owing to the sediment transport threshold of stream power. Above this point, sediment concentration equals that kept in suspension by the balance of splash and deposition rates, Eq. 13.11.

Note, this comparison assumes that for the steady approach the true value of the peak runoff rate, which is difficult to determine in the absence of a time distributed (dynamic) simulation, is known. Since the WEPP steady transport model performs a dynamic simulation to obtain this estimate, it would seem prudent to make a dynamic erosion estimate as well.

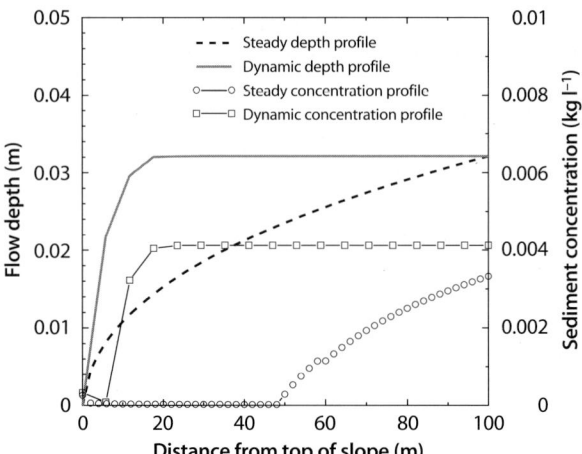

Fig. 13.5. Comparison of actual profile and sediment load conditions at the time of peak flow with a steady flow profile at the same flow rate, as used by the WEPP model

Table 13.1. Effects of plot length on sediment yield from a short storm, exemplified by the Woburn plot case of Fig. 13.4

Plot length	Runoff	Peak flow		Peak time	Time to equil., t_e	Sediment	
(m)	(mm)	(m³ min⁻¹)	(mm h⁻¹)	(min)	(min)	(kg)	(kg ha⁻¹)
5	1.980	0.116	62.10	11.0	1.6	5.0	446
10	1.994	0.216	57.80	11.5	2.6	11.4	509
20	1.971	0.411	55.00	12.0	4.0	28.3	632
40	1.917	0.499	33.40	12.0	6.0	70.7	788
60	1.867	0.500	22.30	12.0	7.6	121.0	898
100	1.764	0.500	13.40	12.0	10.3	230.0	1026
150	1.658	0.500	8.93	12.0	13.2	368.0	1095
200	1.557	0.500	6.69	12.0	15.7	491.0	1095
300	1.354	0.500	4.46	12.0	20.0	680.0	1011

One important reason for dynamic simulation for the treatment of short, intense storms, is to represent the effect of slope length properly. (The USLE and RUSLE approaches both contain factors for length of runoff surface that reflect steady flow assumptions and are based on data from relatively short (<25 m) slopes). The assumption that longer slopes create increasing sediment ignores storm dynamics. Table 13.1 illustrates simulated results for application of the storm of Fig. 13.4 to a range of slope lengths. Unit sediment delivery reaches a peak at some length and then decreases as the length becomes larger. Beyond some length greater than here shown, moreover, the absolute amount of sediment (and water) coming from the hillslope no longer increases, because water and sediment from upslope areas simply cannot reach the outlet on an infiltrating soil. (This condition is more dramatic for storms whose runoff amount is a smaller fraction of the rainfall, and for plots with more moderate slopes than illustrated here (~ 10%)). This fundamental feature of hillslope runoff illustrates the relation between the time scale of the rainstorm and the length scale of the hillslope. t_e is defined as the time needed for runoff from the upstream point to reach the outlet. Sediment yield begins to decrease beyond lengths where t_e exceeds the peak time. For purpose of comparison in this case, t_e was calculated assuming a steady excess rate of 60 mm h^{-1}, although in fact it varies in time during the rainstorm.

Simpler models which seek to avoid dynamic simulation should at least take these scale effects into consideration in some way if they are to be robust and accurate for storm erosion estimation.

13.4 Summary

Erosion processes vary significantly in space and time during an erosive rainfall event. The fundamental features of erosion and sediment transport simulation have been illustrated using two similar dynamic models, EUROSEM and KINEROS2, as examples, with application to experimental situations from plot and catchment studies. There are significant advantages in the dynamic simulation of erosion as compared with the assumption that a single steady profile may be used to estimate sediment transport. While the more simple approach may have relative success in some cases where transport capacity limits the sediment delivery, the results shown here indicate that it may be in error for short storms on long slopes, and where erosion rates are low. The interplay of time and space scales and the time span of erosive storms are important in determining the success of the lumped treatment of erosion processes. Dynamic modeling also can better estimate the spatial distribution of erosion due to the combination of hydraulic and splash detachment.

References

Govers G (1990) Empirical relationships on the transporting capacity of overland flow. Transport and Deposition Processes (Proceedings of the Jerusalem Workshop, March–April 1987). IAHS Publ 189:45–63

Hartley DM, Alonso CV (1991) Numerical study of the maximum boundary shear stress induced by raindrop impact. Water Resources Research 27(8):1819–1826

Lane LJ, Nearing MA (eds) (1989) USDA-Water Erosion Prediction Project: hillslope profile model documentation. NSERL Report No. 2. USDA-ARS NSERL, West Lafayette, Indiana

Meyer LD, Wischmeier WH (1969) Mathematical simulation of the process of soil erosion by water. Transactions of the American Society of Agricultural Engineers 12(6):754–762

Moore IJ (1981) Effect of surface sealing on infiltration. Transactions ASAE 24:1546–1552

Morgan RPC, Quinton JN, Smith RE, Govers G, Poesen JWA, Auerswald K, Chisci G, Torri D, Styczen M (1997) The european soil erosion model (EUROSEM): A process-based approach for predicting soil loss from fields and small catchments. Earth Surface Processes and Landforms

Moss AJ, Walker PH, Hutka P (1979) Raindrop stimulated transportation in shallow water flows: an experimental study. Sedimentary Geology 22:165–184

Renard KG, Laflen JM, Foster JR, McCool DK (1994) The Revised Universal Soil Loss Equation. Chapter 5. Lal R (ed) Soil Erosion Research Methods. Soil and Water Conservation Society, Ankeny, IA, 105–124

Smith RE (1992) Opus, An integrated simulation model for transport of nonpoint-source pollutants at the field scale: Vol I, Documentation. US Department of Agriculture, ARS-98

Smith RE, Parlange J-Y (1978) A parameter-efficient hydrologic infiltration model. Water Resources Research 14(3):533–538

Smith RE, Goodrich DC, Woolhiser DA, Unkrich CL (1995a) Kineros: A KINEmatic Runoff and EROSion model. In: Singh VP (ed) Computer models of watershed hydrology. Water Resources Publications, Highlands Ranch, CO, Chapter 20:697–732

Smith RE, Goodrich DC, Quinton JN (1995b) Dyniamic, distributed simulation of watershed erosion: The KINEROS2 and EUROSEM models. J of Soil and Water Conservation 50(5): 517–520

Smith RE, Goodrich DC, Unkrich CL (1997) Simulation of selected events on the Catsop catchment by KINEROS2, Proceedings of the GCTE Workshop on Catchment-scale Erosion Models. 14–18 April 1997

Wischmeier WH, Smith DD (1978) Predicting rainfall erosion losses – A guide to conservation planning. Agricultural Handbook No. 537, USDA, Washington, DC

Woolhiser DA, Smith RE, Goodrich DC (1990) KINEROS, A Kinematic Runoff and Erosion Model: Documentation and User Manual. US Department of Agriculture, Agricultural Research Service, ARS-77

Woolhiser DA, Smith RE, Giraldez J-V (1996) Effects of spatial variability of saturated hydraulic conductivity on Hortonian overland flow. Water Resources Research 32(3):671–678

Young RA, Onstad CA, Bosch DD, Anderson WP (1987) AGNPS, Agricultural Non-point source pollution model; a large watershed analysis tool. USDA-Agricultural Research Service, Conservation Research Report 35, Washington DC, USA

Chapter 14
Modeling Surface Runoff

St. Hergarten · G. Paul · H.J. Neugebauer

14.1 Introduction

In most climates, soil erosion is governed by the surface runoff of water during excessive rainfalls. Thus, numerical models for the simulation of surface runoff provide a key to quantitative modeling of soil erosion.

In the previous decades there has been much progress on this field. Beside some completely empirical approaches, several physically based models have been established, e.g. CREAMS (Knisel 1980), WEPP (Lane and Nearing 1989), KINEROS (Woolhiser et al. 1990), EROSION-2D/3D (Schmidt 1991; von Werner 1996), OPUS (Smith 1992), EUROSEM (Morgan et al. 1992), LISEM (De Roo et al. 1994), and PEPP (Schramm 1994). In many cases, these models enable us to reproduce, and sometimes even to predict erosion events fairly well. So, is the problem of modeling soil erosion solved now? In fact, it is not – mostly, the prediction does not work satisfactorily without calibration. The reason for this problem lies in the fact that these physically based models regard some physical principles like mass conservation, but the physical knowledge is not enough to describe the phenomena completely. Thus, some empirical approaches have to be integrated, which are not valid universally and require additional parameters. Often, these parameters cannot be determined in the absence of erosion events; additional measurements under artificial precipitation, like the discharge, the runoff velocity or even the soil loss itself, are necessary. Often the parameter values are only valid in the considered situation, so that every change in soil properties, relief or climate requires new measurements. In general, this procedure is called calibration. It is obvious that the need for calibration makes the model application costly and limits the quality of the prediction. So the long term aim of modeling soil erosion should be avoiding calibration wherever it is possible.

There are two general approaches towards this aim. First, the data basis can be enlarged, so that it finally covers a wide field of applications; alternatively, we can try to replace empirical approaches by physics step by step, so that less empirical data are required. Actually, the first way may be more feasible, but sooner or later the data shall be transferred into more general physical laws, which not only help us to predict, but also to understand the phenomena.

Every physically based water erosion model should include three components:

- infiltration
- surface runoff, and
- detachment and transport of soil particles

In some sense, the infiltration process is the most important part. Occurrence and intensity of runoff are governed by the difference between precipitation and infiltration rate. Thus, all the errors and inaccuracies of the infiltration model propagate to the runoff. Furthermore, the errors and inaccuracies of the calculated runoff cause errors in erosion and deposition.

From the view of physics, the infiltration process is quite clear. The Richards equation (Richards 1931), that is a generalization of Darcy's law, describes the flow of water in soils well down to scales of centimeters, even if effects like fingering are not completely understood. The progress in numerical methods and computer capacity enables us to solve this equation in three dimensions, including lateral currents. Adaptive discretization techniques allow resolution of sharp saturation fronts. Macropores can be simulated, too. But such a sophisticated approach requires much data such as the spatial distribution of soil properties and initial soil moisture as well as position and behaviour of the groundwater level. Moreover, a stochastical description of the macropores is required. For the purpose of understanding processes on a basic level, these points are not crucial. On the other hand, a quantitative prediction definitely requires these data. Thus, if a quantitative prediction is needed, the gain of such a theoretically sophisticated approach may not be obvious to the user in comparison with established simpler approaches, mostly based on that of Green and Ampt (1911).

Opposite problems occur if we look at particle detachment and transport. From the view of physics, these processes are not clear at all. So it is not surprising that the most striking differences between the established soil erosion models occur here. The most common approach has been taken from sediment transport in rivers and is based on an estimated shear stress acting on the soil. In contrast, in EROSION-2D/3D (Schmidt 1991; von Werner 1996) the momentum flux of runoff and raindrop impact is assumed to be the driving force. As a third idea, the dissipation of turbulent energy (Paul 1998) seems to be promising. All these approaches have one thing in common: they are mainly empirically based, and thus limited in their validity. From the view of physics, the shear stress approach appears to be most appropriate for particle detachment, the energy dissipation approach most for particle transport. The advantage of the momentum flux approach consists of its ability to combine runoff and raindrop impact. But it should be noted that this requires an additional parameter – a characteristic length – so that even the weighting between runoff and raindrop impact is completely empirical. Currently, physical approaches seem not to be practical. From our point of view, a physical approach first requires a detailed description of the water flow on the surface.

14.2
Approaches to Surface Runoff

From the theoretical point of view, the flow of water on a surface seems to be at least as clear as the infiltration; the Navier-Stokes equations are appropriate for describing the flow of fluids and gases.

In order to obtain a physical model for particle detachment and transport, calculating a three-dimensional flow field on the scale of the grain sizes would be ideal. In this case, the forces acting on the soil particles could be calculated; this would yield a completely physical approach to detachment and transport. But even if the progress in computer hard- and software enabled us to do this, the data collection on this scale would be impossible. Even if we restrict the resolution to the typical scales of flow depth, i.e., some parts of millimeters, the problem remains.

Thus, a further reduction of the resolution to macroscopic scales is necessary. Because the model scale exceeds the flow depth, the three-dimensional Navier-Stokes equations can be replaced by the two-dimensional shallow-water equations, reducing the numerical effort significantly. On the other hand, small turbulent eddies driving detachment and transport cannot be resolved any more. These phenomena have to be introduced artificially using turbulence models. For applications to rivers, channels or even oceans, there has been much progress on this field. But these applications have got one thing in common: the turbulence is induced by the roughness of some walls or a bed, which is small compared to the whole model region. As an example, the roughness of channel walls has the order of magnitude of some centimeters, while channels themselves are usually some meters wide or deep. In contrast, the depth of surface runoff rarely exceeds some millimeters, except for rill flow. Thus, the size of plants, litter, gravel or even coarse sand exceeds the typical dimensions of flow, so that the roughness of the surface cannot be treated as a boundary effect, but governs the whole flow regime.

Due to this problem and the still high numerical effort, this hydrodynamic approach has not been established widely in applications, although if the results of the small scale model of Zhang and Cundy (1989) are promising. Nevertheless, the development of appropriate turbulence models may be worthwhile in future. But until then, the established, more empirical approaches are more appropriate.

Independent of the approach used and the scales considered, the amount of water is conserved. If we imagine a segment of the slope, the balance of inflow and precipitation versus outflow and infiltration determines the gain or loss of water within this segment. This balance can be written down for each segment considered in the model. On the other hand, from theoretical aspects it is better to formulate the balance using a differential equation and introduce segments of finite size later. In the differential equation formalism the mass balance is expressed in a continuity equation:

$$\frac{\partial \delta}{\partial t} + \frac{\partial}{\partial x}(\delta v) = R - I \tag{14.1}$$

where
- δ = flow depth
- v = vertically averaged lateral flow velocity
- R = precipitation rate, and
- I = infiltration rate

This is the one-dimensional case, only applicable to slope profiles. All variables mentioned depend on the spatial coordinate x and on the time t.

The two-dimensional case looks a little more complicated. The variables depend on the spatial coordinates x_1 and x_2 and on the time t; the lateral velocity $\vec{v}$ is a vector with two components. The two-dimensional continuity equation reads:

$$\frac{\partial \delta}{\partial t} + div(\delta \vec{v}) = R - I \qquad (14.2)$$

where the divergence operator applied to any vector field $\vec{u}$ means:

$$div\vec{u} = \frac{\partial u_1}{\partial x_1} + \frac{\partial u_2}{\partial x_2} \qquad (14.3)$$

In principle, this is the trivial part of the problem. The other part consists of determining the flow velocity v, respectively $\vec{v}$.

As mentioned above, empirical constitutive laws are established here. In general, these laws determine the vertically averaged flow velocity as a function of the slope gradient Δ, the flow depth δ, and the soil roughness.

The most common relations are power laws of the form

$$v = C\delta^{\alpha}\Delta^{\beta} \qquad (14.4)$$

where the constant C depends on the soil roughness. For the reason that turbulent friction forces increase quadratically with the velocity, the exponent β shall be one half. Inserting this relation into the one-dimensional continuity Eq. 14.1 leads to a kinematic wave equation:

$$\frac{\partial \delta}{\partial t} + \frac{\partial}{\partial x}(C\delta^{(\alpha+1)}\Delta^{\frac{1}{2}}) = R - I \qquad (14.5)$$

Analytical solutions of this equation, obtained using the method of characteristic curves, have been published (Giraldez and Woolhiser 1996).

If not obtained from measurements directly, the constitutive laws are based on Manning's formula (Giles 1962), which was originally developed for channel construction. Field measurements (Emmett 1970; Pearce 1976) have stated that this equation may be applied to turbulent sheet flow, too. In this case, it reads:

$$v = \frac{1}{n}\delta^{\frac{2}{3}}\Delta^{\frac{1}{2}} \qquad (14.6)$$

which means that the exponent α is two third for sheet flow. The coefficient n expresses the roughness of the soil. Thus, the one-dimensional kinematic wave equation for sheet flow reads:

$$\frac{\partial \delta}{\partial t} + \frac{\partial}{\partial x}(\frac{1}{n}\delta^{\frac{5}{3}}\Delta^{\frac{1}{2}}) = R - I \qquad (14.7)$$

The most doubtful point is the assumption of sheet flow. In most cases relevant for soil erosion, the flow regime is governed by rills. Field measurements (Govers 1992; Abrahams et al. 1996) have shown that the exponent of rill flow is significantly lower than two third. Geometrical investigations (Hergarten and Neugebauer 1997) show that in general the δ-dependence for rill flow is no power law at all. If nevertheless a power law is assumed to be valid over a limited range, they propose an exponent between one third and two thirds. The definition of the coefficient of roughness has to be changed according to the exponent, so that its physical units change, too. There are attempts to correct the effects of rills by modification of the roughness coefficient, but keeping the sheet flow formula. But this causes a dependency of the coefficient from the flow depth, and thus, the modified value is only valid over a strongly limited range, so that this procedure cannot be recommended from a theoretical point of view.

But even if a correct law is used according to the rills present on the site, this only reflects a snapshot taken from a complex evolution. The rill pattern may change significantly during an erosion event, even a complete transition from sheet flow to rill flow may occur. There are promising approaches to include changes in rill geometry into large-scale models (Schramm 1994). But from our point of view, we should first consider the small scale, where effects like rill formation can be resolved, and then try to find a way back to the large scale.

On small scales, the surface runoff cannot be considered as a one-dimensional process. Thus, a generalization to two dimensions is necessary. This point is not new; recent large-scale models such as EROSION-3D are not restricted to slope profiles, too. If a constitutive law like Eq. 14.6 shall be extended to two dimensions, an assumption concerning the direction of flow has to be introduced. The most simple and plausible assumption is a flow downslope at every point of the surface. If the model is directly based on discrete cells of finite size, the realization of this idea is quite complicated, e.g., a flow to the lowest neighbour may lead to unreasonable results. In contrast, in the differential equation formalism the direction downslope is uniquely given by the negative gradient of the surface height, defined by

$$-\vec{\nabla}H = -\begin{pmatrix} \frac{\partial H}{\partial x_1} \\ \frac{\partial H}{\partial x_2} \end{pmatrix} \tag{14.8}$$

The length of the gradient vector equals the value of the slope gradient Δ. Thus, the two-dimensional generalization of Eq. 14.6 reads:

$$\vec{v} = -\frac{1}{n}\delta^{\frac{2}{3}}\frac{\vec{\nabla}H}{|\vec{\nabla}H|^{\frac{1}{2}}} \tag{14.9}$$

Inserting this into the two-dimensional continuity Eq. 14.2 leads to the generalization of Eq. 14.5:

$$\frac{\partial \delta}{\partial t} - \operatorname{div}\left(\frac{1}{n}\delta^{\frac{5}{3}}\frac{\vec{\nabla}H}{|\vec{\nabla}H|^{\frac{1}{2}}}\right) = R - I \tag{14.10}$$

Although this procedure appears to be almost plausible, even in the one-dimensional case the results may be strange: Figure 14.1 shows the water level for stationary flow on a slope with a quadratic profile according to the kinematic wave Eq. 14.7. The scales do not matter here. Near the slope foot the flow velocity decreases due to the low surface gradient, so that the outflow at the slope foot is hindered. Continuity of flow (mass conservation) leads to an unrealistic water column.

Another inconsistency occurs if the temporal behaviour of surface runoff is considered. Figure 14.2 shows how a sine-shaped wave on the water surface propagates according to the kinematic wave Eq. 14.7. Again, the scales are irrelevant. Where the flow depth is large, the water moves fastly. Hence the wave front becomes steeper and steeper, until it is vertical somewhere and a shock front moves downslope. For a slope of constant gradient, it can be shown that this effect occurs under any nontrivial initial condition; if at any time a inhomogeneity within the flow depth occurs, this will finally lead to a shock front.

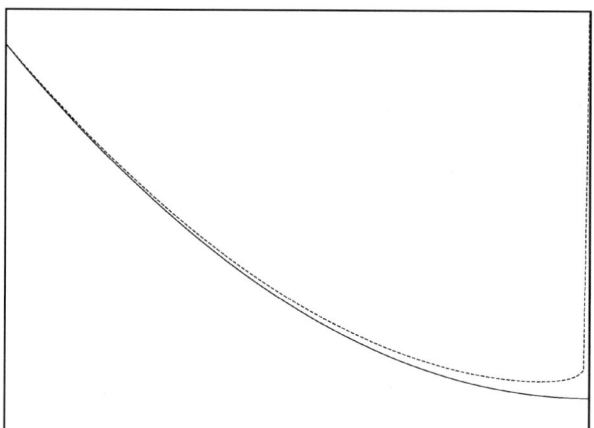

Fig. 14.1. Relief (*solid*) and water level (*dashed*) for stationary flow on a slope with a quadratic profile according to the kinematic wave Eq. 14.7

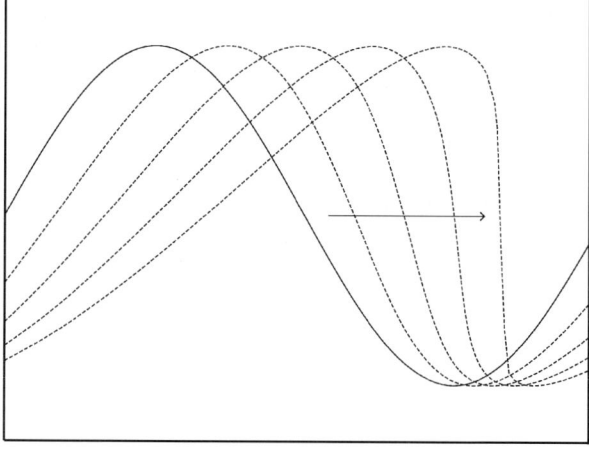

Fig. 14.2. Propagation of a sine-shaped wave on the water surface using a kinematic wave equation. The arrow indicates flow direction and time evolution

Both problems illustrate that in some cases the equations derived from intuition and physical approximations are plausible, but may fail in borderline cases. Often, these problems are obvious only in borderline cases, but may bias the results in general. Thus, before applying approximations such as the kinematic wave we should think about why these problems are not crucial on the scale considered or how they can be avoided.

The inconsistencies illustrated here result from the fact that water flow is not only determined by the slope gradient; water has some kind of diffusive behaviour, too. If the flow depth is spatially inhomogeneous, the water tends to flow from areas with large flow depths to areas with a shallow water film. This effect avoids unrealistic water columns as well as shock fronts.

In order to regard the diffusive behaviour of the water surface, the gradient of the flow depth should be included somehow, so that water flows from higher to lower flow depths. Experience shows us that there is no flow if the water surface is horizontal, i.e., if

$$\vec{\nabla}(H + \delta) = 0 \tag{14.11}$$

Thus, the gradient of the water surface, and not that of the soil surface, should be interpreted as the driving force for surface runoff. In the one-dimensional case this leads to the diffusive wave equation (Govindaraju et al. 1988); in the two-dimensional case Eq. 14.9 should be replaced by (Paul et al. 1995):

$$\vec{v} = -\frac{1}{n}\delta^{\frac{2}{3}} \frac{\vec{\nabla}(H + \delta)}{\left|\vec{\nabla}(H + \delta)\right|^{\frac{1}{2}}} \tag{14.12}$$

Then the differential equation to be solved reads:

$$\frac{\partial \delta}{\partial t} - \mathrm{div}\left(\frac{1}{n}\delta^{\frac{5}{3}} \frac{\vec{\nabla}(H + \delta)}{\left|\vec{\nabla}(H + \delta)\right|^{\frac{1}{2}}}\right) = R - I \tag{14.13}$$

At first sight, this may be a minor correction; but from the mathematical point of view it causes a fundamental change. The type of the differential equation changes from a hyperbolic to a parabolic type. It can be formally be proved that a unique solution exists, so that the evolution does not stop at any time. So the theory confirms the hope that the modification avoids the problem of sharp fronts evolving.

On the other hand, it cannot be denied that this equation requires a completely different and more complicated numerical treatment than that of the hyperbolic Eq. 14.10. But this is the price we have to pay for describing the runoff more precisely without any obvious inconsistencies.

14.3
Some Examples

Let us now take a look at some artificial examples, which illustrate the advantages of small scale models.

First, we show a part of a slope, approximately 36 square meters large, with a shallow hollow in its middle. For the sake off better illustration, we have switched of precipitation and infiltration and start with a dry surface. Then we introduce an instantaneous inflow of 1.6 litres per second at the upper edge. Figure 14.3 shows how the water flows downslope, the grayscale indicates the flow depth. The hydrograph (Fig. 14.4) shows the total outflow at the lower boundary. After 10 seconds, the water reaches the hollow. The sharp front – caused by the instantaneous inflow – has been smoothed significantly. The water passing by the hollow needs about 25 seconds to reach the lower boundary. After 60 seconds, an overflow out of the hollow begins, so that the hydrograph rises again. The fact that this increase takes place very slowly may be a little surprising; from the first overflow it takes two minutes until the outflow nearly reaches its maximum.

The behaviour can easily be understood looking again at Fig. 14.3. After 60 seconds – immediately after the storage capacity of the hollow has been reached, the outflow only affects a minor part of the hollow's perimeter with a small flow depth. So the amount of water stored within the hollow rises still although its static storage capacity is exceeded. Thus, there is a significant difference between static and dynamic storage capacity; the latter one strongly depends on the inflow. As this example shows, the delay caused by the hollow would be underestimated significantly if the hollow was considered as a simple storage which stores water until it is full, and then lets additional water pass through. Thus, the probability of surface runoff becoming sta-

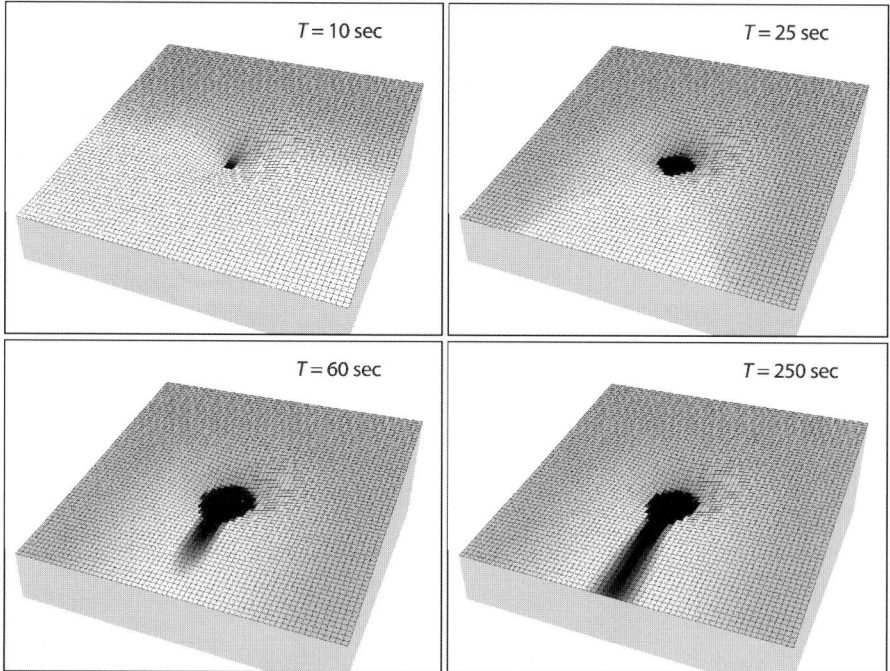

Fig. 14.3. Surface runoff through a hollow

tionary during a short storm may be smaller than it usually is assumed to be, at least if there are some hollows.

On the other hand, the water flowing out of the depression is focused; the dry (light gray) regions in the nearly stationary flow pattern (250 sec) indicate that the hollow collects water from an area that is several square meters large and focuses it on a narrow channel. Thus, the outflow may be as effective for erosion as rill flow is, so that the effects of depressions on erosion may be complex: first, they delay and even reduce runoff, but if overflow occurs, the resulting focused flow may assist erosion significantly.

The second example (Fig. 14.5 and 14.6) illustrates how our approach works on a slope with a complex microtopography. We consider a 100 m long and 5 m wide section of a loamy slope. The slope gradient is 30%; the microtopography is given by a pink noise with a variable standard deviation: a standard deviation of 4 cm constitutes a quite rough surface with several small depressions, whereas a standard devia-

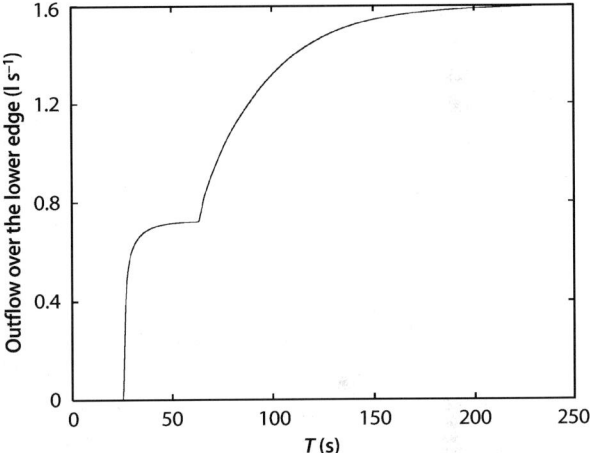

Fig. 14.4. Total outflow at the lower boundary

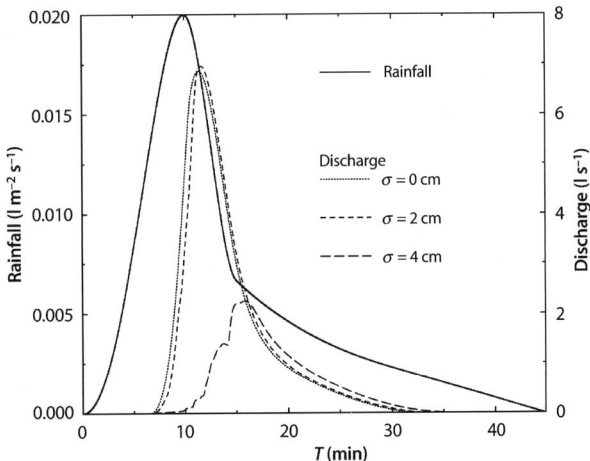

Fig. 14.5. Rainfall intensity and discharge at the slope foot for three slopes with different microtopographies

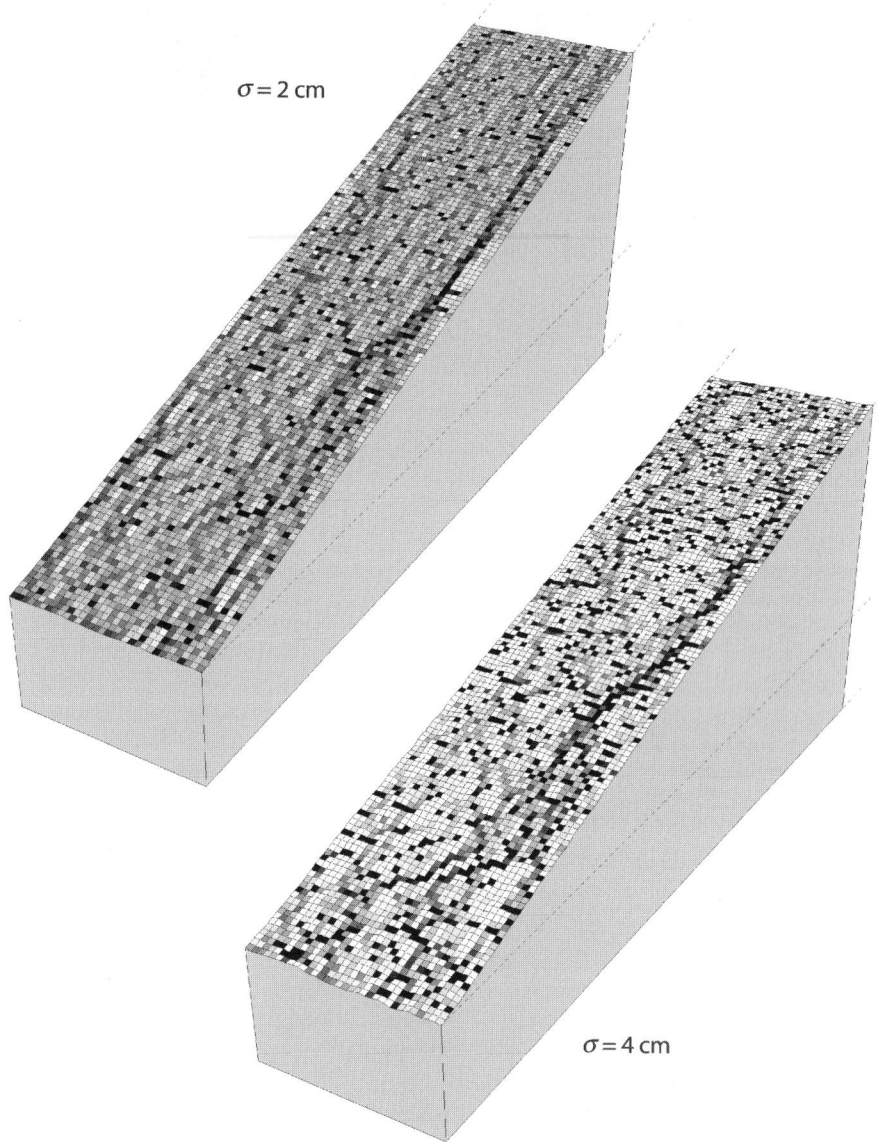

Fig. 14.6. Surface runoff at the time when peak discharge occurs on the moderately rough and the rough slope. The grayscale indicates flow depth. For clarity, only the lowest 20 meters of the slopes are shown

tion of 2 cm leads to a moderate roughness. For comparison, a slope without microtopography is considered, too. The slopes are exposed to a rainstorm observed by the group of G. Heinemann (Institute of Meteorology, University of Bonn) on 4 July 1996, near Bonn (Fig. 14.5).

The hydrographs at the slope foot show that the moderately rough surface behaves similarly to the smooth one. In contrast, the peak discharge on the rough slope is reduced by a factor three and occurs five minutes later compared to the smoother slopes. As discussed before, this effect arises from the water retention of the hollows. Obviously, the precipitation event is not sufficient to achieve the state of stationary overflow through the depressions.

However, the behaviour becomes more complex if we are interested in soil erosion. With increasing surface roughness the flow becomes more channelized (Fig. 14.6), so that it may be more effective for erosion. In this example, the reader may doubt if this effect is able to compensate the significantly reduced discharge. But if the rainstorm is longer or more intense, it is able because then the asymptotic stationary flow through the depressions will finally be reached. In this case the total discharge approaches the discharge of the smooth slope, so that the channelization of the flow is the most significant effect of the microtopography.

Thus, the effects of microtopography on runoff induced soil erosion are significant and quite complex.

14.4
Conclusions

Our examples have shown that the effects of microtopography on soil erosion are significant and quite complex. Especially, they depend on the considered precipitation events. For moderate rainstorms, the microtopography can be expected to reduce the erosivity of the runoff; in contrast, for very heavy events the channelization of flow may drive erosion as significantly as rills do. For this reason, these effects cannot be integrated into soil erosion models by simply modifying parameters like Manning's roughness coefficient.

Thus, it is worthwhile taking a look at the processes on the small scale, even if it requires much theoretical and numerical effort, and it is not possible to cover the field scale with a resolution of a few centimeters. Thus, small scale models should be seen as a tool for understanding the processes in detail. The long term aim should be integrating the knowledge obtained from the small scale into large-scale models.

References

Abrahams AD, Li G, Parsons AJ (1996) Rill hydraulics on a semiarid hillslope, Southern Arizona. Earth Surface Processes and Landforms 21:35–47

De Roo APJ, Wesseling CG, Cremers NHDT, Offermans RJE, Ritsema CJ, Oostindie K van (1994) LISEM: a new physically-based hydrological and soil erosion model in a GIS-environment, theory and implementation. In: Olive LJ, Loughran RJ, Kesby JA (eds) Variability in stream erosion and sediment transport. Int Assoc Hydrol Sci Publ 224, Int Assoc Hydrol Sci Press, Wallingford

Emmett WW (1970) The hydraulics of overland flow on hillslopes. US Geol. Survey Paper 662A

Giles R (1962) Theory and problems of fluid mechnics and hydraulics. McGraw-Hill Inc

Giraldez JV, Woolhiser DA (1996) Analytical integration of the kinematic equation for runoff on a plane under constant rainfall rate and Smith and parlange infiltration. Water Resour Res 32(11):3385–3389

Govers G (1992) Relationship between discharge, velocity, and flow area for rills eroding in loose, non-layered material. Earth Surface Processes and Landforms 17:515–528

Govindaraju RS, Jones SE, Kavvas ML (1988) On the diffusion wave modeling for overland flow. Water Resour Res 24(5):734–744

Green WH, Ampt GA (1911) Studies on soil physics 1: the flow of air and water through soils. J Agr Sci 4:1–24

Hergarten S, Neugebauer HJ (1997) Homogenization of Manning's formula for modeling surface runoff. Geophys Res Letters 24(8): 877–880

Knisel WG (1980) CREAMS: a field scale model for chemicals, runoff and erosion from agricultural management systems. USDA-Conserv Res Report 26, Washington, DC

Lane LJ, Nearing MA (1989) USDA-water erosion prediction project: hillslope profile model documentation. NSERL Report 2 (USDA-ARS), West Lafayette

Morgan RPC, Quinton JN, Rickson RJ (1992) EUROSEM documentation manual. Silsoe College, Silsoe, UK

Paul G (1998) Numerische Simulation der Bodenerosion durch Oberflächenabfluss von Wasser auf einem komplexen Relief. Berichte aus der Geowissenschaft, D98, Shaler Verlag, Aachen

Paul G, Hergarten S, Neugebauer HJ (1995) Numerische Simulation des Oberflächenabflusses und der Versickerung von Wasser auf einem komplexen Relief. Technical Reports of the SFB350, No. 5, University of Bonn

Pearce AJ (1976) Magnitude and frequency of erosion by Hortonian overland flow. J Geol 84:65–80

Richards LA (1931) Capillary conduction of liquids in porous mediums. Physics 1:318–333

Schmidt J (1991) A mathematical model to simulate rainfall erosion. In: Bork HR, De Ploey J, Schick AP (eds) Erosion, transport and deposition processes – theory and models. Catena Suppl 19:101–109

Schramm M (1994) Ein Bodenerosionsmodell mit räumlich und zeitlich veränderlicher Rillenmorphologie. Mitt Inst Wasserbau und Kulturtechnik, 190, TH Karlsruhe

Smith RE (1992) OPUS: an integrated simulation model for transport of nonpoint-source pollutants at the field scale. I: documentation, USDA-Soil Conserv Service 98, Washington, DC

Werner M von (1996) GIS-orientierte Methoden der digitalen Reliefanalyse zur Modellierung von Bodenerosion in kleinen Einzugsgebieten. Dissertation, FU Berlin

Woolhiser DA, Smith RE, Goodrich DC (1990) KINEROS, a kinematic runoff and erosion model: documentation and user manual. USDA-ARS-77, Washington, DC

Zhang W, Cundy WC (1989) Modeling of two-dimensional overland flow. Water Resour Res 25(9):2019–2035

Index

A

ABAG (see *Allgemeine Bodenabtragsgleichung*)
adaptive discretization technique 298
aerial photograph 60, 67, 276
 –, stereoscopic 60
afforestation 59
aggradation 93
aggregate
 –, size distribution 268
 –, stability 36, 268, 270, 279
AGNPS (see *Agricultural Non-Point Source Pollution Model*)
AGNPSm (see *modified Agricultural Non-Point Source Pollution Model*)
Agricultural Non-Point Source Pollution Model (AGNPS) 43–46, 54, 56, 165, 170–173, 199, 285
 –, data sources and input variable 46
agriculture 33, 34, 76
Algeria 59, 77
Allgemeine Bodenabtragsgleichung (ABAG) 120
Alte Schwentine 91
analysis
 –, discriminant a. 69
 –, empirical a. 199
ANSWERS (see *Areal Nonpoint Source Watershed Environmental Response Simulation model*)
approach
 –, empirical 297, 299
 –, hydrodynamic 299
Arc/Info 163, 165
Areal Nonpoint Source Watershed Environmental Response Simulation model (ANSWERS) 35, 166, 199
Arrhenius 3
aspect 167, 171, 173, 202

B

base flow 44
bed
 –, material load 65–67
 –, slope 254
biodiversity 24
biomass 169

BMDP software 69
Bornhoeved Lake 80, 91
Brazil 3, 10ff.
buffer zone 79, 91, 92
bulk density 94, 117, 167, 208, 224, 226, 229, 232, 234, 236–238, 240, 241, 246, 247
Bureau of Land Management of the U.S. Department of the Interior 200
business as usual
 –, scenario 7, 19
 –, strategy 23, 27

C

cadmium 101, 104
calcic brown soil 68
calibration 15, 27, 47, 50, 51, 54, 75, 165, 194, 297
cambisols 80
canopy
 –, coefficient 205
 –, height 202, 207
capillary
 –, drive
  –, effective 185, 209
  –, net 185, 209
 –, tension, effective 208
carbon, cycling in soil 168
carbon dioxide 3–5, 7, 9, 15, 18–20, 27, 28
 –, atmospheric concentration 3–5, 15
 –, future emissions 7
 –, global emissions 7
case study 3
catchment 9, 34–36, 40, 59, 60, 63, 68, 75, 80, 82, 84, 86, 88, 90, 109, 110, 113, 115, 117–119, 123–125, 128, 152, 163–166, 168–175, 177, 201, 203, 253, 267, 270, 273, 275–279, 285, 286, 291, 292, 295
 –, agricultural 166, 170
 –, area 59, 63, 80, 82, 84, 86, 88, 90, 109–119, 123–125, 128
 –, outlets 173
Centre for Agricultural Landscape and Land Use Research (ZALF) 229, 232, 247
cerrado 14
channel 63, 106, 110, 115, 117, 118, 120, 168, 170, 182, 200, 275, 276, 286, 287, 299
 –, bedload 218
 –, flow 220, 223, 287

–, runoff 62, 63
–, slope 44
chemical transport 165
chemical oxygen demand 165
chemicals
  –, agricultural 12, 166
  –, polluting 253
Chemicals, Runoff and Erosion from Agricultural Management Systems (CREAMS) 165, 166, 199, 297
Chezy coefficient 211
chlorofluorocarbons 3
classification 68–70, 72, 75, 226
  –, method 69
  –, of input parameter 227
  –, supervised 69
clay content 11, 99, 268–270, 275, 276, 279
CLIGEN (see *Climate Generator*)
climate 3–6, 8, 9, 15, 19, 21, 23, 26, 27, 43, 45, 47, 59, 76, 125, 164, 169, 201, 297
  –, change 3, 4, 6
  –, data 21, 45, 169
  –, future 7
  –, parameter 19
  –, regional 4
Climate Generator (CLIGEN) 7, 27, 28
coefficients of efficiency 49, 52, 53
cohesion 36, 64, 94, 189, 194, 216, 228, 240, 242, 256
colluvic gleysol 80
commercial area 117
compaction 75, 202
concavity 117
concentration 65, 66, 75, 89–91, 105, 124, 135, 151, 165, 199, 204, 214–216, 218, 221, 237–239, 241, 246, 264, 268–270, 278, 287–289, 291–293
concept of partial watershed area 44
conductivity
  –, effective saturated 208
  –, hydraulic 148
  –, parameter 14, 236
  –, saturated and unsaturated 36
  –, saturated hydraulic 140, 142, 168, 209, 234, 236, 263
conservation
  –, measures 148, 150
    –, efficiency 202
  –, planning for agricultural land 148
  –, practice 164, 199, 228, 240
  –, technique 14
  –, tillage 106, 107, 242
contour farming 40
convexity 117
corn 47, 120, 123, 128, 130, 154, 158, 276
cover crop 36, 39, 106
CREAMS (see *Chemicals, Runoff and Erosion from Agricultural Management Systems*)
critical
  –, flow velocity 45
  –, momentum flux 216, 258
  –, shear stress 213, 214
  –, shear velocity 193, 194

–, slope length 135, 144, 146, 148, 150, 152, 156, 158, 160
–, stream power 219, 238
crop
  –, green cover c. 36
  –, management measures 151
  –, management system 228
  –, rotation 106, 120, 123, 151, 153, 199, 201
  –, yield 33, 38, 106
    –, reduced 33, 38
crusting 9, 12, 63, 75, 269, 270, 272, 276
crusts 12, 35
Czech Republic 135, 139, 149, 150, 152

D

damage
  –, ecological 109
  –, economic 109
Darcy's law 298
data
  –, editing 243
  –, vectorial 126, 127
DEDNM (see *Digital Elevation Drainage Network Model*)
deductive spatial analysis 112
deforestation 3, 10, 12
degradation 12, 68, 166
denitrification 166
deposition
  –, by flow 191, 258
  –, coefficient 66, 67, 258–260, 279
  –, rate 22, 258, 260, 288, 289, 293
  –, spatial distribution 175
desorption 166
detachment 9, 93, 94, 112, 120, 124, 138, 166, 167, 169, 181, 185, 189, 191, 192, 194, 196, 199, 213–217, 219, 220, 238–240, 242, 245, 258, 259, 272, 279, 287–290, 292, 293, 295, 298, 299
  –, by overland flow 166, 194
  –, by raindrop impact 191, 215
  –, capacity 256
    –, of rill flow 214
  –, of soil particles 93, 94, 124, 138, 166, 167
  –, physical approach 299
  –, rates 181, 194, 238, 289
differentiating universal soil loss equation 115
Digital Elevation Drainage Network Model (DEDNM) 45
digital elevation model 60, 67, 94
digital landscape model 113
digital relief analysis 124
digital terrain model 96, 115, 124, 127, 128, 167, 173
discharge 44, 76, 116, 125, 154, 167, 185, 189, 201, 204, 210–212, 215, 217, 218, 220, 221, 239, 254, 263, 269, 275, 277, 287, 292, 293, 297, 307
  –, sensitivity parameter 263
  –, specific 220
  –, volume 263
discretization scheme, spatial 202
discriminant analysis 69
ditches 81, 106, 107, 117–119, 126, 128, 129, 150–156,

Index

158, 201
-, diversion d. 106, 107, 154, 156
-, drainage d. 151, 154, 156
-, spatial pattern 152, 155, 156
downscaling process 112, 120, 132
drainage
-, area 44
-, ditches 151, 154, 156
-, network 45
drinking water 59
-, reservoir 93
dump 157-160
duration 9, 44, 83, 90, 94, 138, 142, 167, 211, 225, 234
Dutch Agricultural Board 33
dynamic model 285, 286

E

earthworm burrow 106
ECHAM3TR 16, 19, 21-23
ecosystem 79, 92, 125, 169
-, semiterrestrial 79
ecotones 79, 84, 92
effective
-, capillary drive 185, 209
-, capillary tension 208
-, height of plant canopy 207
-, hydraulic conductivity 14, 208
-, porosity 208
-, rainfall 136, 207, 212, 261, 262
-, rainfall intensity 207, 212
-, saturated conductivity 208
efficiency measure 236
element retention 80
elementary area 110, 112
emissions 3, 7, 16, 21
-, policies 7
-, scenario IS92a 17, 18, 20, 22
-, scenario IS92d 16, 18, 19, 21
empirical
-, analysis 199
-, approach 297, 299
-, model 112, 113, 164
-, regression model 285
EPIC (see *Erosion Productivity Impact Calculator*)
eroded soil, transport processes 253
erodibility 14, 86, 131, 138, 141, 154, 156, 160, 164, 167, 205, 213, 216, 226, 228, 239, 242, 247, 259, 264, 266, 272
-, factor 164
-, interrill e. 242
erosion 4-7, 9ff.
-, annual 21, 89
-, areal e. 110
-, budget 203
-, calculation 214, 219, 238
-, equation 94
-, event 23, 120, 156, 297
-, hazard 113, 115, 148, 158
-, assessment 148

-, interrill e. 170, 191, 213, 214, 217
-, limiting conditions 291
-, linear 110
-, mechanics 9
-, model 4, 7, 9, 25, 27, 34, 35, 67, 79, 84, 86, 109, 120, 123, 124, 132, 163, 164, 170, 171, 181, 191, 196, 199, 200-202, 224, 234, 237, 238, 245, 253, 264, 267, 279, 298, 307
-, application 279
-, pattern 291
-, pavement 290
-, processes, model 135
-, rate 5, 22, 23, 27, 109, 128, 164, 166, 181, 189, 192, 194, 213-215, 226, 229, 238, 256, 258, 259, 279, 288, 290, 293, 295
-, potential 256, 279
-, rates, modeling 15
-, resistance 94, 106, 110, 114, 118, 125, 226, 228, 242, 258, 261, 264-270, 272, 275, 276, 279, 288, 291
-, determination 268
-, temporal variability 264
-, variation 266
-, sensitivity 113
-, simulation 96
-, spatial distribution 175
Erosion Productivity Impact Calculator (EPIC) 9, 27, 28, 164, 170, 199
EROSION-2D model 199-202, 207, 208, 210, 212, 216, 217, 221, 225-229, 231, 234, 236-240, 242, 244-247
EROSION-3D model 96, 297, 298
Erosion-Productivity Impact Calculator (EPIC) 9, 27, 28, 164, 170, 199
estimation errors 224
European Soil Erosion Model (EUROSEM) 35, 170-173, 181, 185, 191, 194, 196, 199, 200-202, 204, 209-211, 214, 216, 217, 223, 225-229, 231, 234, 236-239, 242-247, 286, 288-290, 293, 295, 297
-, application 228
-, hillslope file 202
-, input parameter 202
eutrophication 91
evaporation 61
evapotranspiration 262, 279
event simulation 205, 244
Everaert transport capacity equation 237, 239, 246
external forces 256, 279

F

FAO system 68
feedback 4
fertilizer 56, 109, 253
-, transport processes 253
field
-, capacity 166, 167
-, measurement 34, 110, 131, 183, 194, 253, 275
-, strip 39
flood prevention scenario 33
flooding 33, 34, 36, 38, 39, 100, 156

–, costs 38
flow
   –, depth 136, 137, 191, 194, 211–213, 216, 220, 223, 259, 270, 289, 299–304
   –, direction 45, 173
   –, erosion 215, 216, 291, 293
   –, impact 206
   –, interrill f. 189, 193, 211, 214, 219
   –, mass rate of 212
   –, path, maximum 44
   –, rate, spatial variation 285
   –, shear stress 214
   –, simulation 292
   –, transport 291
   –, velocity 43, 45, 53, 54, 66, 67, 80, 140, 212, 216, 220, 254, 264, 300, 302
   –, velocity, critical 45
   –, volumetric sediment concentration 215
fluid density 207, 212, 256
flume 185, 221
foliar interception 166
forage fodder 47
forces, external 256, 279
forecast 15, 22, 23, 26
forest 47, 106
fossil fuels 3
friction
   –, internal f. 64, 256
   –, slope 254

G

General Circulation Model (GCM) 3, 6, 7, 8, 9, 15, 16, 19, 21, 22, 23, 26, 28
Geoecological Information System (GOEKIS) 115, 124–126, 132
Geographic Information System (GIS) 35, 36, 40, 45, 56, 68, 90, 109, 113, 115, 117, 118, 122, 123, 125, 126, 128, 129, 132, 163
GLEAMS 166, 170
gleysols 80
Glonn Creek 45
Glonn watershed 46
GOEKIS (see *Geoecological Information System*)
Govers transport capacity equation 239
gradient, spatial 79
grass, buffer strips 36, 40
grassland 33, 36, 40, 117, 128, 153, 154, 169, 183
   –, hydrology 169
gravity 64, 94, 194, 220, 256
   –, specific 220
Green and Ampt equation 9, 36
green cover crop 36
greenhouse
   –, effect, anthropogenic 7
   –, gases 3
   –, warming 3, 6
grid 45
   –, area 112
   –, cell 55, 117, 127, 129, 165, 166
   –, element 65, 123
   –, global 8

growing season 36, 264–266, 273
gully, ephemeral 13

H

HADCM2 9, 16, 19, 21–23, 26
harvesting 169
hazard level 122
heavy metals 93, 95, 96, 100, 101, 103–105
   –, transport 102
   –, yield 93
height of plant canopy
   –, effective 207
HILLFLOW model 253, 255, 257, 259, 261–263, 272, 273, 275–277, 279
   –, input parameters 263
hillslope 74, 201, 262, 279
   –, element 200
   –, profile 21, 23, 202, 217
histosols 80
Holtan 35, 36
   –, equation 35, 36
Hölzelbach 100, 104, 105
   –, watershed, mean heavy metal concentrations 105
Horton
   –, equation, modified 63
   –, overland flow' concept 35, 206
humus content 115, 117, 270
hydraulic
   –, conditions 285
   –, conductivity 9, 14, 34, 124, 140, 142, 147, 148, 159, 160, 167, 168, 208, 209, 226, 234, 236, 241, 263
   –, correction factor 167
   –, effective 14, 208
   –, erosion 288, 290
   –, properties 202
   –, radius 254
   –, roughness (see also *Manning's roughness coefficient*) 94, 191, 194, 167, 212, 270, 276
   –, shear 218, 226, 242, 245, 288
   –, transport 94
hydrograph 76, 152, 181, 189, 191, 194, 204, 211, 234–236, 239, 242, 246, 264, 289, 304, 307
hydrological
   –, data 68, 173
   –, process 40

I

ice cores, Antarctic 3
IDRISI 45
IHDM4 275
Ilde 122
impacts
   –, economic 24
   –, environmental 24
infiltration
   –, calculation 287
   –, capacity 36, 106, 157, 160, 167, 209, 236, 247
   –, maximum 209

–, characteristics 63, 191
–, component 262
–, excess volume 211
–, maximum 209
–, model 124
–, parameters 263, 276
–, rate 61–64, 88, 110, 140, 141, 206–212, 235, 236, 255, 262, 263, 269, 287, 298
   –, temporal change 208
–, relative 64
inflow 60–63, 212, 254, 255, 287, 293, 299, 304
–, lateral 60–62, 254, 255, 287
initial
   –, conditions 62, 244, 265–267, 289, 302
   –, moisture 234, 236
   –, saturation 208
input
   –, data, spatial 45
   –, parameter 82, 83, 86, 88–90, 92, 96, 112, 123–125, 131, 144, 169, 171, 173, 182, 185, 200–202, 224–229, 244, 245, 253, 261, 263
      –, classification 227
      –, determination 171
      –, sensitivity of 228
      –, spatial variability 224
integral management scenario 36
interception 34, 136, 139, 140, 147, 166, 167, 185, 202, 262, 279
   –, characteristics 202
   –, foliar 166
   –, potential 139, 140, 147
   –, storage
      –, calculation 202
      –, maximum 185
Intergovernmental Panel on Climate Change (IPCC) 3, 6
interrill 169, 170, 182, 189, 191–194, 196, 211, 213–215, 217, 219, 220, 223, 242, 291
   –, erodibility 242
   –, erosion 170, 191, 213, 214, 217
   –, sediment delivery rate 213
   –, soil erodibility 213
   –, soil erodibility parameter 213
   –, surface 182
investigation, multiscale 110, 111
irrigation 26, 59, 72

K

KINEmatic Runoff and EROSion model (KINEROS) 35, 168, 170, 201, 209, 275, 293, 297
kinematic wave 62, 125, 135, 204, 211, 212, 245, 255, 287, 300, 302, 303
   –, approach 135
   –, equation 125, 204, 211, 212, 255, 287, 300, 302
   –, velocity 62
KINEROS (see *KINematic Runoff and EROSion model*)
KINEROS2 286–290, 292, 295
kinetic energy 63, 64, 138, 184, 191, 192, 207, 291
   –, of rainfall 138, 191, 192

L

lake
   –, sediments 5
   –, shore zones 79
   –, siltation 24
Lake Belau 79–83, 86, 88, 90, 91
   –, phosphorus input rates 91
Lamme river 119
land
   –, agricultural 33, 47, 80, 93, 95, 107, 135, 148, 150, 156, 165, 199, 220, 247, 253
   –, arable 33, 34, 39, 40, 80, 84, 91, 129, 130, 132
   –, consolidation plan 148
   –, cover conditions 170, 182
   –, management 34, 36, 123, 163, 164, 177
   –, use
      –, agricultural 33, 199
      –, data 117
      –, future 26
      –, pattern 36
      –, scenario 122, 123
land conservation planning, agricultural 148
landcover map 173
Landsat 67, 69, 75
landscape, rehabilitation project 148
Landscape Care Fund 149
leaching 166, 239
   –, of nutrients and pesticides 239
leaf
   –, area
      –, index 36
      –, relative 139, 147
   –, cover 264, 265
   –, drainage, energy of 207
Leierfass 273, 275, 276
   –, natural erosive rainfall events 276
Limburg Soil Erosion Model (LISEM) 33, 34, 35, 36, 37, 38, 39, 40, 41, 45, 47, 297
linear erosion 110
LS factor 120
Lutz
   –, factor 44
   –, method 47, 49, 50
   –, parameter 49
lynchets 33

M

macropore 106, 241, 262, 263, 264, 275, 276, 278, 279, 298
macroscale 115
Maghreb 59
MAGICC (see *Model for the Assessment Of Greenhouse-Gas Induced Climate Change*)
maize 15, 33, 36, 84, 151–154, 264, 265, 266
Malter reservoir, mean annual sediment delivery 98, 99
Malter watershed, heavy metals 96
management 12, 14, 26, 34, 36, 59, 66, 94, 106, 107, 110, 112, 115, 120, 123, 138, 141, 151–154, 163–166, 169, 177, 201, 228, 267

–, file 201
–, practices, agricultural 94, 153, 154
–, scenario, integral 36
–, strategy 34, 59, 94, 106, 123, 153, 154, 164, 165, 267
–, system 164, 166, 228
management system 228
Manning's
  –, equation 43
  –, hydraulic coefficient 140, 148
  –, hydraulic roughness 43, 45, 137–140, 147, 148, 167, 193, 254, 255, 261, 263, 264, 270, 272, 275, 276, 300, 307
  –, roughness coefficient 45, 137, 138, 147, 167, 193, 210, 254, 261, 263, 264, 270, 272, 275, 276, 307
Mato Grosso 10–23, 27
matter, suspended 66, 292
measures
  –, active 106
  –, crop management 151
  –, efficiency m. 236
  –, for conservation 148, 150
    –, assessment 240
  –, passive 106, 107
Mehle 128
mesoscale 110, 123
methane 3
Methau 224, 229–247
  –, experimental site 230
  –, rainfall experiment 247
Methau plot
  –, initial conditions 231
  –, runoff and soil loss values 232
micropore 262
microscale 110–113, 120, 123
microtopography 184, 191, 305, 307
mid infrared (MIR) 67, 73
military training area 163, 173, 175
mineralisation 166
miniature flume 185
mitigation
  –, scenario 7, 19, 22
  –, strategy 23, 27
model
  –, application 228–253
  –, assessment 228, 243
  –, behaviour 200, 226, 228, 234, 261
  –, comparison 200–251
  –, components, temporal resolution 203
  –, deterministic 224
  –, empirical 112, 113, 164
    –, regression m. 285
  –, equations 243
  –, erosion m., overview 163–177
  –, input
    –, spatial resolution 203
    –, temporal resolution 204
  –, performance 200, 229, 244
  –, physically-based 112, 120, 123, 132, 253, 278, 297
  –, predictions 105, 200, 224, 226, 228, 231,
236, 244, 246
  –, accuracy 228, 231, 236, 244
  –, structure 204
  –, type 6
  –, variables 228
Model for the Assessment Of Greenhouse-Gas Induced Climate Change (MAGICC) 6, 8, 28
modeling
  –, concept 45, 46
  –, future erosion rates 15
modified Agricultural Non-Point Source Pollution Model (AGNPSm) 43, 45, 51, 53–56
Modified Universal Soil Loss Equation (MUSLE) 164, 165, 170
moisture
  –, content 34, 131, 167, 177, 185, 234, 236, 263, 265–270, 275, 276, 279
  –, initial 234, 236
momentum flux 94, 124, 207, 212, 216, 221, 256–258, 279, 298
  –, approach 94, 298
  –, critical 216, 258
  –, of flow 212
  –, of overland flow 256, 257
  –, of runoff 298
morphography 125
morphometric data 117
mulch seeding 106
mulching 36, 39, 123
Müncheberg plot 224, 229–240, 244–247
  –, experimental site 230
  –, initial conditions 231
  –, measured and predicted runoff hydrographs 234
  –, runoff and soil loss values 232
MUSLE (see *Modified Universal Soil Loss Equation*)

N

nanoscale 110, 112
National Soil Erosion Research Laboratory (NSERL) 200
near the infrared (NIR) 73
Nette river 119
nitrate 166
nitrification 166
nitrogen 55, 165–169, 279
  –, cycling 169
    –, in soil 168
nitrous oxide 3
numerical effort 299, 307
nutrient 11, 55, 56, 112, 129, 165, 166, 168, 239, 253
  –, adsorbed transport 168
  –, dissolved and absorbed 165
  –, transport 165
  –, uptake 168
  –, yield 43

O

optimum scenario 38

OPUS 168, 170, 199, 286, 297
organic matter 11
 –, content 11, 94, 161, 167, 213, 247, 268, 275
Osterzgebirge 93
Oued Mina 59, 72, 74
 –, catchment 74
outflow, total 305
output, parameter 201
overland erosion process 291
overland flow 34, 35, 40, 93, 94, 124, 135–148, 158, 165–168, 170, 181, 194, 206, 210–213, 216, 218, 220–223, 245, 254, 256, 257, 269, 279, 286, 291
 –, calculations 210
 –, depth 136, 137, 194, 213
 –, detachment by 166, 194
 –, maximum transport capacity 170
 –, model 135
 –, numerical simulation 157, 159
 –, rate 136–138
 –, velocity 206
 –, volume 206
Overton infiltration equation 35

P

parameter
 –, input p. 201
 –, number of 56, 244
 –, output p. 201
particle
 –, class
   –, erosion rate 290
   –, of eroded soil 219
 –, classes 45, 221
 –, deposition 199
 –, detachment 199, 240, 298
 –, diameter 103, 219, 223
 –, size 129, 193, 221, 223, 255, 256, 287, 288, 290
 –, size distribution 268
 –, transport 167, 240, 298
pathway 128
PCA (see *Principal Component Analysis*)
peak flow
 –, algorithm 50
 –, calculation 50
 –, discharge 156
 –, rate 43
   –, measured and predicted 50
pedotransfer function 209, 229, 245
PEPP (see *Process orientated Erosion Prediction Program*)
PEPP-HILLFLOW model
 –, parameter determination 253
percolation 165, 268
pesticide 109, 165, 166, 168, 239, 253, 279
 –, adsorbed transport 168
Philip equation 140
phosphorus 54, 55, 79, 89–92, 165, 166, 168, 253, 279
 –, cycling in soil 168
 –, in sediment 54

 –, retention 79
 –, soluble 54
plane runoff 211
plant
 –, available water content 68
 –, canopy 192, 202, 207
 –, canopy, effective height of 207
 –, cover 125, 126, 205, 285
   –, coefficient 205
 –, growth 4, 5, 9, 168, 169, 202
 –, nutrients 165, 166, 253
 –, parameter 201
 –, stems 185
 –, uptake 56, 166
plasticity limits 268
plot
 –, length 285
 –, rainfall simulation 101, 102
pollutant 253
pollution 12, 253
pond 286
porosity 167, 185, 208, 236, 247
 –, effective 208
pot experiment 5, 27
potatoes 36, 47
Prague 135, 151
precipitation 4, 8, 9, 19, 23, 26, 44, 45, 47, 55, 61–65, 125, 126, 128, 159, 167, 169, 173, 175, 262, 297–299, 304, 307
 –, daily data 169
 –, data 125, 167, 173
 –, kinetic energy 63
prediction error 224
Principal Component Analysis (PCA) 69, 73–75
process
 –, erosional 253
 –, hydrological 253
Process orientated Erosion Prediction Program (PEPP) 253, 255, 257–263, 270, 273, 275, 279, 297
process simulation, sources of error 228
production capacity, potential 34, 39
productivity, agricultural 23
protection of urban areas 148
Province of Limburg 34, 40

R

Radovesice Waste Dump (North Bohemia) 157, 159–161
rain, depth 209
raindrop
 –, detachment 185, 216
 –, erosion 215
 –, erosive impact of 205
 –, fall velocity 207, 256
 –, impact 94, 166, 191, 192, 206, 215–217, 298
   –, detachment by (see also *splash*) 166, 191, 215
 –, splash 164
rainfall
 –, actual intensity 206

-, artificial 184
-, depressional type low intensity r. 36
-, depth 136, 142
-, detachment 289
-, duration 9, 94, 138, 167, 225, 234
-, effective 136, 207, 212, 261, 262
-, energy 164, 207, 216, 289, 291
-, erodibility 242
-, erosivity 199, 207, 245
-, event 34, 40, 47, 82, 88, 89, 94, 96, 98, 107, 112, 125, 128, 131, 136, 163, 164, 173, 202, 228, 245, 263, 276, 295
  -, simulation 228, 245, 273, 275
  -, single 40, 164
-, excess 206, 211, 287, 292
-, excess rate 211
-, experiment 67, 101, 229, 237, 247, 258, 264-272, 279
-, fluid density of 207
-, gross r. 136, 139
-, impact 205
-, input data 142
-, intensity 4, 61, 64, 86, 94, 112, 142, 157, 165, 167, 206, 207, 210-213, 216, 218, 225, 230, 255, 256, 264, 272
  -, effective 207, 212
-, intensity of net r. 262
-, kinetic energy 138, 191, 192
-, net 136, 140, 210, 262
-, parameters 94, 96
-, probabilities 158, 160
-, simulation 101, 102, 159, 184, 213, 224, 229, 235, 264, 265
-, simulation experiment 184, 200, 213
-, simulator 110, 200, 229, 257, 264, 267, 279
-, spatial variability 203
rainsplash 228, 287, 290, 291
  -, energy 291
random roughness 36
rating equation 211
relief parameter 94, 96
remote sensing 67
  -, data 72
reservoir 59, 93-100, 105-107, 149
  -, sedimentation 59
residue, buried 208
resolution, spatial and temporal 202
resource, natural 59
retention 36-40, 44, 47, 62, 63, 70, 75, 79, 80, 86, 88, 92, 99, 100, 105-107, 151, 167, 199, 307
  -, basin 38-40, 99, 100, 105-107
  -, capacity 62, 70, 75, 151
  -, kinematic 63
  -, potential maximum r. 44, 47
return period 36, 107, 152
Revised Universal Soil Loss Equation (RUSLE) 164, 165, 170, 285, 294
Reynolds number 194
rhythm, annual 125
Richards equation 35, 36, 263, 298
rill 9, 12, 63, 116, 158, 164, 169, 170, 211-217, 219, 220, 223, 226, 229, 236, 237-246, 253, 257, 261,
264, 275, 291, 292, 2992301, 307
-, erosion 169, 170, 213, 214, 217, 226, 238
  -, processes 125
-, flow 213-217, 219, 223, 226, 275, 299-301
-, soil erodibility 213
rillanz 63
riparian
-, buffer 106, 107
-, forest 106
river 12, 24, 91, 109, 110, 117-119, 149, 298, 299
-, points of intersection of 117
-, siltation 24
road 33, 36, 38, 117, 120, 126, 128-132, 149, 150, 201
-, incised ('hollow') r. 33
root
-, channel 106
-, development depth 115, 117, 122
Rote Weißeritz 95, 99
roughness 36, 45, 72, 73, 84, 86, 94, 125, 137, 138, 147, 163, 191, 194, 210-214, 220, 226, 236, 245, 254, 264, 267, 270, 279, 299-301, 306
-, coefficient 45, 137, 138, 147, 210, 211, 254, 264, 272, 275, 276, 301, 307
Rouse number 65, 66
runoff
-, characteristics 202
-, direction 117, 118
-, erosive 81
-, excess rate 211
-, generation 153, 199, 201, 225, 226, 245
-, hydraulics 181, 189, 194
-, hydrograph 152, 181, 189, 191, 234-236, 246
-, linear channel 117
-, peak r. 33
-, prediction 194, 225, 231, 236
  -, accuracy 235
-, rate 151, 211, 213, 236, 238, 292, 293
-, simulation 286
-, total r. 33
-, velocity 151, 288, 297
-, volume 43, 44, 47-50, 55, 56, 151, 153, 156, 165, 211, 214, 218, 232, 233, 240, 241, 246
  -, calculation 43, 47
  -, measured and predicted 48, 49
  -, predicted 233, 246
RUSLE (see *Revised Universal Soil Loss Equation*)

S

Saidenbach 95, 96, 98, 99, 100, 104, 105
-, reservoir 99, 104
  -, measured and simulated 98
-, watershed, heavy metals 96
satellite data 69
saturated conductivity, effective 208
saturation, initial 208
Saxonian Research Institute for Agriculture 201
Saxonian State Office for Agriculture (SLfL) 229
Saxonian State Office for Environment and Geology 201
Scenario Generator (SCENGEN) 6, 8, 9, 15, 23, 28

Index   315

sedigraph 201, 242, 244
sediment 14–23, 51–59, 88–120, 163–170, 213–223, 237–241, 257–259, 266–270, 285–287, 290–295
  –, concentration
    –, maximum 221
    –, simulation 278
  –, delivery 52, 53, 55
    –, coefficients of efficiency 52, 53
    –, measured and predicted 51–53
    –, minimizing 105, 107
    –, ratio (SDR) 165, 214
  –, deposition 43
  –, discharge 201, 204, 217, 221, 239, 292, 293
  –, in runoff 241
  –, input 79
    –, erosive 80
  –, load 33, 84, 88, 109, 128, 185, 213, 257–259, 266, 273, 294
  –, movement 213
  –, production 93, 95, 177, 292
  –, texture 104, 239
  –, transport 34, 43, 59, 76, 90, 93, 99, 100, 105, 163–167, 246, 286, 290–295, 298
    –, capacity 218
    –, on slopes 167
    –, simulation 286
  –, yield 14, 15, 18–21, 23, 43, 98–103, 106, 107, 109, 112, 113, 117–120, 165, 166, 177, 181, 185
    –, estimated by WEPP 19
    –, estimation 101
    –, probability 20
sedimentation, map 37
sensitivity 8, 15, 23, 84, 113, 144, 145, 200, 223–228, 236, 238, 244–246, 260–263, 277
  –, analysis 144–161, 224–251, 263
    –, HILLFLOW model 263
    –, PEPP model 262
  –, concept, deterministic 224
  –, measure 224, 244
  –, parameter 225, 260, 261, 263
    –, definition 224
  –, value 225, 236
settling
  –, rate 66
  –, velocity 259
sewage plant 79
shear
  –, strength 268
  –, stress 213, 214, 218, 219, 220, 298
    –, critical 213, 214
  –, velocity, critical 193, 194
sheet
  –, flow 62, 63, 79, 220, 223, 237, 253–259, 275, 300, 301
  –, runoff 164
Shields critical shear velocity 193, 194
shore zone 79, 82, 90, 91
  –, element retention 80
siltation 12, 24
Simulating Production and Utilization of Range Land (SPUR) 169
Simulation Model of Overland Flow and Erosion Processes (SMODERP) 135, 137, 139, 141–152, 157–159, 161
simulation, small scale 164
Simulator for Water Resources in Rural Basins (SWRRB) 9
single rainstorm, simulation 204
slope
  –, angle 255, 263
  –, geometry 142, 173
  –, grade 199
  –, gradient 34, 39, 115, 117, 124, 137, 216, 218, 219, 255, 300, 301, 303, 305
  –, inclination 65, 207
  –, input data 143
  –, length 10, 27, 120, 135, 142, 144, 146–150, 152, 156, 158, 160, 164, 199, 255, 294
    –, critical 135, 144, 146, 148, 150, 152, 156, 158, 160
    –, permissible 144, 147
  –, length and grade 199
  –, profile 123, 301
  –, spatial variation 285
SMODERP (see: Simulation Model of Overland Flow and Erosion Processes)
soil
  –, aggregate 106
  –, alcaline 68
  –, alluvial 68, 72
  –, bare 73, 137, 147, 192
  –, biota 11
  –, bulk density 208
  –, cohesion 36, 189, 194, 216, 228, 240, 242
  –, conservation 40, 109, 113, 129, 132, 135, 148
  –, conservation scenario 33
  –, consolidation 9
  –, critical momentum flux 258
  –, crusted 35
  –, cultivation, traditional 59
  –, degradation 68
  –, detachability 192, 196, 216, 237
    –, index 216
  –, detachment 120, 169, 185, 191, 192, 215, 238, 288
  –, detachment by flow 191
  –, detachment by raindrop impact 191, 215
  –, disturbance 202
  –, eroded, particle class 219
  –, fertility 120
  –, file 202
  –, humidity 72, 74
    –, mapping 72
  –, hydraulic parameter 277
  –, hydraulic properties 202
  –, irrigated 72
  –, matrix 208, 241, 262, 263
  –, maximum volumetric moisture content 185
  –, organic matter 11, 169
  –, parameter 96, 268
  –, particle 94, 100, 106, 124, 138, 141, 149, 166, 167, 205, 216, 245, 256, 258, 270, 291
  –, particles

-, detachment 106, 138, 141
-, mobilization 106
-, pores 244
-, porosity 185, 236
-, productivity 135
-, profile 27, 125, 166, 202
-, propertiy 72, 115, 153, 156, 168, 202, 224, 241, 268, 279, 297, 298
-, protection measure 153
-, roughness 300
-, sorptivity 140
-, structural stability 106
-, structure 36, 161, 228, 242, 244
-, surface 4, 60, 73, 94, 106, 120, 124, 136, 146, 147, 184, 185, 191, 192, 202, 205, 216, 220, 226, 243, 246, 256, 270, 290, 303
  -, geometric roughness 270
-, surface depression storage 136, 146, 147
-, suspended 93
-, texture 36, 167, 168, 191, 199, 213, 216, 224, 234, 239, 242, 273
-, tilled loess 269
-, uncrusted 35
-, units 125
soil cover 36, 39, 65, 69, 94, 106, 168, 199, 256, 267, 276
-, changes 168
-, crop 39
soil erodibility 167
-, parameters 213, 228
-, relative 138, 141
soil erodibility factor 206
soil erosion 3, 4, 28, 34, 39, 59, 109–111, 163, 170, 181, 200, 201, 245
-, costs 38
-, future, potential and actual increases 4
-, map 37
-, model 67, 109, 120, 123, 132
  -, ranking of 234, 237, 238
-, modeling 110, 111
-, modeling system 199
-, multiscale approach for predicting 109
-, process 33, 40, 109, 110, 200
-, quantitative estimate 163
-, quantitative modeling 297
-, resistance 110
soil loss 38, 39, 90, 115, 118–123, 128, 135, 144, 156, 163–166, 181, 199, 200, 225–233, 236–242, 246, 261, 265–269, 273, 275, 278, 279, 297
-, annual 122, 164
-, areal prediction 120–129
-, predicted 200, 226, 236–239
-, prediction 225, 236
-, spatial and temporal distribution 181
soil moisture 34, 64, 67, 73, 94, 131, 132, 140, 167, 176, 177, 208, 209, 226, 263, 265–272, 275, 276, 279, 298
-, budget model 131
-, content 34, 131, 177, 263, 265–268, 270, 275, 276, 279
-, data 131, 132
-, deficit 208, 209

-, estimation 131
-, mapping 67
-, simulation 175, 176, 177
soil nutrients, depletion 11
soil particle
-, detachment 93, 94, 124, 138, 166, 167
-, detachment of 166
-, transport by overland flow 166
soil type 64, 70, 75, 83, 115, 117, 125, 137, 140–142, 147, 164, 185, 202, 208, 229, 273–277, 279
soil water 169
-, content 73, 83, 208, 209, 226
-, movement 199
solar radiation 5, 19, 169
solifluction 70
solution characteristics 289
sorptivity 140, 147, 159, 160
Sorriso
-, area 12, 23, 27
-, monthly minimum and maximum temperatures 11
-, monthly rainfall 11
-, simulated annual rainfall and sediment yield 20
-, site data 14, 15
sowing, costs 38
soya bean 12–15, 26, 28
-, production in Brazil 13, 14
SPANS 45
splash 12, 164, 166, 169, 185, 192, 217, 240, 241, 288–295
-, detachment 217, 289, 292, 293, 295
-, energy 288
-, erosion 12, 289, 291, 292
-, suspension 291
SPOT 67
sprinkler system 184
SPUR (see *Simulating Production and Utilization of Range Land*)
SPUR I model 169
SPUR II model 169
steady flow 293, 294
-, erosion equation 293
steepness 120, 164, 245
stillwater reach 93
stoniness 68, 70
storage capacity 47, 126, 241, 304
straw 39
stream, network 124, 173, 173
stream power 44, 45, 193, 219, 220, 238, 288, 292, 293
-, critical 219, 238
-, modified 193
-, theory 44, 45
subcatchment 173–177
subsurface flow 159
suction potential 208
sugarbeet 33, 36
surface
-, cover 192
-, depression storage 140
-, flow 159

Index

-, flow depth 216
-, layer 157, 160
-, roughness 86, 125, 167, 211, 214, 226, 236, 267, 307
-, runoff 9, 33, 34, 60, 63, 66, 67, 80, 82, 90, 109, 115, 118, 128, 129, 137, 149-153, 156, 166-168, 214, 262, 272-275, 279, 287, 297, 299, 301, 302, 304
  -, approach 298-307
  -, modeling 60, 297
  -, model 67, 297
  -, routing 210
  -, temporal behaviour 302
  -, storage 34, 206, 210
SWATRE 36
SWRRB (see *Simulator for Water Resources in Rural Basins*)

T

temperature 3, 5, 9, 15, 16, 23, 26, 27, 169
temporal variability 264, 266
terracing 14, 59
texture 36, 68, 70, 72, 104, 167, 168, 191, 199, 213, 216, 221, 224, 234, 236, 239, 242, 273
throughfall
  -, calculation 202
  -, direct 140, 207
thunderstorm 36, 269
tillage 9, 106, 107, 115, 120, 123, 131, 132, 169, 201, 214, 228, 236, 241, 242, 245, 265-268
tillage effects, assessment 240
topography 45, 75, 152, 167-169, 262, 275, 276
topology grid 173
transfer point 117-120
transpiration 4, 5
transport
  -, by flow 191, 217
  -, coefficient 218, 219
  -, physical approach 299
transport capacity 128, 138, 165, 170, 199, 206, 214-221, 223, 226, 229, 237-239, 242, 244-246, 255-260, 275, 279, 288, 290-292, 295
  -, calculation 219-221
  -, equation 223, 239
  -, estimate 285
  -, formula 238, 244
  -, net 290
  -, overland flow 138, 165, 170, 218
  -, rill flow 219
  -, volumetric 215
transport deficit approach 258, 259
transport processes 43, 253, 290
transport/erosion rate approach 258, 259
tributary stream 106
tropospheric ozone 3
turbulence, model 299

U

U.S. Department of Agriculture 200
unit discharge 218
Universal Soil Loss Equation (USLE) 27, 43, 51-53, 90, 92, 113, 115, 138, 141, 163-167, 170, 199, 207, 253, 285, 294
urban area 33, 148, 165
USDA-Water Erosion Prediction Project model (WEPP) 3, 7, 9, 14, 15, 18-23, 27, 28, 35, 164, 169-171, 199-202, 208-211, 213, 215-218, 221, 225-232, 236-246, 253, 285, 293, 294, 297
user interface 243
USLE (see *Universal Soil Loss Equation*)
USLE-DABAG (see *Modified United Soil Loss Equation*) 90

V

validation 228
  -, event 47, 50, 51
Van Genuchten parameter 273
vegetation 4, 36, 72, 81, 83, 135, 137-139, 141, 147, 153, 167, 168, 184, 185, 262, 265, 288
  -, change 4
  -, cover 137, 138, 141, 153, 164, 167, 168, 184
velocity 43, 45, 53, 54, 62, 65-67, 80, 135, 140, 151, 169, 189, 193, 194, 206, 207, 212, 216, 219, 220, 245, 254, 256, 259, 260, 264, 288, 297, 299, 300, 302
  -, wind 169
vertisols 72
Vinare 151, 153, 156
  -, watershed 153

W

Walnut Gulch Experimental Watershed 183
wash load 65-67
waste
  -, dump 148, 157, 161
    -, restoration 148
  -, land 47
  -, water 79
    -, municipal 91
water
  -, budget model 131, 132
  -, conservation 40, 109, 113, 129, 132, 135, 148
  -, content 67, 68, 72, 73, 83, 110, 112, 124, 166, 208, 209, 224, 226, 267, 270, 276
    -, plant available 68
  -, erosion 135, 149, 285
  -, holding capacity 68
  -, interaction of surface and soil water 168
  -, pollution 115
  -, quality 24, 43, 149, 165, 290
  -, resources, mobilisation 59
  -, retention curve 36
  -, runoff 68
  -, storage capacity 47
  -, transport process 253
  -, use, human 168
  -, vapour 3
  -, vertical movement 34
  -, vertical transport in soil 35
Water Erosion Prediction Project (WEPP) 3, 7, 9,

14, 15, 18–23, 27, 28, 35, 164, 169–171, 199–202, 208ff., 253, 285, 293, 294, 297
–, application 228
–, input parameter 202
Waterboard 'Roer en Overmaas' 34, 40
watercourse 117, 118, 128, 129, 131
watershed 43–56, 59, 76, 93, 95–101, 104–107, 123, 151–156, 201, 275, 276, 285, 291
–, management 59
waterway, grassed 36, 39, 106, 107
weather data 28, 201
Weiherbach
    –, catchment 264, 267
    –, research project 253, 278
    –, soil types 277
WEPP (see *Water Erosion Prediction Project*)
wetness 72–75

wetting front 208
wheel track 35, 36
–, eraser 39
wind velocity 169
winter wheat 33, 36, 120, 123
Woburn, catchment 292
wormholes 208

Y

Yalin equation 218, 220, 223
Yalin transport capacity 218

Z

ZALF (see *Centre for Agricultural Landscape and Land Use Research*)

Printing (Computer to Film): Saladruck, Berlin
Binding: Stürtz AG, Würzburg